AF361873

Emerging Protein Sources for Food Production and Human Nutrition

Emerging Protein Sources for Food Production and Human Nutrition

Editors

Przemyslaw Lukasz Kowalczewski
Anubhav Pratap Singh
David Kitts

MDPI • Basel • Beijing • Wuhan • Barcelona • Belgrade • Manchester • Tokyo • Cluj • Tianjin

Editors

Przemyslaw Lukasz
Kowalczewski
Department of Food
Technology of Plant Origin
Poznań University of Life
Sciences
Poznań
Poland

Anubhav Pratap Singh
Food, Nutrition, and Health
The University of British
Columbia
Vancouver
Canada

David Kitts
Food, Nutrition, and Health
The University of British
Columbia
Vancouver
Canada

Editorial Office
MDPI
St. Alban-Anlage 66
4052 Basel, Switzerland

This is a reprint of articles from the Special Issue published online in the open access journal *Molecules* (ISSN 1420-3049) (available at: www.mdpi.com/journal/molecules/special_issues/protein_food).

For citation purposes, cite each article independently as indicated on the article page online and as indicated below:

LastName, A.A.; LastName, B.B.; LastName, C.C. Article Title. *Journal Name* **Year**, *Volume Number*, Page Range.

Contents

Editorial

Emerging Protein Sources for Food Production and Human Nutrition

Przemysław Łukasz Kowalczewski [1,*], Anubhav Pratap-Singh [2,*] and David D. Kitts [2]

1   Department of Food Technology of Plant Origin, Faculty of Food Science and Nutrition, Poznań University of Life Sciences, 60-637 Poznań, Poland
2   Food, Nutrition, and Health, Faculty of Land & Food Systems, The University of British Columbia, 2205 East Mall, Vancouver, BC V6T 1Z4, Canada; david.kitts@ubc.ca
*   Correspondence: przemyslaw.kowalczewski@up.poznan.pl (P.Ł.K.); anubhav.singh@ubc.ca (A.P.-S.)

1. Introduction

It is estimated that by 2050, the world's population will be up to 9 billion. Feeding the rising population with the appropriate amounts of food, and in particular, with adequate protein supply is an emergent focus of current research activities in the field of food science. Global environmental changes, rapidly changing socio-economic conditions, and geopolitical issues endanger food sustainability, and generate the need to search for new, unconventional sources of protein for human nutrition. Considering the above-mentioned worldwide circumstances, many scientists around the world are dealing with the subject of alternative protein sources, as well as examining their properties and safety. Ensuring an adequate supply of protein, but also its appropriate quality and nutritional value, is one of the main challenges facing the world of science.

With this mission in mind, the scope of this Special Issue of Molecules, entitled "Emerging Protein Sources for Food Production and Human Nutrition", is devoted to the latest advances in analytics and the application of new methods in the processing of proteins, both plant and animal, as well as methods of production and testing foods enriched with these proteins. In this SI, 10 original research manuscripts were collected and published. The articles included in this collection are briefly described below.

2. Plant-Based Protein Sources

Miedzianka et al. analyzed the effect of the acetylation of plant proteins on their functional and nutritional properties. In their first paper, an attempt was made to increase the value of rice protein concentrate [1] by improving the properties of a commercial protein preparation. After the acetylation process, the influence of this chemical modification on the chemical composition, digestibility and protein modeling was analyzed using SDS-PAGE, electrophoresis and FT-IR spectroscopy. Electrophoresis showed that the content of the main fractions of rice proteins (prolamine and glutelin) decreased with the increase in the concentration of the modifying reagent. Using spectroscopic analysis, wavenumbers corresponding to the presence of proteins or lipids, aromatic systems and carbohydrates were observed. However, the use of acetic anhydride did not significantly affect the digestibility of the modified rice protein concentrate, while acetylation resulted in a significant increase in its emulsifying properties and water-binding capacity. A slight increase in protein solubility and a decrease in foaming capacity were also observed in the modified rice protein concentrate.

In the second publication, the effect of acetylation with different doses of acetic anhydride on the chemical composition and selected functional properties of a commercial pumpkin protein concentrate was determined [2]. The electrophoretic analysis showed that in the acetylated pumpkin protein, the content of the heaviest protein (with a molecular weight of 35 kDa) decreased with the increase in concentration of the modifying reagent. The acetylation of pumpkin protein caused a significant increase in the water-binding

Citation: Kowalczewski, P.Ł.; Pratap-Singh, A.; Kitts, D.D. Emerging Protein Sources for Food Production and Human Nutrition. *Molecules* **2023**, *28*, 2676. https://doi.org/10.3390/molecules28062676

Received: 13 March 2023
Accepted: 14 March 2023
Published: 16 March 2023

capacity, oil absorption and emulsifying properties that were already at a dose of 0.4 mL/g. In addition, an increase in the foaming capacity of preparations obtained with 2.0 mL/g of acetic anhydride was demonstrated, while the acetylation of 0.4 and 1.0 mL/g caused a decrease in protein solubility compared to native pumpkin protein. It can, therefore, be concluded that acetylation improves the functional properties of commercial protein preparations, which broadens the possibilities of their use in the food industry.

Potato protein is considered to be one of the most valuable plant-based proteins due to the high content of essential amino acids and biological activity [3]. Numerous literature data indicate, however, that the enzymatic hydrolysis of proteins that has been properly carried out can significantly improve both the nutritional value and biological activity. The study of Kowalczewski et al. [4] is a report on the effect of the enzymatic hydrolysis of potato juice proteins, combined with membrane filtration. The obtained hydrolysate was characterized in terms of nutritional value and biological activity, including the evaluation of the amino acid profile and score, the content of mineral compounds, as well as antioxidant and cytotoxic activity in vitro. It was also found that the enzymatic hydrolysis of potato juice in the reactor with the ultrafiltration membrane separation system increased the cytotoxic activity of the processed material. IC_{50} toxic doses of the hydrolysate for cancer cells were significantly lower than those of fresh potato juice. Moreover, IC_{50} toxic doses of the concentrate were lower for cancer cells than for normal cells. Cytotoxicity against the human gastric cancer cell line (Hs 746T), the human colon cancer cell line (Caco-2), the human colorectal adenocarcinoma cell line (HT-29) and the normal human colon mucosa cell line (CCD 841 CoN) showed cytotoxic activity specifically directed against cancer cells. Therefore, it can be concluded that the membrane filtration-assisted enzymatic hydrolysis of potato juice proteins may increase their biological activity and allow for potato juice to be used in the production of medicinal preparations.

3. Whey and Insects as a Protein Source

Another important article in the Special Issue is that of Smułek et al. [5] on hemp-seed-oil-based emulsions. The protein used to stabilize the emulsions was whey protein, which is an important raw material in the food industry. A novel solution proposed by the authors of the paper was the addition of an extract from *Aesculus hippocastanom*, which contains natural compounds—for example, saponins, as a co-surfactant. The coexistence of these two emulsifiers made it possible to obtain stable emulsions with an average droplet size of 200–300 nm. Moreover, the obtained emulsions were characterized by good rheological properties. In addition, microbiological studies showed that the tested emulsion systems positively influenced the activity of a probiotic strain of the genus *Lactobacillus*. Thus, the described emulsions represent a good solution with high application values.

Edible insects are commonly used as food in many parts of the world, mainly in Africa, Latin America and Asia [6]. According to the FAO, more than 1900 different species of insects are eaten worldwide. These include mealworms and crickets [7].

Dion-Poulin et al. [8], in their paper, described studies on the evaluation of the functional properties of two commercial insect meals (obtained from *Gryllodes sigillatus* and *Tenebrio molitor*) and their respective hydrolysates obtained using Alcalase®, conventionally and after the pressurization pretreatment of insect powders. It was observed that water-binding capacity, foaming and gelling properties were not improved after enzymatic hydrolysis. The pre-pressure treatment of mealworm meals probably caused protein denaturation and aggregation, which reduced the degree of hydrolysis. As expected, enzymatic digestion (with and without pressure) increased the solubility, reaching values close to 100%. The pre-pressure treatment of mealworm flour further improved its solubility compared to the control hydrolysate, while pressure treatment reduced the solubility of cricket meals. The oil-binding capacity was also improved after enzymatic hydrolysis.

In the research of Boukil et al. [9], pressure treatment was also used. The ability of high hydrostatic pressure (HHP) in combination with enzymatic hydrolysis by Alcalase® or pepsin has been studied to improve the in vitro digestion of mealworm proteins, particularly allergy-causing proteins. The effect of the in vitro digestion of the main allergenic proteins of mealybug was enhanced by the use of HPP; therefore, HHP-assisted enzymatic hydrolysis is an alternative strategy to conventional hydrolysis to produce a large amount of peptide derived from allergenic insect proteins and reduce their immunoreactivity in food, nutraceuticals and pharmaceuticals.

The analysis of the properties of mealworm proteins was also undertaken by a group of scientists led by Gravel [10]. The effect of defatting powder from mealworm (*Tenebrio molitor*) with hexane on protein profiles and the techno-functionality of the obtained preparations was analyzed. Major protein profiles were shown to be similar between hexane defatted and non-defatted samples; however, some specific differences in content (e.g., hexamerin 2) were observed and characterized using proteomic tools. Protein solubility was significantly lower in the case of *Tenebrio molitor* meals in comparison to protein extracts defatted with hexane. A significant increase in the foaming capacity of the defatted fractions was also observed.

Consumer acceptance of insects as food is a necessary step to expand their presence in the market. The most popular way to eat insects in Europe is to use them to enrich traditional products. Zielińska and Pankiewicz [11] studied the characteristics of shortcake biscuits enriched with *Tenebrio molitor* flour and they examined properties such as nutrient composition, color, physical and antioxidant properties, starch digestibility and in vitro glycemic index. They showed that the substitution of wheat flour with mealworm flour changed the nutritional value of the products—a progressive increase in the protein and ash contents of biscuits as the concentration of mealworm flour increased. Moreover, mealworms were found to have high antioxidant potential, as evident from the higher free-radical scavenging activity of biscuits enriched in mealworm flour compared to the control. Additionally, the supplementation of mealworm flour to biscuits caused an increase in slowly digested starch, with a decrease in rapidly digested starch. This is important information for consumers because the dietary benefits attributed to SDS are associated with a slower postprandial rise in blood glucose and glycemia maintenance for longer periods compared to RDS, which results in a rapid rise and then a rapid fall in blood glucose. Thus, the authors emphasized that edible insects are a source of valuable nutrients and manifest health-promoting properties; thus, using them in designing health-promoting foods seems to be justified.

A group of researchers led by Smarzyński [12] analyzed the impact of using cricket powder (CP) on the molecular properties of water in model shortcake biscuits, and characterized their nutritional properties. The partial replacement of wheat flour with CP in biscuits increased their nutritional value, but also affected the analyzed physical properties. In addition, a small addition of CP improved the taste, texture, appearance and overall attractiveness ratings of the biscuits. Changes in the physical properties of the biscuits were also observed. The more wheat flour was replaced by CP, the lower the hardness of the biscuits. Analysis of the molecular dynamics of water, measured by LF NMR, indicated a decrease in the value of the short components of the spin–spin (T_{21}) relaxation times, which indicates a decrease in the dynamics of water molecules bound to the polymer matrix.

The analysis of the impact of the use of CP in another product—gluten-free bread—was also dealt with by a group of scientists led by Kowalczewski [13]. The nutritional value as well as antioxidant and β-glucuronidase activities were assessed after the simulated in vitro digestion of gluten-free breads enriched with 2%, 6% and 10% of house cricket (*Acheta domesticus*) powder. The addition of CP significantly increased the nutritional value, both in terms of protein and, above all, minerals. A significant increase in the content of polyphenolic compounds and antioxidant activity in enriched bread was also demonstrated. The use of CP also reduced the undesirable activity of β-glucuronidase by 65.9% (compared to the control bread) in the small intestine, and by as much as 78.9% in the large intestine. The effect of bread on the intestinal microflora was also assessed and no inhibitory effect on the growth of the intestinal microflora (Bifidobacterium and Lactobacillus) was found.

The presented results indicate the benefits of using CP to increase the nutritional value and biological activity of gluten-free food products.

Author Contributions: All authors have contributed to the conceptualization, writing, review and editing of this manuscript. All authors have read and agreed to the published version of the manuscript.

Institutional Review Board Statement: Not applicable.

Informed Consent Statement: Not applicable.

Data Availability Statement: Not applicable.

Acknowledgments: This Special Issue has been made possible by the contributions of several authors, reviewers and editorial team members. Their efforts are acknowledged, and thanks are due to all of them.

Conflicts of Interest: The authors declare no conflict of interest.

References

1. Miedzianka, J.; Walkowiak, K.; Zielińska-Dawidziak, M.; Zambrowicz, A.; Wolny, S.; Kita, A. The Functional and Physicochemical Properties of Rice Protein Concentrate Subjected to Acetylation. *Molecules* **2023**, *28*, 770. [CrossRef] [PubMed]
2. Miedzianka, J.; Zambrowicz, A.; Zielińska-Dawidziak, M.; Drożdż, W.; Nemś, A. Effect of Acetylation on Physicochemical and Functional Properties of Commercial Pumpkin Protein Concentrate. *Molecules* **2021**, *26*, 1575. [CrossRef] [PubMed]
3. Kowalczewski, P.Ł.; Olejnik, A.; Białas, W.; Rybicka, I.; Zielińska-Dawidziak, M.; Siger, A.; Kubiak, P.; Lewandowicz, G. The Nutritional Value and Biological Activity of Concentrated Protein Fraction of Potato Juice. *Nutrients* **2019**, *11*, 1523. [CrossRef] [PubMed]
4. Kowalczewski, P.Ł.; Olejnik, A.; Rybicka, I.; Zielińska-Dawidziak, M.; Białas, W.; Lewandowicz, G. Membrane Filtration-Assisted Enzymatic Hydrolysis Affects the Biological Activity of Potato Juice. *Molecules* **2021**, *26*, 852. [CrossRef] [PubMed]
5. Smułek, W.; Siejak, P.; Fathordoobady, F.; Masewicz, Ł.; Guo, Y.; Jarzębska, M.; Kitts, D.D.; Kowalczewski, P.Ł.; Baranowska, H.M.; Stangierski, J.; et al. Whey Proteins as a Potential Co-Surfactant with *Aesculus hippocastanum* L. as a Stabilizer in Nanoemulsions Derived from Hempseed Oil. *Molecules* **2021**, *26*, 5856. [CrossRef] [PubMed]
6. Raheem, D.; Carrascosa, C.; Oluwole, O.B.; Nieuwland, M.; Saraiva, A.; Millán, R.; Raposo, A. Traditional consumption of and rearing edible insects in Africa, Asia and Europe. *Crit. Rev. Food Sci. Nutr.* **2019**, *59*, 2169–2188. [CrossRef] [PubMed]
7. van Huis, A. Potential of insects as food and feed in assuring food security. *Annu. Rev. Entomol.* **2013**, *58*, 563–583. [CrossRef] [PubMed]
8. Dion-Poulin, A.; Laroche, M.; Doyen, A.; Turgeon, S.L. Functionality of cricket and mealworm hydrolysates generated after pretreatment of meals with high hydrostatic pressures. *Molecules* **2020**, *25*, 5366. [CrossRef] [PubMed]
9. Boukil, A.; Perreault, V.; Chamberland, J.; Mezdour, S.; Pouliot, Y.; Doyen, A. High hydrostatic pressure-assisted enzymatic hydrolysis affect mealworm allergenic proteins. *Molecules* **2020**, *25*, 2685. [CrossRef] [PubMed]
10. Gravel, A.; Marciniak, A.; Couture, M.; Doyen, A. Effects of Hexane on Protein Profile, Solubility and Foaming Properties of Defatted Proteins Extracted from Tenebrio molitor Larvae. *Molecules* **2021**, *26*, 351. [CrossRef] [PubMed]
11. Zielińska, E.; Pankiewicz, U. Nutritional, Physiochemical, and Antioxidative Characteristics of Shortcake Biscuits Enriched with Tenebrio molitor Flour. *Molecules* **2020**, *25*, 5629. [CrossRef] [PubMed]
12. Smarzyński, K.; Sarbak, P.; Kowalczewski, P.Ł.; Różańska, M.B.; Rybicka, I.; Polanowska, K.; Fedko, M.; Kmiecik, D.; Masewicz, Ł.; Nowicki, M.; et al. Low-field NMR study of shortcake biscuits with cricket powder, and their nutritional and physical characteristics. *Molecules* **2021**, *26*, 5417. [CrossRef] [PubMed]
13. Kowalczewski, P.Ł.; Gumienna, M.; Rybicka, I.; Górna, B.; Sarbak, P.; Dziedzic, K.; Kmiecik, D. Nutritional Value and Biological Activity of Gluten-Free Bread Enriched with Cricket Powder. *Molecules* **2021**, *26*, 1184. [CrossRef] [PubMed]

 *molecules*

Article

The Functional and Physicochemical Properties of Rice Protein Concentrate Subjected to Acetylation

Joanna Miedzianka [1,*], Katarzyna Walkowiak [2], Magdalena Zielińska-Dawidziak [3], Aleksandra Zambrowicz [4], Szymon Wolny [1] and Agnieszka Kita [1]

[1] Department of Food Storage and Technology, Wroclaw University of Environmental and Life Sciences, 51-630 Wrocław, Poland
[2] Department of Physics and Biophysics, Poznań University of Life Sciences, 60-637 Poznan, Poland
[3] Department of Biochemistry and Food Analysis, Poznań University of Life Sciences, 60-623 Poznań, Poland
[4] Department of Functional Products Development, Wroclaw University of Environmental and Life Sciences, 51-630 Wrocław, Poland
* Correspondence: joanna.miedzianka@upwr.edu.pl

Abstract: The aim of the present study was to increase the value of rice protein concentrate (RPC) by improving the functional properties of a preparation subjected to acetylation and analyze the impact of this chemical modification on chemical composition, digestibility, and protein patterning using SDS-PAGE electrophoresis and FT-IR spectroscopy. In the modified samples, the protein content increased (80.90–83.10 g/100 g cf. 74.20 g/100 g in the control). Electrophoresis revealed that the content of the main rice protein fractions (prolamin and glutelin) decreased as the concentration of the modifying reagent increased. Through spectroscopic analysis, wavenumbers, corresponding to the presence of proteins or lipids, aromatic systems, and carbohydrates, were observed. The use of acetic anhydride did not change the digestibility of the modified RPC significantly when compared to that of the control sample. The acetylation of the RPC caused a significant increase in its emulsifying properties at pH 8 (1.83–14.74%) and its water-binding capacity but did not have a statistically significant impact on the oil-absorption capacity. There was a slight increase in protein solubility and a decrease in foaming capacity in the modified RPC.

Keywords: commercial rice protein preparation; acetic anhydride; amino acid composition; digestibility; FT-IR spectroscopy; functional properties

Citation: Miedzianka, J.; Walkowiak, K.; Zielińska-Dawidziak, M.; Zambrowicz, A.; Wolny, S.; Kita, A. The Functional and Physicochemical Properties of Rice Protein Concentrate Subjected to Acetylation. *Molecules* **2023**, *28*, 770. https://doi.org/10.3390/molecules28020770

Academic Editor: Susy Piovesana

Received: 13 November 2022
Revised: 27 December 2022
Accepted: 9 January 2023
Published: 12 January 2023

1. Introduction

Plant-based proteins are innovative ingredients with fast-growing applications in the food industry. They are sources of important bioactive compounds and ingredients that are used in the production of functional foods [1]. Sustainability, ethical implications, population growth, variety, and the formulation of healthier products are among their main advantages over animal proteins. However, on the other hand, most plant-based protein preparations obtained by aqueous extraction methods are characterized by poor aqueous solubility, weak functionality (i.e., gelling, emulsification, and foaming), and a high degree of complexity, as well as susceptibility to pH, ionic strength, and temperature, which limits their applications in the industrial sector [2]. Moreover, they are less digestible and have less ability to transport other important nutrients, such as calcium and iron. Therefore, efficient modification processes are needed to improve the value of plant-based proteins.

One of the methods of protein extraction from plant-based materials that retain their original functional properties is the dry tribo-electrostatic separation method. This technology relies on the efficiency of milling to mechanically dissociate the proteins, which are subsequently separated by an air stream based on particle size and density [3,4]. As a result, the dry fractionation method has many advantages, including having no chemical residues, a minimal impact on the techno-functional properties, and the loss of insoluble protein, as well as having low energy consumption. The low protein extraction yield is the main disadvantage [3]. Based on the studies of Tabtabaei et al. [5], dry tribo-electrostatic separation can produce protein-rich plant fractions from navy bean flour which were characterized by superior solubility, emulsifying, and foaming when compared to wet process isolates. Vitelli et al. [6] investigated the performance of electrostatic separation to produce protein- and carbohydrate-rich fractions from hammer-milled navy bean flour. The results revealed that the samples collected from the middle and top part of the plate had a significantly higher protein content and, therefore, a lower starch content than the samples collected from the bottom of the plate.

Some of the functional properties can also be changed, inter alia, by chemical modifications [7]. Chemically modified proteins are produced by the addition of new functional moieties or the elimination of the components from the protein's structure. In proteins, the most reactive nucleophilic groups are the *E*-amino groups of lysyl residues and the SH group of cysteine. Therefore, these amino acids undergo the most modification. Examples of chemical modification include acetylation and succinylation based on the use of acetic or succinic anhydride, respectively, as the acylating agent. Acylation is a type of modification that has been widely used due to its efficiency, low cost, and ease of operation. Hydrophilic functional groups, such as hydroxyls, ε-amino groups, and phenols, are susceptible to acetylation [8]. Acetylation is advantageous from a technological point of view because the protein concentrates and isolates obtained can have improved solubility, emulsifying ability, and foaming and water-binding capacity and also stabilize the emulsions and determine their color. Additionally, it has been scientifically proven that modification through acetylation may cause changes in the chemical composition of the obtained preparations without adversely affecting their nutritional value [8–10]. However, the chemical modification methods (including acetylation with acetic anhydride) are not favorable for food applications as they use chemicals and produce chemical byproducts. Additionally, the modified proteins can be less digestible and are not utilized in animal feeding tests.

One of the plant-based alternatives to meat or soy protein is rice (*Oryza sativa* L.). It is a staple cereal and is widely consumed around the world. It has high nutritional value, being a source of starch, mineral salts, vitamins, and dietary fiber. The protein content of rice is relatively small (about 3%). However, rice proteins have been recognized as highly nutritious, hypoallergenic, and particularly healthful for human consumption. Additionally, the antioxidant and nutraceutical properties of rice are associated with a reduction in the risk of oxidative stress, which contributes to the prevention of hypercholesterolaemia [11,12].

Commercial rice protein concentrate is first subjected to alkaline extraction, and then the proteins are precipitated by adjusting the pH to their isoelectric point, where the nonprotein components are isolated by enzymatic processes [12,13]. Depending on factors such as the rice cultivar and the degree of milling of the rice, the protein content of the products of these treatments ranges from 65% to 90%. Rice protein concentrate (RPC) produced on an industrial scale is popular in the food and pharmaceutical industries, mainly in protein supplements and also in drink shakes, bars, or gels. Additionally, studies indicate that RPC can be used as a value-added ingredient in the production of bread [14] or biscuits [15]. It is considered as a substitute for soy protein as it has no beany flavor [16]. However, its powder presents fine particles and poor dissolution properties, limiting its use.

So far, rice protein has been subjected to chemical modification by phosphorylation with sodium trimetaphosphate (STMP) [17,18] and alkaline deamidation [19]. As for acetylation using acetic acid anhydride, authors have analyzed the physicochemical and functional properties of rice bran protein concentrate [17] and rice protein successively subjected to trypsin hydrolysis [20]. The novelty of our research is the analysis of commercial RPC, and it is known that among the factors which have a great influence on the properties of modified plant protein preparations are the protein's origin and method of isolation, as well as the conditions of the modification process [21]. The poor functional properties of RPC limit its use, but due to the high biological value of these proteins, it is justified to search for modification methods that might lead to the improvement of its functional properties. Therefore, the aim of this work was to try to improve the functional properties of commercial RPC by acetylation using acetic anhydride and to analyze the influence of the modification on the physicochemical properties (chemical composition, structure, protein profile, and digestibility).

2. Results and Discussion

2.1. Functional Properties

2.1.1. Protein Solubility Index (PSI)

Solubility is often used as an indicator of other functional properties. It is defined as the ability of a substance to form homogeneous mixtures with other substances. The solubility of the control and modified proteins in this study was pH-dependent, as indicated in Figure 1. Over the pH range studied [2,12,16], both the unmodified and acetylated RPC were soluble almost exclusively at pH 12; however, even under this condition, the solubility was very low (below 0.5%). The acetylation of the RPC at pH 12 with a 0.4 mL/g dose of acetic anhydride caused an almost two-fold increase in PSI to 0.45%, as compared to the control sample (0.28%). This could have resulted from the covalent incorporation of the acetyl groups into the amino group of the protein, which results in a certain degree of protein unfolding (electrostatic attraction is diminished). The hydrophilic groups can be exposed, and therefore hydrophilicity increases, which improves solubility [17].

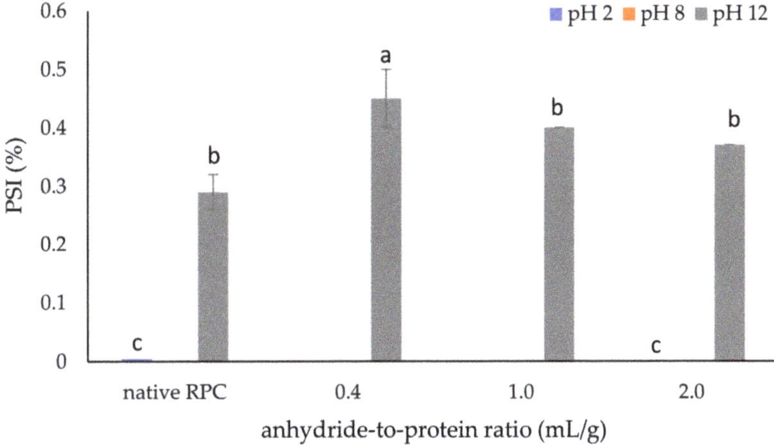

Figure 1. Effect of anhydride-to-protein ratio on the protein solubility index (PSI) of native and acetylated RPC. a, b, c—the same letters within the same analysis indicate values that are not significantly different; RPC—rice protein concentrate; 0.4, 1.0, 2.0—rice protein preparations after acetylation conducted with different concentrations of acetic anhydride (mL/g).

As the dose of acetic anhydride increased (from 1.0 mL/g (with a degree of acetylation of 92.24%) to 2.0 mL/g (with a degree of acetylation of 99.98%), data presented in Table 1), the protein solubility decreased, which is in line with the reports by other authors [8,22,23]. The data provided by Heredia-Leza et al. [21] show that when the degree of acetylation does not exceed 78%, the solubility of proteins increases. Thus, the solubility depends on the degree of acetylation, and, in some cases, a deep degree of acetylation is not desired. Similar results have been found for pumpkin proteins wherein a dose of 2 mL of acetic anhydride/g protein caused a decrease in the solubility from 62.44% to 49.52% [21]. Moreover, hydrogen bonds play a fundamental role in regulating protein and water interactions and rely on polarity. RPC is characterized by low polarity; thus, it cannot interact and form hydrogen bonds; therefore, the solubility of the control and acetylated samples was very low [21].

Table 1. Characteristics of native and acetylated RPC.

Anhydride-to-Protein Ratio (mL/g)	Dry Matter	Protein	Fat	Ash	Degree of N-Acylation	Amount of Protein Released into the Intestinal Fluid in Two-Step Digestion
	(g/100 g)					%
native RPC	94.49 ± 0.10 [a]	74.20 ± 1.15 [b]	6.32 ± 0.09 [a]	2.15 ± 0.04 [a]	-	70.00 ± 6.6 [a]
0.4	97.32 ± 0.16 [a]	83.10 ± 0.09 [a]	4.99 ± 0.26 [b]	1.93 ± 0.00 [ab]	71.23 ± 0.15 [c]	69.00 ± 10.4 [a]
1.0	96.96 ± 0.54 [a]	82.30 ± 3.69 [a]	1.58 ± 0.21 [c]	1.69 ± 0.14 [b]	92.24 ± 0.17 [b]	66.80 ± 1.9 [a]
2.0	96.16 ± 2.21 [a]	80.90 ± 0.42 [a]	1.94 ± 0.31 [c]	1.56 ± 0.27 [b]	99.88 ± 0.11 [a]	67.00 ± 5.8 [a]

Values are means ± standard deviation; n = 3; a, b, c—the same letters within the same column indicate values that are not significantly different; RPC—rice protein concentrate; 0.4, 1.0, 2.0—rice protein preparations after acetylation conducted with different concentrations of acetic anhydride.

In an alkaline environment, protein molecules may unfold, increasing the exposure of the hydrophobic groups and thus changing the solubility of the protein. However, considering the low solubility of the RPC and acetylated samples, this change may have resulted from the increased degradation of the commercial protein, which resulted in fewer interactions between the protein molecules. Additionally, the high insolubility of rice protein in water could be due to extensive aggregation and crosslinking through the disulphides of one of the main fractions, glutelin [8]. Rice proteins are composed of two major hydrophobic fractions: glutelins, which are high-molecular-weight proteins composed of subunits bound by disulphide linkages and are soluble only in dilute acid or alkali, and prolamins, which are soluble in 70% ethanol [11,13]. Also, according to the results of these studies [11,13], RPC displays poor solubility in the pH range of 2–10, with a minimum at pH 5. When considering the amino acid profile (Table S2), it also can be stated that rice proteins have an acidic character; that is, the sum of the aspartic acid and glutamic acid residues (25.95 g/100 g protein) is greater than the sum of the lysine, arginine, and histidine residues (12.51 g/100 g protein). Therefore, they exhibit maximum solubility at alkaline pH. Additionally, the very low solubility of the RPC samples can be explained by the protein isolation method used and, therefore, by the enhanced hydrophobic interactions [10]. Generally, protein preparations with low solubility in water can potentially be applied to products, e.g., meat analogs, baked goods, breakfast cereals, protein bars, and pet food.

2.1.2. Water- and Oil-Absorption Capacity

Water-binding capacity (WBC), defined as the ability of a protein to retain water in a physical or physicochemical way regardless of the forces of gravity or heating, is a common property of all proteins and protein products. This property is dependent on amino acid composition, hydrophobicity, pH, temperature, and ionic strength [24]. The results indicate that WBC increased after acetylation, already at the dose of 0.4 mL/g and almost two-fold at the dose of 2.0 mL/g (3.76 g/g), as compared to the control rice protein (2.07 mL/g) (Figure 2). No statistically significant difference was found between the acetylated samples. However, we noted that, contrary to the trend observed for solubility, that WBC increased slightly as the degree of acetylation increased. This is in line with the results of Lawal [25], who also observed that water absorption capacity increases with an increase in the level of modification. In his study on African locust bean protein isolate, the highest water absorption capacity of 6.20 mL/g was recorded with 73.6% acetylation. WBC increases after acetylation because, during the reaction, the hydrophobic properties of the control protein are reduced after the modification through the incorporation of additional hydrophilic groups from acetic anhydride. After chemical modification, the high-molecular-weight protein dissociates, which facilitates enhanced water absorption, which is also observed. Similar results have also been noted for jack bean protein [22], African bean protein isolate [25], and acylated mucuna protein [26]. However, the acetylated rice protein showed a low WBC when compared to other plant-based proteins subjected to the same chemical modification. This observation can be explained by differences in the chemical characteristics of the different plant-based protein preparations, such as the total protein content and the changes in their structure resulting from different isolation and modification conditions. The high WBC of a rice protein is related, in part, to its amino acid composition—the greater the number of charged residues (glutamic and aspartic acids), the greater the WBC (Table S2). Such proteins with a high WBC can be used in meat sausages, cakes, or bread. An acetylated preparation of RPC can be used in the processing of meat, fish, and plant products because it increases their juiciness, improves their rheological properties, and reduces weight loss during heating [10].

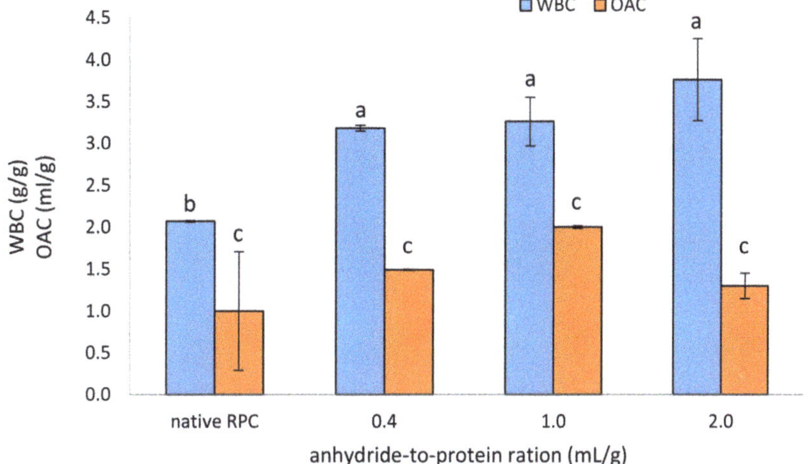

Figure 2. Effect of anhydride-to-protein ratio on water-binding capacity (WBC) and oil-absorption capacity (OAC) of native and acetylated RPC. a, b—the same letters within the WBC analysis indicate values that are not significantly different; c—the same letter within the OAC analysis indicates values that are not significantly different; RPC—rice protein concentrate; 0.4, 1.0, 2.0—rice protein preparations after acetylation conducted with different concentrations of acetic anhydride (mL/g).

Oil-absorption capacity (OAC) is the physical entrapment of fat molecules and is influenced by the protein concentration in the preparation, particle size, and porosity, availability of hydrophobic amino acid groups, and protein–fat–carbohydrate interactions. This functional property improves the mouthfeel and flavor retention of certain food products [12]. The control RPC sample was characterized by a low OAC (amounted 1.0 mL/g), which could be related to its high density and particle size (Figure 2). The acetylation using 1.0 mL/g of acetic anhydride resulted in a two-fold increase in the OAC of the modified RPC. In contrast to the WBC, a decrease in the ability to absorb oil was observed as the degree of acetylation increased. The same results were obtained by Lawal [25]. In his study, OAC reduced progressively with an increase in the level of modification, and the lowest OAC of 0.80 mL/g was recorded for African locust bean protein isolate, with 73.6% acetylation compared with the 1.90 mL/g recorded for the unmodified protein isolate. This could be related to the increased net negative charge and consequent decrease in hydrophilicity of the control protein after acetylation.

2.1.3. Foaming Properties

The ability of a compound to create a foam is called the foaming capacity (FC). Foam is a dispersion mixture in which the dispersing medium is a liquid or a solid, and the dispersed phase is a gas [21]. The effect of acetylation on the FC measured at different pH values is presented in Figure 3. Generally, the acetylation reaction led to a deterioration in the foaming measured at different pHs. However, as the dose of the modifying reagent increased (from 0.4 to 2.0 mL/g), there was a decrease in FC under the same conditions, except at pH 12 for the 2.0 mL/g dose, where the FC was comparable to that of the FC of the unmodified RPC. Similar observations were noted for the foam stability (FS) of rice preparations subjected to modification by acetylation (Table S1). A decrease in the FC of plant-based protein preparations subjected to acetylation with an increasing degree of acetylation was also observed by Bora [27] and Miedzianka et al. [28]. Furthermore, from the review prepared by Heredia-Leza et al. [21], it can be concluded that an increase in the FC of different plant-based proteins can be obtained up to a certain level of acetylation. For example, the 26% acetylation of canola protein leads to a 4.38-times increase in FC, while for 62.5% acetylated lentil and 78% acetylated Bambara groundnut, the FC does not increase significantly [21]. In the studies presented here, RPC acetylation ranged from 71.23% to 99.88%, which could even cause the deterioration in FC. The results disagree with previous observations by El-Adawy [8] and Bora [27]. They noted that the acetylation of canola, mung bean and lentil protein samples, respectively, improves FC because this reaction introduces acetyl groups for the ε-amino groups, lowering the number of positive charges. Additionally, the molecular size of acetylated proteins diminishes, which allows them to move faster toward the air–water interface [21]. A low FC can also be related to low protein solubility. Acetylated RPC, characterized by a low FC, cannot be used in the production of, for example, ice cream.

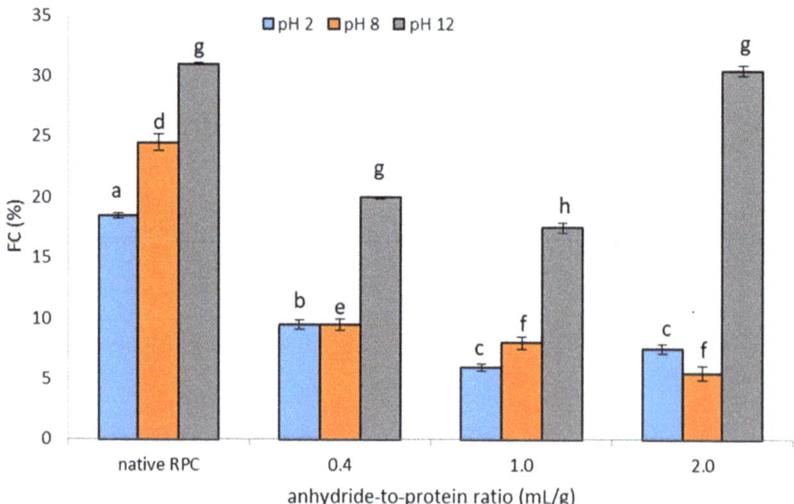

Figure 3. Effect of anhydride-to-protein ratio on foam capacity (FC) at different pH of native and acetylated RPC. a, b, c—the same letters within the pH 2 indicate values that are not significantly different; d, e, f—the same letters within the pH 8 indicate values that are not significantly different; g, h—the same letters within the pH 12 indicate values that are not significantly different. PRC—rice protein concentrate; 0.4, 1.0, 2.0—rice protein preparations after acetylation conducted with different concentrations of acetic anhydride (mL/g).

2.1.4. Emulsifying Properties

Emulsifying properties are among the most important properties in the manufacturing of many formulated foods. Emulsifying activity (EA) is defined as the ability of a protein solution or suspension to emulsify oil, whereas emulsion stability (ES) is the ability to remain unchanged over a period of time at a specific temperature and gravitational force [12,21]. The type of protein, its concentration, and pH are factors that influence emulsifying capacity [21]. The EA and ES of the control and acetylated RPC were strongly pH-dependent, as indicated in Figure 4A,B, respectively. The acetylation reaction had a significant influence on the EA of the commercial rice sample at pH 8.0. At a dose of 1.0 mL/g and 2 mL of acetic anhydride/g, the EA increased by 7.31 and 8.03 times, respectively. Here, under this condition, the modification enhanced the protein–oil interaction and thus the exposure of the hydrophobic and lipophobic residues, which results in better emulsifying properties. Furthermore, according to [13], the EA of rice protein isolate was relatively low, especially at pH < 6. As observed in Figure 4A, the EA of RPC acetylated with the lowest dose of acetic anhydride improved by more than seven times compared to its control counterpart. This result agrees with previous reports on mung bean isolate [8] and African locust bean [25]. At pH 2 and 12, a decrease was observed in EA and ES with an increasing dose of acetic anhydride. A minimal EA of the control and acetylated RPC was observed at pH 2 due to the limited solubility in this isoelectric region. These acetylated preparations, characterized by strong emulsifying properties (pH 8), may be of importance in the production of finely ground meat products, coffee whiteners, milk, or frozen desserts.

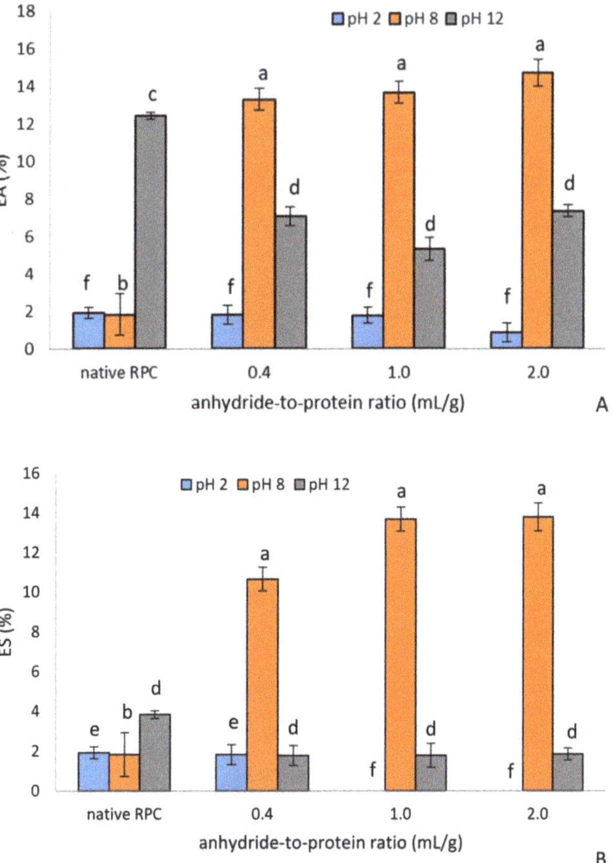

Figure 4. Effect of anhydride-to-protein ratio on emulsion activity, EA (**A**), and emulsion stability, ES (**B**), at different pH for the native and acetylated RPC. a, b—the same letters within pH 8 indicate values that are not significantly different; c, d—the same letters within pH 12 indicate values that are not significantly different; e, f—the same letters within pH 2 indicate values that are not significantly different; RPC—rice protein concentrate; 0.4, 1.0, 2.0—rice protein preparations after acetylation conducted with different concentrations of acetic anhydride (mL/g).

2.2. Chemical Composition

The characteristics of control and acetylated RPC are presented in Table 1. The analyzed samples differed statistically significantly in their protein, fat, and ash contents. No differences in dry matter content were observed. All of the analyzed samples met the condition that the water content in the protein preparation cannot exceed 10% [28].

The protein content of the nonacetylated RPC reached a value of 74.20 g/100 g, which resulted from the protein content of the rice seeds (approximately 3% of dried seeds). This commercial product was obtained under harsh processing conditions, and it was recovered in a denatured state with poor functional properties. Along with an increase in the acetic anhydride concentration from 0.4 to 2.0 mL/g, the preparations were characterized by a higher protein content than that of the control sample not subjected to chemical modification. However, the protein content in the acetylated samples was not statistically significantly different and ranged from 80.90 to 83.10 g/100 g. However, it can be seen that as the degree of acetylation increased, the protein content decreased. The higher protein content in the acetylated rice samples could be related to a series of operations performed during the preparation of the modifications. Similar results were noted by Khader et al. [29] upon the acetylation of whey protein concentrate. It is worth highlighting that the acetylation of plant-based protein preparations affects their protein content differently.

According to the data presented in Table 1, the fat content decreased by up to 75% after the acetylation of the protein with a dose of 1.0 mL/g. This could be related to washing during the precipitation of the modified proteins. Similar results have been noted for acetylated pumpkin protein concentrate, with the fat content ranging from an average of 9.17 g/100 g before chemical modification to an average of 8.89 g/100 g in the modified pumpkin protein concentrate [28].

Plant-based protein preparations are characterized by an ash content of less than 10%, which is mainly affected by the type of raw material. The control RPC, not subjected to chemical modification, contained only 2.15 g of ash/100 g. Along with an increase in the anhydride-to-protein-ratio, the ash content decreased, so that at a dose of 2.0 mL/g an almost 28% decrease in the analyzed compound was observed. This may be related to the removal of excess modifying reagent (because the acetylated samples were washed 3–5 times). Similar results have been noted for acetylated jack bean protein [22] and acetylated pumpkin protein concentrate [28].

2.3. Amino Acid Composition

The amino acid composition of a protein significantly affects its functional properties. As presented in Table S2, glutamic acid (17.05 g/100 g protein) was the dominant amino acid among those in the control sample, followed by aspartic acid (8.90 g/100 g protein), valine (7.14 g/100 g protein), and leucine (6.53 g/100 g protein). Proline (1.50 g/100 g protein), histidine (2.21 g/100 g protein), and lysine (3.43 g/100 g protein) were the least noted in the nonacetylated RPC. Other authors have found rice protein to be a source of methionine as well as branched-chain amino acids (BCAA), i.e., leucine, isoleucine, and valine, with lysine as the limiting amino acid. Similar levels of amino acids in a rice protein preparation have been reported by other authors [12,13,16]. The acetylation of the rice protein preparation did not statistically significantly affect the amino acid profile, except for the histidine content, which decreased at higher doses of acetic anhydride (1.0 and 2.0 mL/g). We proved that with an increase in the concentration of the modifying reagent, the content of the main rice protein fractions, i.e., prolamin and glutelin, decreases (Figure 5). Most likely, the change in the protein profile slightly changed the histidine content.

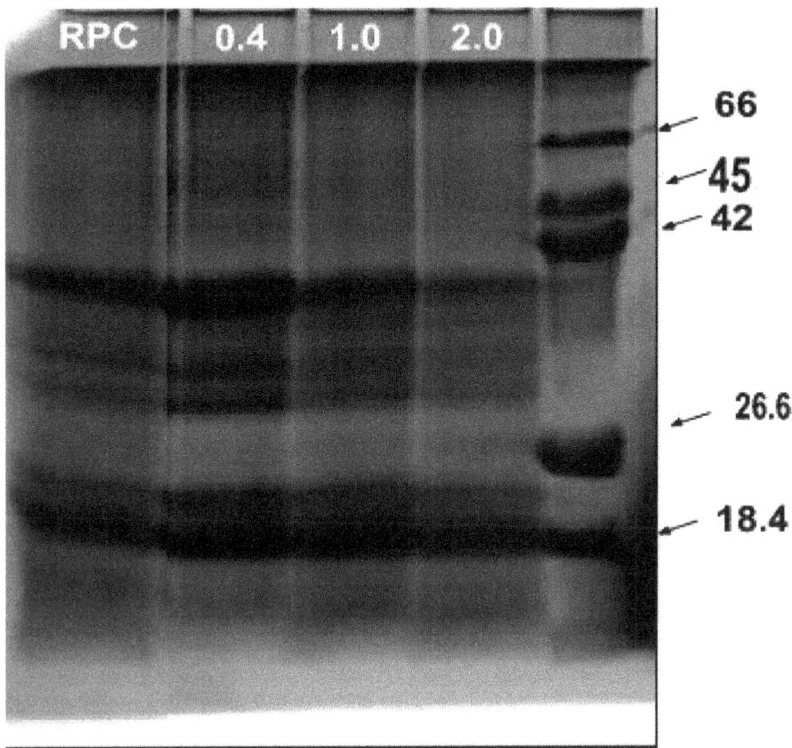

Figure 5. SDS-PAGE of the native and acetylated RPC. RPC—rice protein concentrate; 0.4, 1.0, 2.0—rice protein preparations after acetylation conducted with different concentrations of acetic anhydride (mL/g); MW (kDa)—molecular weight marker prepared in the laboratory (18.4 kDa (β-lactoglobulin), 26.6 kDa (β-casein), 42, 45 kDa (ovalbumin), 66 kDa (bovine serum albumin).

2.4. Degree of Acetylation and In Vitro Digestibility

In the analyzed samples, an increase in the degree of N-acetylation was observed along with an increase in the concentration of acetic acid anhydride (Table 1), but only when using the highest dose of acetic anhydride (2.0 mL/g), resulting in an almost complete blockage of the amino acid residues observed (99.88%). This means that there was a decrease in the number of ε-amino groups involved in the acetylation reaction, leading to an increase in the degree of N-acetylation. The degree of acetylation after the chemical modification of the rice protein was higher than that of other plant preparations, which confirms that the source of the protein and the conditions of its acetylation have the greatest influence on them. Some modified plant-based preparations reach only 70–80% acetylation [28].

The amount of protein ingested by an organism in proportion to the amount consumed is called digestibility, and this depends on the structure of the protein, the pre-processing, and the antinutritional compounds present [21]. Following the digestion of the analyzed commercial and acetylated rice preparations, the determined protein digestibility ranged from 66.80% (for the samples acetylated with a dose of 1.0 mL/g) up to 70.00% (for the samples not subjected to a modification) (Table 1). Additionally, Amagliani et al. [12] confirmed that the protein digestibility of rice is higher than that of other major cereals (i.e., wheat or corn). The use of acetic anhydride did not change the digestibility of the modified RPC significantly compared to that of the control sample. Similar observations were noted by Bergner et al. [30] and are in accordance with the effect of the process on the amino acid profile discussed above. The influence of the acetylation of lysine on its bioavailability has not been presented in the literature, and thus it is expected that the accessibility of this amino acid to trypsin was not modified. The effectiveness of digestive enzymes (both pepsin and serine proteases) should not be influenced by decreasing histidine content because they do not act on the bond formed with the amino acid [31,32]. On the other hand, acetylation may have a positive effect on the digestibility not related to the protein profile; it may contribute to a significant increase in the digestibility attributed to the destruction of antinutritional factors through chemical modifications [21].

2.5. Sodium Dodecyl Sulfate–Polyacrylamide Gel Electrophoresis (SDS-PAGE)

The control and resulting acetylated RPC were analyzed by SDS-PAGE (Figure 5). Based on the electrophoretic analyses, it was revealed that the rice protein not subjected to acetylation was composed of glutelin, with a molecular mass of about 22 kDa and 35–39 kDa, and prolamin, with a molecular mass of about 18 kDa. Similar observations were presented by Wang et al. [11], who additionally observed globulin and proglutelin fractions with molecular masses of 26 kDa and about 57 kDa, respectively. Based on the data in the literature, the major component of rice protein is glutelin, constituting about 80% of the total endosperm protein: prolamin, a minor storage protein, accounts for 5% to 10%. Other proteins include fractions such as globulin and albumin [11].

Based on the presented studies, it was shown that with an increase in the concentration of the modifying reagent (from 0.4 to 2.0 mL/g), the content of the main rice protein fractions, i.e., prolamin and glutelin, decreased. The applied acetylation process contributed to the removal of those proteins with a mass of about 18 kDa and about 35 kDa. This also confirmed a decrease in the protein content of the acetylated preparations. Similar results were observed in previous studies [28] when pumpkin protein concentrate was subjected to the same chemical modification.

2.6. Fourier Transform Infrared Spectroscopy (FT-IR)

FT-IR is a rapid and noninvasive spectroscopic technique for characterizing various biomaterials. It shows the changes that occur in the tested material at the molecular level, visualized on the basis of the absorption bands of various functional groups present in a given material. By analyzing the spectra of the samples of commercial plant-derived protein concentrate from rice (*Oryza sativa* L.) (RPC) shown in Figure 6, we can observe the maxima at a wavenumber of 1200–1700 cm^{-1}, which corresponds to the presence of proteins or lipids. The enhanced absorption bands at wavenumbers of 1630 and 1517 cm^{-1} can be attributed to the presence of the C=C vibrations characteristic of aromatic systems. In all samples, the presence of a region corresponding to carbohydrates (1200–900 cm^{-1}) was also apparent [33]. For the first sample, we observed the absorption maximum in the wavenumber area of 2929 cm^{-1}, which corresponds to the C-H stretching vibrations. These vibrations did not occur in trials 2, 3, and 4, in which the RPC was acetylated. When analyzing the course of the absorption changes at higher wave numbers, we observed a band at 3300 cm^{-1}, which indicates the presence of -O-H- functional groups for each of the tested samples, as well as for C-H and O-H stretching vibrations [34].

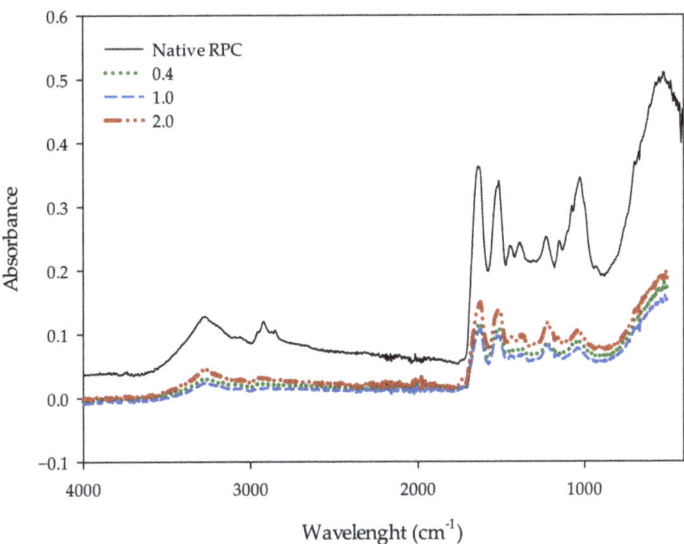

Figure 6. FTIR spectrum of the native and acetylated RPC.

3. Materials and Methods

3.1. Materials and Chemicals

Commercial RPC was purchased from the Diet Food company (Opatówek, Poland). The producer recommends direct consumption in the amount of 2–3 tablespoons a day, depending on the body's needs and physical effort. Acetic anhydride, 2,4,6-trinitrobenzenesulfonic acid (TNBS) were obtained from Sigma-Aldrich (St. Louis, MO, USA). All chemicals used in the experiment were of an analytical grade.

3.2. Acetylation of RPC

Acetylated RPC samples were prepared using the method of Miedzianka et al. [28]. Briefly, the commercial RPC was acetylated with acetic anhydride by adding different concentrations of modifying reagent (0.4, 1.0, and 2.0 mL of anhydride per 1 g of protein contained in the preparation) to the 1% aqueous suspension. After 60 min, the precipitate was centrifuged using an MPW-351 centrifuge (3720× *g* force for 15 min, Heraeus Sepatech, Osteorode, Germany), freeze-dried using Christ Alpha 1–4 LSCplus lyophilizer (Osterode am Hatz, Germany), and then the preparation was sieved and stored at about −20 °C until further analysis. Not subjected to the acetylation process, the native, commercial RPC was used as the control. Acetylation was performed over two technological repetitions.

3.3. Determination of Functional Properties

3.3.1. Effect of pH on Protein Solubility

The pH-solubility profile index (PSI) of the native and acetylated RPC was determined according to the method of Achouri et al. [35] with slight modifications. Briefly, 750 mg of preparation was weighed in the tube, and 15 mL of distilled water was added, and then the pH was adjusted (2, 8 or 12) using either 0.5 M NaOH or 0.5 M HCl. The protein solutions were shaken at room temperature for 30 min and successively centrifuged at 4500× *g* for 15 min (Rotofix 32A by Hettich, Tuttlingen, Germany). The protein content of the supernatants was determined by the Lowry method. Protein solubility was calculated as:

$$PS = (PCS/TPC) \times 100\ (\%)$$

where PCS is the protein content in the supernatant after centrifugation and TPC is the total protein content present in the protein sample.

3.3.2. Water-Binding Capacity and Oil-Absorption Capacity

Water-binding capacity (WBC) of the native and acetylated RPC was determined according to the method described by Jeżowski et al. [36]. For this, 1 g of powdered sample was weighed in a test tube containing 20 mL of distilled water. It was allowed to mix in a laboratory mixer for 60 s and was allowed to stand for 15 min at ambient temperature. This slurry was centrifuged at $4500\times g$ for 15 min (Rotofix 32A by Hettich, Tuttlingen, Germany). The separated solid was oven-dried. WBC was expressed as the amount of water (g) absorbed by 1 g of the preparation.

Oil-absorption capacity (OAC) was determined using the method of Wu et al. [37]. Briefly, 1 g of protein preparation was weighed in the test tube and mixed with 15 mL of rapeseed oil using a Vortex mixer. Samples were allowed to stand for 30 min. The resulting protein–oil mixture was separated using a centrifuge ($4000\times g$; Rotofix 32A by Hettich, Tuttlingen, Germany) for 10 min. Immediately after centrifugation, the supernatant was carefully poured into a 15 mL graduated cylinder, and the volumes were recorded. OAC was expressed as the amount of oil (mL) absorbed by 1 g of the preparation.

3.3.3. Effect of pH on Foaming Capacity and Stability

Foam capacity (FC) and stability (FS) were measured according to the method of Jeżowski et al. [36]. Briefly, 1 g of the preparation was weighed into the tube, and 200 mL of distilled water was added to it. The resulting mixture was adjusted to the appropriate pH (2, 8, or 12) using 0.5 M NaOH or 0.5 HCl. The sample was then homogenized for 2 min at 16,000 rpm (T25 basic ULTRA-TURRAX®; IKA Werke, Staufen im Breisgau, Germany). The beaten sample was immediately transferred to a measuring cylinder where the total foam volume was determined after 0, 5, 10, 30, and 60 min. FC and FS were calculated according to the following equations:

$$FC = (VA/VB) \times 100\ (\%)$$

where VA denotes the volume after whipping (mL) and VB is the volume before whipping (mL).

$$FS = (VC/VT) \times 100\ (\%)$$

where VC denotes the volume before whipping (cm^3) and VT is the volume after a certain time (mL).

3.3.4. Effect of pH on Emulsifying Properties

Emulsification activity (EA) and stability (ES) were estimated by the method described by Miedzianka et al. [28]. Briefly, the protein suspensions with added oil were mixed using a T25 basic ULTRA-TURRAX® homogenizer (IKA-Werke GmbH & Co. KG, Staufen im Breisgau, Germany) to produce a crude emulsion for 1 min at 20,000 rpm. Then, they were centrifuged at $3000\times g$ for 10 min. Emulsion stability was determined by centrifugation after heating at a temperature of 80 °C for 30 min.

$$EA = a/b \times 100\ (\%)$$

where a is the height of the emulsified layer in the tube and b is the height of the total contents in the tube.

$$ES = c/d \times 100\ (\%)$$

where c is the height of emulsified layer after heating and d is the height of the total contents after heating.

3.4. Basic Chemical Composition

The moisture and ash content were measured by the constant mass method [38]. The total protein content was calculated from the amount of nitrogen by a 5.95 factor, evaluated according to the Kjeldahl method, using a Büchi Distillation Unit K-355 (Athens, Greece) (https://www.fao.org/3/y5022e/y5022e03.htm#bm3.1 (accessed on 20 April 2022)). Fat content was determined according to the standard method [38].

3.5. Amino Acid Composition

The amino acid composition of the native and acetylated RPC was determined by ion-exchange chromatography after 23 h of hydrolysis with 6 N HCl at 110 °C, according to the method described previously [28]. The hydrolyzed amino acids were determined using an AAA-400 analyzer (INGOS, Prague, Czech Republic). A photometric detector was used, working at two wavelengths: 440 nm and 570 nm. A column of 350×3.7 mm, packed with ion exchanger Ostion LG ANB (INGOS) was utilized. Column temperature was kept at 60–74 °C and the detector at 121 °C. The prepared samples were analyzed using the ninhydrine method. No analysis of tryptophan was carried out.

3.6. Measurement of Degree of N-Acylation

The measurement of the degree of N-acylation was prepared, as described by Habeeb [39]. Briefly, to the protein suspensions, 1 cm^3 of 4% $NaHCO_3$ solution and 1 cm^3 of 0.1% trinitrobenzenesulfonic acid (TNBS) solution were added. After heating at 60 °C for 2 h, the samples were cooled down to room temperature. Next, 1 cm^3 of 10% SDS and 0.5 cm^3 of 0.1 N HCl were added. The absorbance of the solutions was read at 335 nm in a Rayleigh UV-2601 PC spectrophotometer (Beijing, China) against a reagent blank. The absorbance of the control protein concentrate was set equal to 100% free amino groups, and the extent of acetylation of the modified samples was calculated based on the decrease in absorbance because fewer amino groups were able to react with the TNBS reagent.

3.7. Digestibility of Protein Preparations

The protein digestibility of the native RPC and its acetylated preparation was determined using a method that simulates two-stage digestion [40], omitting the oral cavity and large intestine steps, as these are irrelevant to protein digestion. Gastric digestion was carried out at 37 °C for 2 h. About 0.5 g of the sample was introduced into the water, and the stomach acidic environment was achieved by decreasing the pH down to 2.0 by 1 N HCl and pepsin addition (60,000 U). The first step was stopped by increasing the pH up to 7.4 by 0.1 M $NaHCO_3$. The mixture was enriched in bile salts (0.03 g) and porcine pancreatin (0.005 g), which contains the proteases (trypsin, protease A, ribonuclease)amylase and lipase. The intestine digestion was performed in the same conditions (37 °C; 2 h). The digested sample was centrifuged. The remaining proteins present in the supernatant—not digested but extracted from the sample—were removed by trichloroacetic acid precipitation. In the prepared supernatant, with the use of the Kjeldahl method, protein nitrogen was determined [38] and was related in percentage to the amount of protein nitrogen content in the nondigested sample.

3.8. Sodium Dodecyl Sulfate–Polyacrylamide Gel Electrophoresis (SDS–PAGE)

SDS–PAGE analysis was performed according to the method of Laemmli [41]. The protein samples were diluted (10 mg/mL) with the buffer containing SDS and β-mercaptoethanol as the reducing reagents and were denatured. Then, the samples were loaded (10 µL) onto gel slabs (12%). At the end of the analysis, the gel slabs were stained with Coomassie Brilliant G-250 dye. Protein molecular weights were analyzed using the Infinity Capt (BioRad) program for electropherogram analysis.

3.9. Fourier Transform Infrared (FT-IR) Spectroscopy

FTIR spectra were obtained by a spectrophotometer from the Perkin Elmer company (Waltham, MA, USA) equipped with an ATR device with diamond as the internal reflection element [42]. Data were collected over a spectral range of 4000–800 cm^{-1}.

3.10. Statistical Analysis

Statistical analysis of all data was performed using one-way analysis of variance (ANOVA). The analysis of the chemical composition and functional properties were performed in duplicate and in triplicate, respectively. Duncan's range test was used to determine the differences among the samples with a probability level of 0.05. Statistical analysis and standard deviations were determined using Statistica v. 13.3 software (Dell Software Inc., Round Rock, TX, USA).

4. Conclusions

Based on the literature data, the acetylation of rice proteins with acetic anhydride, which was used in the presented work, is one of the most effective chemical methods for improving some functional properties without negatively affecting a change in chemical composition. However, it is still difficult to find an explanation for the relationship between the structure of the modified proteins and their function. The acetylation of a commercial rice protein preparation using a dose of 0.4 mL of acetic anhydride/g of protein improved the WBC and emulsifying properties at pH 8, which extends the possibilities of using the modified rice samples in the processing of meat, fish, and plant products, as well as in finely ground meat products, coffee whiteners, milk, or frozen desserts.

Supplementary Materials: The following supporting information can be downloaded at: https://www.mdpi.com/article/10.3390/molecules28020770/s1. Table S1: Amino acid composition, Table S2: Foam stability results.

Author Contributions: Conceptualization, J.M.; methodology, J.M., K.W., M.Z.-D. and A.Z.; software, S.W.; validation, J.M., K.W., M.Z.-D. and A.Z.; formal analysis, J.M., K.W., M.Z.-D. and A.Z.; investigation, J.M., resources, J.M.; data curation, A.K.; writing—original draft preparation, J.M.; writing—review and editing, A.K.; visualization, J.M., supervision, A.K.; project administration, A.K.; funding acquisition, A.K. All authors have read and agreed to the published version of the manuscript.

Funding: This research was financially supported by the National Science Centre of Poland (2019/03/X/N29/01823). The APC/BPC was financed/co-financed by the Wroclaw University of Environmental and Life Sciences.

Institutional Review Board Statement: Not applicable.

Informed Consent Statement: Not applicable.

Data Availability Statement: Data are contained within the article and Supplementary Materials.

Acknowledgments: J.M. would like to thank Przemysław Łukasz Kowalczewski for his help with editing the manuscript (Poznań University of Life Sciences).

Conflicts of Interest: The authors declare no conflict of interest.

Sample Availability: Samples of the compounds are available from the Authors.

References

1.   Kowalczewski, P.Ł.; Olejnik, A.; Świtek, S.; Bzducha-Wróbel, A.; Kubiak, P.; Kujawska, M.; Lewandowicz, G. Bioactive compounds of potato (*Solanum tuberosum* L.) juice: From industry waste to food and medical applications. *CRC Crit. Rev. Plant Sci.* **2022**, *41*, 52–89. [CrossRef]
2.   Day, L.; Cakebread, J.A.; Loveday, S.M. Food proteins from animals and plants: Differences in the nutritional and functional properties. *Trends Food Sci. Technol.* **2022**, *119*, 428–442. [CrossRef]
3.   Schutyser, M.A.I.; Pelgrom, P.J.M.; van der Goot, A.J.; Boom, R.M. Dry fractionation for sustainable production of functional legume protein concentrates. *Trends Food Sci. Technol.* **2015**, *45*, 327–335. [CrossRef]

4. Assatory, A.; Vitelli, M.; Rajabzadeh, A.R.; Legge, R.L. Dry fractionation methods for plant protein, starch and fiber enrichment: A review. *Trends Food Sci. Technol.* **2019**, *86*, 340–351. [CrossRef]

5. Tabtabaei, S.; Konakbayeva, D.; Rajabzadeh, A.R.; Legge, R.L. Functional properties of navy bean (*Phaseolus vulgaris*) protein concentrates obtained by pneumatic tribo-electrostatic separation. *Food Chem.* **2019**, *283*, 101–110. [CrossRef]

6. Vitelli, M.; Rajabzadeh, A.R.; Tabtabaei, S.; Assatory, A.; Shahnam, E.; Legge, R.L. Effect of hammer and pin milling on troboelectrostatic separation of legume flour. *Powder Technol.* **2020**, *372*, 317–324. [CrossRef]

7. Nikbakht Nasrabadi, M.; Sedaghat Doost, A.; Mezzenga, R. Modification approaches of plant-based proteins to improve their techno-functionality and use in food products. *Food Hydrocoll.* **2021**, *118*, 106789. [CrossRef]

8. El-Adawy, T.A. Functional properties and nutritional quality of acetylated and succinylated mung bean protein isolate. *Food Chem.* **2000**, *70*, 83–91. [CrossRef]

9. Zhao, C.B.; Zhang, H.; Xu, X.Y.; Cao, Y.; Zheng, M.Z.; Liu, J.S.; Wu, F. Effect of acetylation and succinylation on physicochemical properties and structural characteristics of oat protein isolate. *Process Biochem.* **2017**, *57*, 117–123. [CrossRef]

10. Damodaran, S.; Parkin, K.L. (Eds.) *Fennema's Food Chemistry*, 5th ed.; CRC Press: Boca Raton, FL, USA, 2017. Available online: https://www.taylorfrancis.com/books/edit/10.1201/9781315372914/fennema-food-chemistry-srinivasan-damodaran-kirk-parkin?refId=bc906e60-da21-4a53-827a-5198971cd958&context=ubx (accessed on 26 August 2022).

11. Wang, Z.; Li, H.; Liang, M.; Yang, L. Glutelin and prolamin, different components of rice protein, exert differently in vitro antioxidant activities. *J. Cereal Sci.* **2016**, *72*, 108–116. [CrossRef]

12. Amagliani, L.; O'Regan, J.; Kelly, A.L.; O'Mahony, J.A. The composition, extraction, functionality and applications of rice proteins: A review. *Trends Food Sci. Technol.* **2017**, *64*, 1–12. [CrossRef]

13. Shih, F.F.; Daigle, K.W. Preparation and characterization of rice protein isolates. *JAOCS J. Am. Oil Chem. Soc.* **2000**, *77*, 885–889. [CrossRef]

14. Jiamyangyuen, S.; Srijesdaruk, V.; Harper, W.J.; Jiamyangyuen, A.; Srijesdaruk, V.; Harper, W.J.; Sci, S.J. Extraction of rice bran protein concentrate and its application in bread. *Songklanakarin J. Sci. Technol.* **2005**, *27*, 55–64.

15. Yadav, R.B.; Yadav, B.S.; Chaudhary, D. Extraction, characterization and utilization of rice bran protein concentrate for biscuit making. *Br. Food J.* **2011**, *113*, 1173–1182. [CrossRef]

16. Lee, J.S.; Oh, H.; Choi, I.; Yoon, C.S.; Han, J. Physico-chemical characteristics of rice protein-based novel textured vegetable proteins as meat analogues produced by low-moisture extrusion cooking technology. *LWT* **2022**, *157*, 113056. [CrossRef]

17. Lee, S.K.; Jang, I.S.; Kim, K.M.; Park, S.K.; Lee, W.Y.; Youn, K.S.; Bae, D.H. Changes in Functional Properties of Rice Bran and Sesame Meal Proteins through Chemical Modifications. Available online: https://agris.fao.org/agris-search/search.do?recordID=KR2005012450 (accessed on 25 August 2022).

18. Wang, Y.R.; Yang, Q.; Fan, J.L.; Zhang, B.; Chen, H.Q. The effects of phosphorylation modification on the structure, interactions and rheological properties of rice glutelin during heat treatment. *Food Chem.* **2019**, *297*, 124978. [CrossRef]

19. Guan, J.; Takai, R.; Toraya, K.; Ogawa, T.; Muramoto, K.; Mohri, S.; Ishikawa, D.; Fujii, T.; Chi, H.; Cho, S.J. Effects of alkaline deamidation on the chemical properties of rice bran protein. *Food Sci. Technol. Res.* **2017**, *23*, 697–704. [CrossRef]

20. Li, Y.Y.; Wang, L.; Qian, H.F.; Zhang, H.; Qi, X.G. Study on physicochemical properties and trypsin hydrolysis of acetylated rice protein. *Mod. Food Sci. Technol.* **2015**, *31*, 81–86. [CrossRef]

21. Heredia-Leza, G.L.; Martínez, L.M.; Chuck-Hernandez, C. Impact of Hydrolysis, Acetylation or Succinylation on Functional Properties of Plant-Based Proteins: Patents, Regulations, and Future Trends. *Processes* **2022**, *10*, 283. [CrossRef]

22. Lawal, O.S.; Adebowale, K.O. The acylated protein derivatives of Canavalia ensiformis (jack bean): A study of functional characteristics. *LWT-Food Sci. Technol.* **2006**, *39*, 918–929. [CrossRef]

23. Miedzianka, J.; Pęksa, A.; Aniołowska, M. Properties of acetylated potato protein preparations. *Food Chem.* **2012**, *133*, 1283–1291. [CrossRef]

24. Moure, A.; Sineiro, J.; Domínguez, H.; Parajó, J.C. Functionality of oilseed protein products: A review. *Food Res. Int.* **2006**, *39*, 945–963. [CrossRef]

25. Lawal, O.S. Functionality of African locust bean (*Parkia biglobossa*) protein isolate: Effects of pH, ionic strength and various protein concentrations. *Food Chem.* **2004**, *86*, 345–355. [CrossRef]

26. Lawal, O.S.; Adebowale, K.O. Effect of acetylation and succinylation on solubility profile, water absorption capacity, oil absorption capacity and emulsifying properties of mucuna bean (*Mucuna pruriens*) protein concentrate. *Food* **2004**, *48*, 129–136. [CrossRef]

27. Bora, P.S. Effect of acetylation on the functional properties of lentil (*Lens culinaris*) globulin. *J. Sci. Food Agric.* **2003**, *83*, 139–141. [CrossRef]

28. Miedzianka, J.; Zambrowicz, A.; Zielińska-Dawidziak, M.; Drożdż, W.; Nemś, A. Effect of Acetylation on Physicochemical and Functional Properties of Commercial Pumpkin Protein Concentrate. *Molecules* **2021**, *26*, 1575. [CrossRef]

29. Khader, A.E.; Salem, O.M.; Zedan, M.A.; Mahmoud, S.F. Impact of substituting nonfat dry milk with acetylated whey protein concentrates on the quality of chocolate ice milk. *Egypt. J. Dairy Sci.* **2001**, *29*, 299–312.

30. Bergner, H.; Seidler, W.; Simon, O.; Schmandke, H. Digestibility and dietary quality of nonacetylated and acetylated Vicia faba proteins in maintenance: Studies on 15N-labelled adult rats. *Ann. Nutr. Metab.* **1984**, *28*, 156–163. [CrossRef]

31. Gráf, L.A.; Szilágyi, L.A.; Venekei, I.A. Chymotrypsin. *Handb. Proteolytic Enzym.* **2013**, *3*, 2626–2633. [CrossRef]

32. Zielińska-Dawidziak, M.; Tomczak, A.; Burzyńska, M.; Rokosik, E.; Dwiecki, K.; Piasecka-Kwiatkowska, D. Comparison of Lupinus angustifolius protein digestibility in dependence on protein, amino acids, trypsin inhibitors and polyphenolic compounds content. *Int. J. Food Sci. Technol.* **2020**, *55*, 2029–2040. [CrossRef]
33. Ying, D.Y.; Hlaing, M.M.; Lerisson, J.; Pitts, K.; Cheng, L.; Sanguansri, L.; Augustin, M.A. Physical properties and FTIR analysis of rice-oat flour and maize-oat flour based extruded food products containing olive pomace. *Food Res. Int.* **2017**, *100*, 665–673. [CrossRef] [PubMed]
34. Amir, R.M.; Anjum, F.M.; Khan, M.I.; Khan, M.R.; Pasha, I.; Nadeem, M. Application of Fourier transform infrared (FTIR) spectroscopy for the identification of wheat varieties. *J. Food Sci. Technol.* **2013**, *50*, 1018–1023. [CrossRef] [PubMed]
35. Achouri, A.; Nail, V.; Boye, J.I. Sesame protein isolate: Fractionation, secondary structure and functional properties. *Food Res. Int.* **2012**, *46*, 360–369. [CrossRef]
36. Jeżowski, P.; Polcyn, K.; Tomkowiak, A.; Rybicka, I.; Radzikowska, D. Technological and antioxidant properties of proteins obtained from waste potato juice. *Open Life Sci.* **2020**, *15*, 379–388. [CrossRef] [PubMed]
37. Wu, H.; Wang, Q.; Ma, T.; Ren, J. Comparative studies on the functional properties of various protein concentrate preparations of peanut protein. *Food Res. Int.* **2009**, *42*, 343–348. [CrossRef]
38. Al-mentafji, H.N. *Official Methods of Analysis of AOAC International*; AOAC: Washington, DC, USA, 2005.
39. Habeebl, A.F.S.A. Determination of Free Amino Groups in Proteins by Trinitrobenzenesulfonic Acid. *Anal. Biochem.* **1966**, *14*, 328–336. [CrossRef]
40. Kowalczewski, P.Ł.; Olejnik, A.; Białas, W.; Rybicka, I.; Zielińska-Dawidziak, M.; Siger, A.; Kubiak, P.; Lewandowicz, G. The nutritional value and biological activity of concentrated protein fraction of potato juice. *Nutrients* **2019**, *11*, 1523. [CrossRef]
41. Laemmli, U.K. Cleavage of structural proteins during the assembly of the head of bacteriophage T4. *Nature* **1970**, *227*, 680–685. [CrossRef]
42. Kowalczewski, P.Ł.; Siejak, P.; Jarzębski, M.; Jakubowicz, J.; Jeżowski, P.; Walkowiak, K.; Smarzyński, K.; Ostrowska-Ligęza, E.; Baranowska, H.M. Comparison of technological and physicochemical properties of cricket powders of different origin. *J. Insects Food Feed* **2022**, 1–10. [CrossRef]

Article

Effect of Acetylation on Physicochemical and Functional Properties of Commercial Pumpkin Protein Concentrate

Joanna Miedzianka [1,*], Aleksandra Zambrowicz [2], Magdalena Zielińska-Dawidziak [3], Wioletta Drożdż [1] and Agnieszka Nemś [1]

1   Department of Food Storage and Technology, Wroclaw University of Environmental and Life Sciences, 37 Chelmonskiego Street, 51-630 Wrocław, Poland; wioletta.drozdz@upwr.edu.pl (W.D.); agnieszka.nems@upwr.edu.pl (A.N.)
2   Department of Functional Products Development, Wroclaw University of Environmental and Life Sciences, 37 Chelmonskiego Street, 51-630 Wrocław, Poland; aleksandra.zambrowicz@upwr.edu.pl
3   Department of Biochemistry and Food Analysis, Poznań University of Life Sciences, 48 Mazowiecka Street, 60-623 Poznań, Poland; magdalena.zielinska-dawidziak@up.poznan.pl
*   Correspondence: joanna.miedzianka@upwr.edu.pl

Abstract: The purpose of the present study was to determine the effects of acetylation with different doses of acetic anhydride on the chemical composition and chosen functional properties of commercial pumpkin protein concentrate (PPC). The total protein content decreased as compared to unmodified samples. Electrophoretic analysis revealed that in the acetylated pumpkin protein, the content of the heaviest protein (35 kDa) decreased in line with increasing concentrations of modifying reagent. Acetylation of PPC caused a significant increase in water-binding and oil-absorption capacity and for emulsifying properties even at the dose of 0.4 mL/g. Additionally, an increase in foaming capacity was demonstrated for preparations obtained with 2.0 mL/g of acetic anhydride, whereas acetylation with 0.4 and 1.0 mL/g caused a decrease in protein solubility as compared to native PPC.

Keywords: pumpkin protein concentrates; acetylation; chemical composition; amino acid profile; in vitro digestibility; SDS-PAGE; functional properties

Citation: Miedzianka, J.; Zambrowicz, A.; Zielińska-Dawidziak, M.; Drożdż, W.; Nemś, A. Effect of Acetylation on Physicochemical and Functional Properties of Commercial Pumpkin Protein Concentrate. *Molecules* **2021**, 26, 1575. https://doi.org/10.3390/molecules26061575

Academic Editors: Anubhav Pratap Singh, Przemyslaw Lukasz Kowalczewski and David Kitts

Received: 31 January 2021
Accepted: 10 March 2021
Published: 12 March 2021

Publisher's Note: MDPI stays neutral with regard to jurisdictional claims in published maps and institutional affiliations.

1. Introduction

Pumpkin seeds are a source of fat (37–45%) and protein (25–37%), which is distinguished by a high content of indispensable amino acids. They also contain about 11% carbohydrates and 6% dietary fibre, as well as many vitamins and minerals [1]. In Europe, pumpkin (*Cucurbita pepo* L.) is cultivated in the middle south region, primarily for oil production from seed, where the main by-product from oil extraction is defatted pumpkin cake, which contains up to 65% protein [2,3]. Due to the high protein content, this by-product left over from pumpkin oil seed extraction is considered by many authors as one of the most attractive and promising sources of vegetable proteins with proven health-promoting and functional effects [4,5].

One of the uses of pumpkin cake, apart from enrichment with amino acids, is protein preparations constituting newly developed functional ingredients that enrich the nutritional properties of food products. Favourable results have been obtained for extraction of functional proteins from pumpkin seed cakes under alkaline conditions. In this technique, pumpkin cakes are treated with an NaOH solution at pH 8–10, after which the solution is centrifuged and the proteins are precipitated therefrom at the isoelectric point. In this way, isolates containing as much as 80–90% protein can be obtained with high yield. Additionally, this method is used to obtain the broad spectrum of proteins found in pumpkin seeds [5].

Widely available pumpkin protein concentrates and isolates are highly appreciated, despite their poor functional properties. During the manufacture of pumpkin protein preparations, the applied heat treatments cause extensive protein denaturation and the resulting preparations are characterized mainly by poor solubility, limiting their incorporation into food systems. Therefore, a roadblock to the large-scale use of plant proteins is their poor solubility under mildly acidic (pH 3–6) conditions. This excludes their use in acidic foods such as coffee whiteners, acidic beverages, yogurts and pourable and non-pourable dressings. The weak solubility of proteins, however, sets limits for their utilization in formulated food systems and so the solubilization of pumpkin protein preparations has been attempted to extend their usefulness in the food industry [6].

One of the most convenient and frequently used methods for altering the functional properties of proteins is acetylation. This chemical modification of proteins involves the reaction of acetic anhydride with the ε-amino group of lysine and other nucleophilic groups such as phenolic (e.g., tyrosine) and aliphatic hydroxyl groups (e.g., serine and threonine) [7]. The acetic anhydride (CAS number 108-24-7) used here is a chemical compound widely used in organic synthesis. It is a colourless liquid that smells strongly of acetic acid. Acetylation of proteins in the range of 0.2–1.0 mL/g protein has been widely known for more than 40 years [8–11]; modified proteins have been applied in the preparation of products such as coffee whiteners [12], flavouring agents for roasted meat [13] and carbonated beverages [14]. Additionally, recently, studies on improving functionality by acylation have been applied to many food proteins including spray-dried egg white [15], oat protein isolates [16] and fish myofibrillar protein [17]. However, in the literature, there is no report available on the functionality of acetylated commercial pumpkin protein concentrate. This chemical modification is considered safe because acetic anhydride is rapidly hydrolysed (half-life 4.4 min) to acetic acid, which is readily biodegradable. In the atmosphere, it is converted to acetic acid, which is subject to photooxidative degradation (half-life 22 days). Toxicity to aquatic organisms is moderate (18 to 3400 mg/L), but it persists only for a short time due to its rapid hydrolysis to acetate/acetic acid. Acetic acid is further used as a food and animal feed additive or a preservative in pickles [18].

For the above-stated reason, this investigation through chemical modification has been undertaken in order to improve the functional properties of pumpkin protein preparations, mainly water-binding, oil-absorption capacities, solubility, foaming and emulsifying properties, while limiting the negative impact on changes in the chemical composition and digestibility of proteins. Therefore, the purpose of the present study was to determine the effect of the anhydride-to-protein ratio of acetylation on the chemical composition and functional properties of commercial pumpkin protein preparations. It is hoped that the data will provide information on acetylated derivatives, which could be useful in experiments concerning the chemical modification of commercial pumpkin protein preparations appropriated for food purposes.

2. Results and Discussion

2.1. Chemical Composition

The chemical composition of native and acetylated pumpkin protein concentrate (PPC) is presented in Table 1. All analysed samples differed in dry matter, fat and ash content. The dry matter content of PPC preparations ranged from 94.90% to 99.01%. This is in line with the results obtained by Rutkowski and Kozłowska [19], who reported a water content for protein preparations not exceeding 10%. In this study, the application of 1.0 and 2.0 mL/g acetic anhydride had a significant impact on the dry matter content of PPC. The protein content in the analysed commercial pumpkin preparation reached a value of 65.11%. Based on the data found in the literature [20], it can be concluded that these proteins were extracted and purified in a neutral or acidic environment. Moreover, according to Ozuna and León-Galván [5], the composition of protein preparations depends on the chemical composition and characteristics of the proteins contained in the raw material. In most cases, acetylation had no effect on the protein content. No statistically significant differences were found between the control and the most experimental samples. Along with an increase in acetic anhydride concentration from 0.0 to 2.0 mL/g, there was no observed tendency to a decrease in the total protein share in the analysed preparations, except for the sample modified with 1.0 mL/g acetic anhydride, where a statistically significant ($p < 0.05$) decrease was noted. Under these conditions, acetylation caused a decrease in total protein content in modified samples, from 63.24 to 53.52 g/100 g. These results are in line with those of other authors who acetylated protein preparations, such as El-Adawy [21], Lawal and Adebowale [22], Lawal et al. [23] and Miedzianka et al. [24]. However, our results disagree with those of Khader et al. [25] and Zedan et al. [26] in which the total protein percentage increased after acetylation. The different research results indicate that the effect of acetylation on protein content is not unequivocal.

Table 1. Characteristics of native and acetylated PPC.

Anhydride-to-Protein Ratio (mL/g)	Dry Matter	Protein Content (g/100g)	Fat	Ash
Native PPC	97.49 ± 0.39 [b]	65.11 ± 0.76 [a]	9.17 ± 0.17 [a]	8.36 ± 0.01 [b]
0.4	94.90 ± 0.01 [c]	63.24 ± 2.68 [a]	8.91 ± 0.15 [b]	8.68 ± 0.02 [a]
1.0	99.01 ± 0.24 [a]	53.52 ± 0.88 [b]	8.88 ± 0.13 [b]	7.99 ± 0.02 [c]
2.0	98.80 ± 0.01 [a]	62.24 ± 2.99 [a]	8.89 ± 0.09 [b]	5.49 ± 0.01 [d]

Values are means ± standard deviation; n = 2; [a,b,c,d]—the same letters within the same row were not significantly different; PPC—pumpkin protein concentrate; 0.4, 1.0, 2.0—pumpkin protein preparations after acetylation conducted with different concentrations of acetic anhydride.

Acetylation reduced the fat content in modified PPC. As shown in Table 1, the fat content ranged from an average of 9.17 g/100 g before chemical modification to an average of 8.89 g/100 g in modified samples. However, the acetylated samples were not significantly differed in fat content. Increasing the amount of anhydride did not affect the level of fat.

The total ash content in native PPC (8.36 g/100 g) was similar to the 9.09 g/100 g found by Zduńczyk et al. [27] in pumpkin oil seed cake. The results are in line with statements that plant-derived protein preparations usually have an ash content of less than 10%, because they contain relatively fewer minerals than preparations of animal origin, and their availability is additionally limited by the presence of chelating compounds. Moreover, the total ash content in protein preparations is mainly influenced by the quality of the raw material, which is largely due to the plant variety, the degree of purification and insulation conditions (pressing technology or the type of press used). The ash content in acetylated samples differed significantly (Table 1). A statistically different decreasing trend was found on the ash content for higher anhydride-to-protein ratios. It ranged from 5.49 g/100 g (in samples modified with 2.0 mL/g) to 8.68 g/100 g (in samples acetylated with 0.4 mL/g). Similar findings were observed by Lawal and Adebowale [22], who acetylated the jack bean protein. This decreasing ash content in acetylated protein preparations can be associated with more frequent removal of an excess modifying reagent (samples modified in the amount of 2.0 mL/g of protein were washed 5 times).

2.2. Amino Acid Composition

One of the reasons proteins exhibit such different functional properties is that their amino acid composition is varied [28]. This also influences the functional properties of a protein depending on how they are arranged in the polypeptide chain. As presented in Table 2, the dominant amino acids found in PPC were aspartic acid (7.04 g/100 g protein), glutamic acid (21.03 g/100 g protein), arginine (16.69 g/100 g protein) and leucine (6.67 g/100 g protein). Similar levels of amino acids in protein preparations obtained from pumpkin oilseed cake were reported by Ozuna and León-Galván [5]. Of the indispensable amino acids, lysine (3.19 g/100 g protein) and sulphur amino acids (2.04 g/100 g protein) were the least noted in PPC. These results provide support for Kotecka-Majchrzak et al.'s [29] theory that the amino acids present at the highest concentrations in oilseed proteins are leucine and valine, whereas sulphur-containing amino acids and hydrophobic tryptophan are present in the lowest amounts. Moreover, the amino acid profile depends on the methods used in both the extraction and coagulation of protein preparations [30].

Acetylation of PPC decreased all amino acids (except lysine and alanine) in preparations. The sum of amino acids decreased from 88.63 to 79.88 g/100 g protein after modification with 2.0 mL/g. A statistically significant decrease was observed in all analysed amino acids, particularly in lysine, tyrosine, methionine with cysteine and phenylalanine with threonine. These results confirm the data presented by other authors [11,24,31].

Table 2. Amino acid concentration (g/100 g protein) in native and acetylated PPC.

Amino Acids	Native PPC	Anhydride-to-Protein Ratio (mL/g)		
		0.4	1.0	2.0
IAA *				
Leucine	6.67 ± 2.43 [a]	6.56 ± 1.01 [b]	6.13 ± 1.00 [c]	6.60 ± 0.72 [ab]
Isoleucine	4.15 ± 0.34 [a]	3.80 ± 1.07 [b]	3.55 ± 0.32 [c]	3.88 ± 0.42 [b]
Methionine	1.62 ± 0.51 [a]	1.27 ± 1.21 [b]	0.97 ± 0.48 [d]	1.16 ± 0.85 [c]
Cysteine	0.42 ± 0.88 [a]	0.32 ± 0.44 [b]	0.23 ± 0.01 [d]	0.27 ± 0.25 [c]
Phenyloalanine	4.51 ± 2.26 [a]	3.40 ± 1.56 [b]	2.95 ± 0.18 [d]	3.05 ± 0.87 [c]
Threonine	1.99 ± 0.07 [a]	1.97 ± 2.19 [b]	0.84 ± 0.34 [d]	1.14 ± 0.26 [c]
Lysine	3.32 ± 0.34 [a]	2.69 ± 1.76 [b]	2.02 ± 0.86 [c]	1.89 ± 0.69 [d]
Tyrosine	1.70 ± 0.26 [a]	1.21 ± 0.45 [b]	0.75 ± 0.51 [d]	1.13 ± 0.30 [c]
Valine	4.10 ± 0.86 [a]	3.15 ± 0.86 [b]	2.72 ± 0.30 [c]	2.56 ± 0.52 [c]
DAA **				
Aspartic acid	7.04 ± 0.72 [a]	6.15 ± 0.72 [d]	6.29 ± 0.79 [b]	6.18 ± 0.95 [c]
Glutamic acid	21.03 ± 4.26 [a]	19.28 ± 2.81 [c]	19.78 ± 0.93 [b]	18.05 ± 1.58 [d]
Serine	3.86 ± 1.07 [a]	3.81 ± 1.27 [b]	3.56 ± 0.23 [d]	3.68 ± 1.92 [c]
Glycine	2.83 ± 2.14 [a]	2.44 ± 0.72 [b]	1.73 ± 0.52 [d]	2.07 ± 1.92 [c]
Alanine	3.64 ± 0.57 [d]	4.07 ± 0.57 [a]	3.89 ± 0.38 [b]	3.84 ± 0.33 [c]
Histidine	1.72 ± 1.31 [a]	1.61 ± 1.31 [b]	1.49 ± 0.21 [c]	1.60 ± 0.19 [b]
Arginine	16.69 ± 0.51 [a]	16.08 ± 0.91 [c]	16.30 ± 2.83 [b]	15.90 ± 0.17 [d]
Proline	3.46 ± 1.90 [b]	3.55 ± 2.24 [a]	3.21 ± 0.97 [d]	3.35 ± 1.05 [c]
Total amino acids	88.63 ± 1.20 [a]	85.98 ± 1.24 [a]	81.07 ± 1.26 [b]	79.88 ± 0.82 [b]

Values are means ± standard deviation; n = 2; [a,b,c,d]—the same letter in verse mean homogenous group; IAA *—indispensable amino acids; DAA **—dispensable amino acids.

2.3. Degree of Acetylation and In Vitro Digestibility

As shown in Table 3, along with an increase in acetic anhydride concentration from 0.0 to 2.0 mL/g, there was a higher degree of N-acetylation in analysed preparations. This phenomenon was due to the number of ε-amino groups involved in the acetylation reaction being reduced severely with the increase in the anhydride level, leading to an increase in the degree of N-acetylation. Moreover, this increase might be correlated with process conditions and the source of protein subjected to chemical modification [32]. Additionally, Achouri and Zhang [33] stated that the extent of amino group acetylation in a polypeptide mixture depends markedly on the amount of acetic anhydride used. However, 78.67% of the ε-amino groups was acetylated with 2.0 mL/g. This means that a complete blockage of the amino acid residues is possible only in the presence of a high dose of acetic anhydride.

Table 3. Degree of N-acetylation and in vitro digestibility of native and acetylated PPC.

Dose of Acetic Anhydride	Degree of N-Acylation	Amount of Protein Released into the Intestinal Fluid in Two-Step Digestion
(mL/g)	(%)	(%)
Native PPC	-	23.27 ± 0.01 [b]
0.4	54.89 ± 0.20 [c]	21.04 ± 0.01 [c]
1.0	62.77 ± 0.33 [b]	25.31 ± 0.01 [a]
2.0	78.67 ± 0.12 [a]	23.09 ± 0.01 [b]

Values are means ± standard deviation; [a,b,c]—the same letters within the same row were not significantly different; PPC—pumpkin protein concentrate; 0.4, 1.0, 2.0—pumpkin protein preparations after acetylation conducted with different concentrations of acetic anhydride.

After digestion of the analysed preparations, protein nitrogen extraction ranged from 21.4% (for samples acetylated with dose 0.4 mL/g) only up to 25.31% (for samples acetylated with dose 1.0 mL/g) (Table 3). The amount of protein nitrogen determined in the supernatant obtained after the two-stage digestion contained nitrogen compounds and not those extracted from the sample proteins. Thus, the determination should reflect the presence of short peptides, possible to the absorption by the intestine enterocytes. The digestion step did not include digestion by the enzymes excreted by enterocytes. The cleavage specificity of porcine pepsin includes peptides with an aromatic amino acid, preferentially at carboxylic groups of the amino acid, especially if the other residue is also an aromatic or a dicarboxylic amino acid and does not destroy bonds containing valine, alanine or glycine linkages [34], while trypsin cleaves the peptide chain on the C-terminal side of lysine and arginine amino acid residues. Similar findings were reported by Shukla [35]. However, the applied method is definitely more precise than the method proposed by Salgó et al. [36] because that determines only peptides, not protein, extracted from the studied material. Thus, the results obtained should not be compared with those obtained for pH changes after placement of the sample in the intestine solutions. We showed that the use of 0.4 mL/g acetic anhydride significantly reduces the digestibility of modified PPC in comparison to native PPC, while acetylation with 1.0 mL/g acetic anhydride significantly improves the digestibility of PPC. Therefore, the effect of acetylation on the digestibility of PPC preparations cannot be clearly determined. However, according to Bergner et al. [37], acetylation of proteins does not reduce either apparent or true N digestibility. Additionally, this modification of food proteins does not influence the nutritive value.

2.4. SDS-PAGE

The resulting native and acetylated PPC were analysed by SDS-PAGE (Figure 1). Electrophoretic analyses revealed that the PPC contained a predominant protein of about 35 kDa·MW. As we can see on the electropherogram, proteins of about 18–22 kDa·MW accompanied this main protein. According to previous research studies, cucurbitin, the major storage protein in pumpkin seed, is a hexamer with a molecular size of about 325 kDa, consisting of two major bands (MW d 22 kDa) corresponding to acidic and basic polypeptides [38–41]. The vast majority of cucurbitins are salt-soluble globulins with a sedimentation index of about 11–12S [42]. They are accompanied by albumin with a sedimentation index of 2S and a mass of about 12.5 kDa, composed of two subunits: a smaller one, about 4.8 kDa and above, and another with a mass of 7.9 kDa [43,44]. Moreover, cucurbitins have proven antimicrobial, antitumour and anti-inflammatory effects [45]. Analysis revealed that in the acetylated pumpkin protein the content of the heaviest protein (35 kDa) decreased in line with increasing concentrations of modifying reagent. The addition of 0.4–2.0 mL acetic anhydride per g of protein resulted in removal of a protein of about 18 kDa·MW and, at the same time, resulted in an increase in the participation of a protein of about 20 kDa·MW in the composition of acetylated pumpkin protein preparations. Additionally, the proteins of about 35 W and 18 kDa·MW were more susceptible than the protein of about 22 kDa·MW to the presence of the modifying reagent. Furthermore, the intensity of the protein patterns decreased in line with increasing concentrations of modifying reagent, which confirms the reduction in protein content in acetylated pumpkin protein preparations. These results are not in line with those of Zhao et al. [16], who found that acetylation of oat protein isolate with different anhydride-to-protein ratios left the molecular weight of the proteins almost unchanged, as compared to native protein isolate.

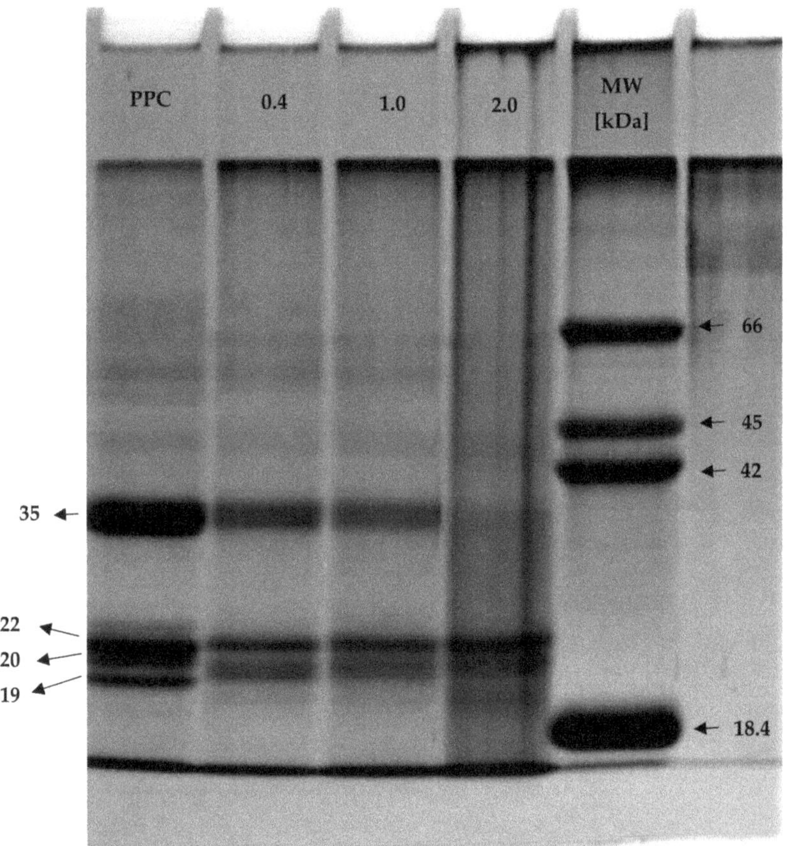

Figure 1. SDS-PAGE of native and acetylated PPC. PPC—pumpkin protein concentrate; 0.4, 1.0, 2.0—pumpkin protein preparations after acetylation conducted with different concentrations of acetic anhydride (mL/g); MW—molecular weight marker prepared at laboratory (18.4 kDa (β-lactoglobulin), 42, 45 kDa (ovalbumin), 66 kDa (bovine serum albumin).

2.5. Functional Properties

2.5.1. Protein Solubility Index

Plant protein preparations used in food production should be characterized not only by a favourable chemical composition, but also by optimal functional properties, of which the most significant for other functional applications of the proteins is solubility (PS). PS depends on the pH value and ionic strength as well as the method of obtaining protein preparations. As shown in Figure 2, the applied chemical modification of the protein caused statistically significant ($p < 0.05$) changes in the PS of analysed preparations. The solubility profile of native PPC indicates that PS reduced as the pH increased from 2 to 4 (from 10.16% to 6.48%), which corresponds to its isoelectric point, after which an increase in pH increased PS progressively (to 62.55%). This solubility profile of pumpkin proteins is U-shaped, common for oil seeds [40,41]. Other authors [46–49] have emphasized that the solubility of the proteins is strongly influenced by pH and ionic strength, with higher values in the alkaline pH regions, and by the type of solvent used to degrease the seeds during preparation of the raw material. Increased PS was observed only for the preparation modified with 2.0 mL/g. This preparation had the highest solubility of proteins at pH = 12 (49.42%) and the lowest at pH = 4 (20.11%). This phenomenon can be explained by higher protein degradation in this sample and by the increased interaction between proteins and water, decreasing protein–protein interactions and, hence, increasing their solubility [48]. These results are in line with the studies presented by Lawal and Adebowale [22] and El-Adawy [21] who analysed jack bean and acylated mung bean protein isolate, respectively. These well-soluble proteins could be used in the production of meat products, improving the stability of the emulsion [19]. Furthermore, acetylation with 0.4 and 1.0 mL/g caused a decrease in PS, as compared to native PPC.

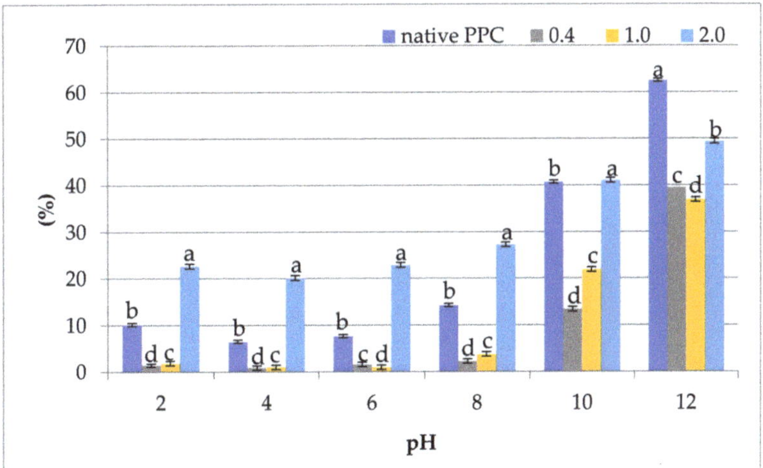

Figure 2. Effect of anhydride-to-protein ratio on protein solubility (PS) at different pH of native and acetylated PPC. Values are means ± standard deviation; [a,b,c,d]—the same letters within the same pH were not significantly different; PPC—pumpkin protein concentrate; 0.4, 1.0, 2.0—pumpkin protein preparations after acetylation conducted with different concentrations of acetic anhydride (mL/g).

2.5.2. Water-Binding Capacity and Oil-Absorption Capacity

The water-binding capacity (WBC) of native and acetylated PPC is shown in Figure 3. Acetylation significantly increased ($p \leq 0.05$) the WBC at all levels of modification compared to native preparation (2.54 g/g). Similar findings were present by El-Adawy and Taha [49] who obtained water absorption of pumpkin flour at the level of 2.51 g/g of preparation. According to Rodriques et al. [50], pumpkin protein preparations show a low WBC, which may be due to the relatively high fat content and low availability of non-polar amino acids (i.e., glycine, proline, valine and tyrosine). Depending on the anhydride-to-protein ratio, the WBC of the modified concentrates was in the range 4.89–5.62 g/g. However, significantly the highest WBC was noted for sample modified with 0.4 mL/g of acetic anhydride. This resulted from sufficient exposure of hydrophilic groups and the higher amounts of polar amino acids present in protein particles. This phenomenon is confirmed in the decreased PS of this sample, since poorly soluble proteins have been found to exhibit good water absorption [6]. Acetylation can cause dissociation and unfolding of the protein; therefore, protein–protein reactions are limited, and protein-water interactions are facilitated [22]. Similar findings presenting an increase in WBC by acetylation have been reported by El-Adawy [21] for mung bean, by Liu and Hung [51] for chickpea proteins, by Dua et al. [52] for rapeseed, by Bora [53] for lentil and by Miedzianka et al. [24] for potato. Due to its high WBC, acetylated PPC could be used in viscous products (dough, processed cheese) [19].

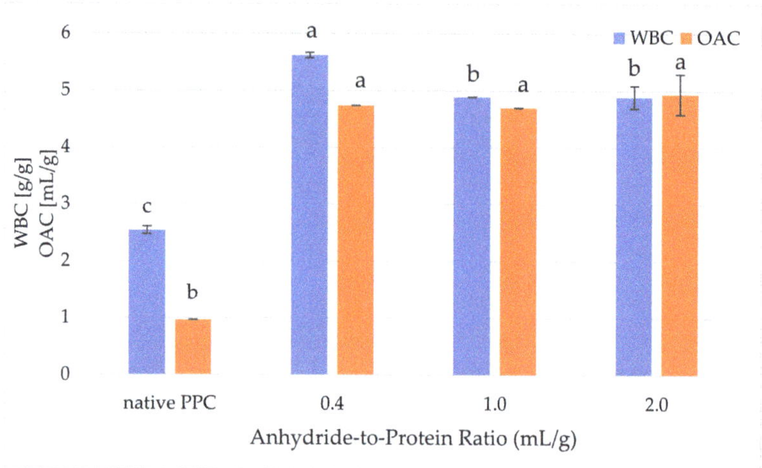

Figure 3. Effect of anhydride-to-protein ratio on water-binding capacity (WBC) and oil-absorption capacity (OAC) of native and acetylated PPC. [a,b,c]—the same letters within the same analysis were not significantly different; PPC—pumpkin protein concentrate; 0.4, 1.0, 2.0—pumpkin protein preparations after acetylation conducted with different concentrations of acetic anhydride (mL/g).

As shown in Figure 3, acetylated pumpkin protein preparations were characterized by a much higher oil-absorption capacity (OAC), in the range 4.7–4.94 mL/g, as compared to native PPC not subjected to the modification process (0.97 mL/g), independent of the anhydride-to-protein ratio. Therefore, it can be assumed that native pumpkin protein is characterized by high density and particle size, which is why it absorbs less oil [48,54]. The results revealed that even at a dose of 0.4 mL/g, the OAC of analysed protein concentrates increased fourfold. This fact may be related to the introduction of lipophilic groups in the pumpkin protein molecules during acetylation [55]. Generally, the OAC of acetylated protein preparations may be attributed to the degree of denaturation due to the method of obtaining them as well as on the type of protein sample used for modification [56]. Similar findings were stated by Lawal and Adebowale [22] and by Miedzianka et al. [24]. The obtained acetylated preparations characterized by high OAC may, therefore, be applied as meat product extenders.

2.5.3. Foaming Properties

The effect of anhydride-to-protein ratio on the foaming capacity (FC) at different pH of native and acetylated PPC is shown in Figure 4. These data reveal that the commercially purchased preparation is characterized by very low FC, regardless of pH, which may be due to the compact, globular structure and high molecular mass of these proteins, which inhibits their ability to reorient effectively. These results are not in line with those of El-Adawy and Taha [49], who found pumpkin flour was characterized by a significantly higher capacity to create foam (18.65%). Additionally, Rezig et al. [46] stated that, in fact, the mechanisms causing the formation and stability of foam in the presence of single pumpkin protein molecules and protein aggregates are not yet fully understood and insufficiently described in the literature. Acetylation with higher doses of acetic anhydride (1.0 and 2.0 mL/g) significantly increased ($p \leq 0.05$) the FC, mainly at pH = 2, pH = 8 and pH = 12. This phenomenon is related to an increase in the net charge of the protein molecules, which weakens hydrophobic interactions and increases protein flexibility [26]. Similar findings were stated by El-Adawy [21] for acetylated mung bean. Foam was not noticeable at pH = 4 and pH = 6 in native and acetylated PPC, which was related to their low solubility at different pH.

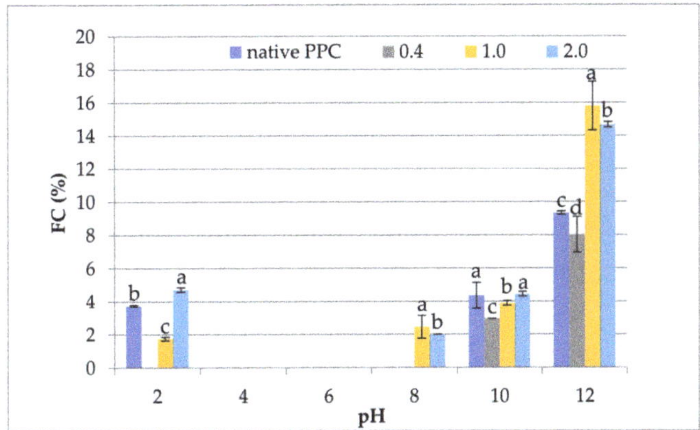

Figure 4. Effect of anhydride-to-protein ratio on foam capacity (FC) at different pH of native and acetylated PPC. [a,b,c]—the same letters within the same pH were not significantly different; PPC—pumpkin protein concentrate; 0.4, 1.0, 2.0—pumpkin protein preparations after acetylation conducted with different concentrations of acetic anhydride (mL/g).

Table 4 presents the effect of anhydride-to-protein ratio on the foam stability (FS) of native and acetylated PPC. Generally, commercial pumpkin protein preparation had weak FS, and the formed foam was stable only at pH = 2, pH = 10 and pH = 12. For foam to be stable, a thick, flexible, continuous and air-permeable protein film is required around each air bubble [46]. Additionally, Wani et al. [57] analysed the FC of albumin, globulin, glutelin and prolamin in two varieties of watermelon belonging to the same botanical family as the pumpkin (Cucurbitaceae). They showed that the FS in both varieties is the highest for the albumin fraction. The differences in stability were attributed to differences in the protein structure. Acetylation with different doses of acetic anhydride had a negative effect on the FS of the analysed samples. After acetylation, the FS was decreased probably due to the negative charges imparted during modification, causing the protein molecule to unfold [6,21,23]. However, stabilization of the formed foam increased only at pH = 12. In general, preparations characterized by low molecular weight, high surface hydrophobicity, good solubility and greater denaturability exhibit improved foaming properties [58].

Table 4. Results of the foam stability (FS, %) at different pH of native and acetylated PPC.

Type of Preparation		Time	Foam Stability (%)					
			pH					
			2	4	6	8	10	12
Native PPC		5 min	1.9	0	0	0	4.5	8.0
		10 min	1.9	0	0	0	4.5	3.7
		30 min	1.4	0	0	0	3.0	1.9
		60 min	1.4	0	0	0	2.0	1.9
Anhydride-to-Protein Ratio (mL/g)	0.4	5 min	0	0	0	0	0	5.8
		10 min	0	0	0	0	0	3.6
		30 min	0	0	0	0	0	2.2
		60 min	0	0	0	0	0	1.7
	1.0	5 min	0	0	0	0	0	4.3
		10 min	0	0	0	0	0	3.5
		30 min	0	0	0	0	0	3.5
		60 min	0	0	0	0	0	3.5
	2.0	5 min	0	0	0	0	0	8.7
		10 min	0	0	0	0	0	8.6
		30 min	0	0	0	0	0	0
		60 min	0	0	0	0	0	0

PPC—pumpkin protein concentrate; 0.4, 1.0, 2.0—pumpkin protein preparations after acetylation conducted with different concentrations of acetic anhydride.

2.5.4. Emulsifying Properties

The effect of anhydride-to-protein ratio on the emulsifying activity (EA) and stability (ES) at different pH levels of native and acetylated PPC is shown in Figures 5 and 6, respectively. Commercially purchased pumpkin protein preparation was characterised by poor EA and ES, mainly at a pH ranging between 4 and 10 (on average, it valued less than 5%). This confirmed the fact that highly insoluble proteins are not good emulsifiers [59]. The better EA for PPC was found at extreme pH, i.e., 2 and 12, where it was noted as 15.62 and 30.61%, respectively. The formed emulsion with PPC retained its properties after heating to 80 °C. Acetylation improved the EA, especially when using the acetic anhydride dose of 0.4 and 2.0 mL/g, in the pH ranging between 6 and 10. One possible explanation for this is the increase in solubility at dose of 2.0 mL acetic anhydride/g, where these changes were found to facilitate the diffusion of protein at oil–water interface [60]. Chemical modification with 1.0 mL/g caused EA similar to PPC, independently from pH range. Similar observations for acetylation the plant-derived proteins were presented by Bora [53] and Miedzianka et al. [24]. In contrast, El-Adawy [21] observed lower EA of preparations obtained as an effect of acetylation of mung bean protein with doses of acetic anhydride higher than 0.6 g/g protein. Furthermore, El-Adawy [21] proved that the best effect for mung bean proteins is obtained using 0.4–0.6 g/g of acetic anhydride. This proves that apart from the conditions of acetylation, the origin (properties) of proteins influences the change in their functional properties.

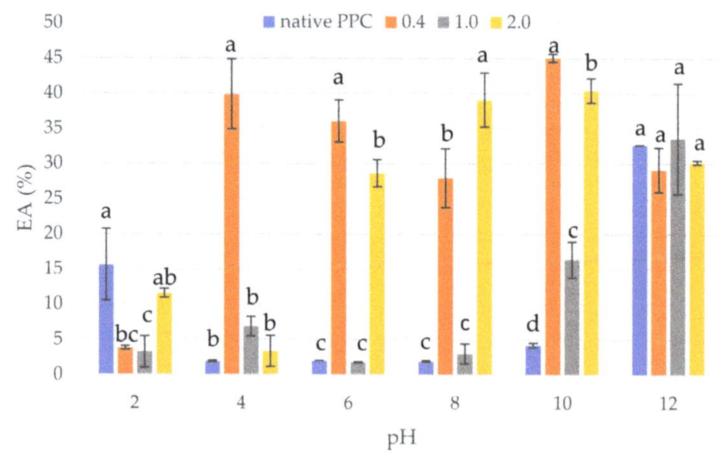

Figure 5. Effect of anhydride-to-protein ratio on emulsion activity (EA) at a different pH of native and acetylated PPC. [a,b,c,d]—the same letters within the same pH were not significantly different; PPC—pumpkin protein concentrate; 0.4, 1.0, 2.0—pumpkin protein preparations after acetylation conducted with different concentrations of acetic anhydride (mL/g).

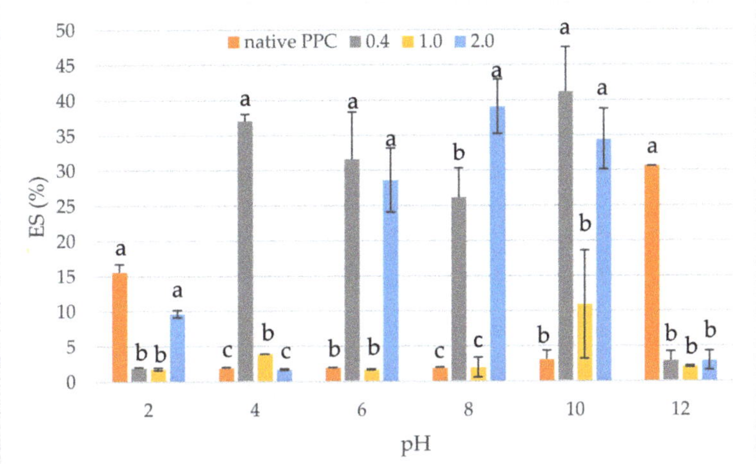

Figure 6. Effect of anhydride-to-protein ratio on emulsion activity (ES) at different pH of native and acetylated PPC. [a,b,c]—the same letters within the same pH were not significantly different; PPC—pumpkin protein concentrate; 0.4, 1.0, 2.0—pumpkin protein preparations after acetylation conducted with different concentrations of acetic anhydride (mL/g).

3. Materials and Methods

3.1. Materials and Chemicals

The material analysed was commercial plant-derived protein concentrate from pumpkin seeds (PPC), purchased from the Diet Food company (Opatówek, Poland). Acetic anhydride, 2,4,6-trinitrobenzenesulfonic acid (TNBS), pepsin (60,000 U), bile salts and pancreatin for porcine pancreas were obtained from Sigma-Aldrich (St. Louis, MO, USA). All chemicals used in the experiment were of analytical grade.

3.2. Acetylation of PPC

Acetylated PPC samples were prepared using the method of Miedzianka et al. [24] with slight modifications. The pumpkin protein preparation was acetylated with acetic anhydride by adding different concentrations of modifying reagent (0.4, 1.0 and 2.0 mL of anhydride per 1 g of protein contained in the preparation) to the 1% aqueous suspension. The reaction took place over 30–90 min, and the pH was established at pH 7.5–8 by dropwise addition of 1 M NaOH. The solution was constantly monitored for changes in pH and maintained at the assumed pH with constant stirring. After this time, the precipitate was separated from the supernatant by centrifugation using a Biofuge 28 RS centrifuge (5260 rpm/min for 15 min, Heraeus Sepatech, Osterode, Germany). Excess modifying reagent was removed from the resulting precipitate, i.e., the precipitate was mixed with water until the liquid's conductance was close to that of distilled water (3–5 times). Then, it was bathed with distilled water until the conductance of liquid was similar to that of distilled water. Then, the protein was freeze-dried at a pressure of 63 Pa, with the heating shelves at 50 °C, for 24 h using a Christ Alpha 1-4 LSCplus (Osterode am Hatz, Germany), sieved through a sieve with a pore size of 420 µm and stored in a sealed plastic container at about −20 °C until further analysis. Untreated native PPC was used as the control. Acetylation was performed in two technological repetitions.

3.3. Basic Chemical Composition

In order to determine the moisture content, approximately 2 g of a sample was placed into a pre-weighed vessel. Samples were dried at 105 °C until a constant weight was reached [61]. The total protein content was evaluated according to the Kjeldahl method of the Association of Analytical Chemists [62]. Approximately 0.5 g of material was hydrolysed with 25 mL concentrated sulfuric acid (H_2SO_4) containing one catalyst tablet in a heat block (Büchi Digestion Unit K-424, Labortechnik AG, Flawil, Switzerland) at 370 °C for 2 h. After cooling, H_2O was added to the hydrolysates before neutralization, using a Büchi Distillation Unit K-355 (Athens, Greece) and titration. The protein content was calculated by multiplying the percentage of nitrogen content by a factor of 6.25 [63]. Fat content was determined according to the standard method of the association of Official Analytical Chemists International [64]. A sample of 2 g of material was hydrolysed using 4 N HCl. Fat extraction and solvent (diethyl ether) removal were performed in an automated Soxhlet apparatus B-811 (Büchi Labortechnik AG, Flawil, Switzerland); the extraction time was 180 min. Samples for determining the ash content were heated gradually to 550 °C, and the residues were weighed [62].

3.4. Amino Acid Composition

The amino acid composition of native and acetylated PPC was determined by ion-exchange chromatography after 23 h hydrolysis with 6 N HCl at 110 °C, according to the method described by Miedzianka et al. [24]. The hydrolysed amino acids were determined using an AAA-400 analyser (INGOS, Prague, Czech Republic). A photometric detector was used, working at two wavelengths, 440 and 570 nm. A column of 350 × 3.7 mm, packed with ion exchanger Ostion LG ANB (INGOS, Prague, Czech Republic), was utilized. Column temperature was kept at 60–74 °C and the detector at 121 °C. The prepared samples were analysed using the ninhydrine method. No analysis of tryptophan was carried out.

3.5. Measurement of Degree of N-Acylation

The measurement of degree of N-acylation was prepared, as described by Habeeb [65]. The procedure involved the addition of 1 cm^3 of 4% $NaHCO_3$ solution and 1 cm^3 of 0.1% TNBS solution to protein suspensions. The samples were heated in a water bath at 60 °C for 2 h. Then, they were cooled down to room temperature. Next, 1 cm^3 of 10% SDS and 0.5 cm^3 of 0.1 N HCl were added to protein solutions. The absorbance of solutions was read at 335 nm in a Rayleigh UV-2601 PC spectrophotometer (Beijing, China) against a reagent blank. The absorbance of the control protein concentrate was set equal to 100% free amino groups, and the extent of acetylation of the modified samples was calculated based on the decrease in absorbance, because fewer amino groups were able to react with the TNBS reagent.

3.6. Digestibility of Protein Preparations

The digestibility of the native and acetylated PPC were determined with the method simulating multienzymatic two-stage digestion [66,67]. The oral digestion and large intestine steps were omitted as irrelevant to protein digestion. The digestion in the stomach was simulated by introducing the sample (~0.5 g) into the medium imitating the stomach acidic environment (by decreasing water pH down to 2.0 by 1 N HCl) and containing pepsin (60,000 U). Gastric digestion was carried out at 37 °C for 2 h. Then, the pH was increased up to 7.4 by 0.1 M $NHCO_3$, and mixture was enriched in bile salts, acting as a surfactant (0.03 g) and porcine pancreatin (0.005 g), which contains proteases (trypsin, protease A, ribonuclease), amylase and lipase. The second digestion step was performed again at 37 °C for 2 h. The digested sample was centrifuged. The remaining proteins present in the supernatant—extracted and not digested—were precipitated with trichloroacetic acid. In this prepared supernatant, with the use of the Kjeldahl method, protein nitrogen was determined [68] and was related in percentage to the amount of protein nitrogen introduced with the sample into the digestion analysis.

3.7. Sodium Dodecyl Sulfate–Polyacrylamide Gel Electrophoresis (SDS-PAGE)

SDS-PAGE analysis was performed according to the method of Laemmli [69]. The protein samples were diluted (10 mg/mL) with the buffer containing SDS and β-mercaptoethanol as reducing reagents and denatured. Then, the samples were loaded (10 µL) onto gel slabs (12%). At the end of analysis, gel slabs were stained with Coomassie Brilliant G-250 dye. The molecular masses of analysed proteins were estimated from the response curve of the mobility of molecular weight standards versus logMW.

3.8. Determination of Functional Properties

3.8.1. Effect of pH on Protein Solubility

The pH-solubility profile (PS) of native and acetylated PPC was determined according to the method of Achouri et al. [70] with slight modifications. Briefly, 750 mg of preparation was weighed in the tube, 15 mL of distilled water was added, and then, the pH was adjusted (2, 4, 6, 8, 10 or 12) using either 0.5 M NaOH or 0.5 M HCl. The protein solutions were shaken at room temperature for 30 min and successively centrifuged at 4.500 g for 15 min (Rotofix 32A by Hettich, Tuttlingen, Germany). The protein content of the supernatants was determined by Kjeldahl method. Protein solubility was calculated as (Equation (1)):

$$PS = \frac{PCS}{TPC} \times 100 \, (\%) \tag{1}$$

where PCS is the protein content in the supernatant after centrifugation, and TPC is the total protein content present in the protein sample.

3.8.2. Water-Binding Capacity and Oil-Absorption Capacity

Water-binding capacity (WBC) of native and acetylated PPC was determined according to the method described by Jeżowski et al. [71]. For this, 1 g of powdered sample was weighed in a test tube containing 20 mL of distilled water. It was allowed to mix in a laboratory mixer for 60 s and was allowed to stand for 15 min at ambient temperature. This slurry was centrifuged at 4500× g for 15 min (Rotofix 32A by Hettich, Tuttlingen, Germany). The separated solid was oven-dried. WBC was expressed as the amount of water (g) absorbed by 1 g of the preparation.

Oil-absorption capacity (OAC) was determined using the method of Wu et al. [72]. Briefly, 1 g of protein preparation was weighed in the test tube and mixed with 15 mL of rapeseed oil using a Vortex mixer. Samples were allowed to stand for 30 min. The resulting protein–oil mixture was separated using a centrifuge (4000 g; Rotofix 32A by Hettich, Tuttlingen, Germany) for 10 min. Immediately after centrifugation, the supernatant was carefully poured into a 15 mL graduated cylinder, and the volumes were recorded. OAC was expressed as the amount of oil (mL) absorbed by 1 g of the preparation.

3.8.3. Effect of pH on Foaming Capacity and Stability

Foam capacity (FC) and stability (FS) were measured according to the method of Jeżowski et al. [71]. Briefly, one gram of the preparation was weighed into the tube and 200 mL of distilled water was added to it. The resulting mixture was adjusted to the appropriate pH (2, 4, 6, 8, 10 or 12) using 0.5 M NaOH or 0.5 HCl. The sample was then homogenized for 2 min at 16,000 rpm (T25 basic ULTRA-TURRAX®; IKA Werke, Staufen, Germany). The beaten sample was immediately transferred to a measuring cylinder, where the total foam volume was determined after 0, 5, 10, 30 and 60 min. FC and FS were calculated according to the following equations (Equations (2) and (3)):

$$FC = \frac{VA}{VB} \times 100 \, (\%) \tag{2}$$

where VA denotes the volume after whipping (mL), and VB is the volume before whipping (mL).

$$FS = \frac{VB}{VT} \times 100 \, (\%) \tag{3}$$

where VB denotes the volume before whipping (cm^3), and VT is the volume after a certain time (mL).

3.8.4. Effect of pH on Emulsifying Properties

Emulsification activity (EA) and stability (ES) were estimated by the method of Miedzianka et al. [24] with slight modifications (Equations (4) and (5)). The protein suspensions with added oil were mixed using an Ultra-turrax T-18 model homogenizer (IKA-Werke GmbH and Co. KG, Staufen, Germany) to produce a crude emulsion for 1 min at 20,000 rpm. Then, they were centrifuged at $3000 \times g$ for 10 min. Emulsion stability was determined by centrifugation after heating at the temperature of 80 °C for 30 min.

$$\text{EA (\%)} = \frac{\text{height of emulsified layer in the tube}}{\text{height of the total contents in the tube}} \times 100 \qquad (4)$$

$$\text{ES (\%)} = \frac{\text{height of emulsified layer after heating}}{\text{height of the total contents before heating}} \times 100 \qquad (5)$$

3.9. Statistical Analysis

Statistical analysis of all data was performed using one-way analysis of variance (ANOVA). The analysis of chemical composition and functional properties were performed in duplicate and in triplicate, respectively. Duncan's range test was used to determine the differences among samples with a probability level of 0.05. Statistical analysis and standard deviations were determined using Statistica v. 13.3 software (StatSoft Inc., Tulsa, OK, USA).

4. Conclusions

Protein acylation with acetic acid anhydride used in this study is one of the methods of shaping the functional properties of proteins. However, despite the large number of publications dealing with the problem of improving the functional properties of protein preparations, it is still difficult to find universal regularities in the explanation of the mechanisms of the relationship between the structure of modified proteins and their function. Acetylation generally improves the functional properties of the obtained commercial pumpkin protein preparations, mainly protein solubility, as well as water-binding and oil-absorption capacity, which extends the possibilities of their use in the food industry.

Author Contributions: Conceptualization, J.M. and A.N.; methodology, J.M., A.Z. and M.Z.-D.; software, J.M.; validation, J.M.; formal analysis, J.M. and A.Z.; investigation, J.M. and A.N.; resources, J.M.; data curation, J.M., W.D. and A.N.; writing—original draft preparation, J.M.; writing—review and editing, A.Z.; visualization, J.M.; supervision, A.Z.; project administration, J.M.; funding acquisition, J.M. All authors have read and agreed to the published version of the manuscript.

Funding: This research was funded by National Science Centre of Poland, grant number 2019/03/X/N29/01823. The publication is financed under the Leading Research Groups support project from the subsidy increased for the period 2020-2025 in the amount of 2% of the subsidy referred to Art. 387 (3) of the Law of 20 July 2018 on Higher Education and Science, obtained in 2019.

Data Availability Statement: Data is contained within the article.

Conflicts of Interest: The authors declare no conflict of interest.

Sample Availability: Samples of the compounds are not available from the authors.

References

1. Antoniewska, A.; Adamska, A.; Rutkowska, J.; Zielińska, M. Olej z nasion dyni jako źródło cennych składników w diecie człowieka. *Probl. Hig. Epidemiol.* **2017**, *98*, 17–22. (In Polish)
2. Murkovic, M. Pumpkin seed oil. In *Gourmet and Health—Promoting Specialty Oils*, 1st ed.; Moreau, R.A., Kamal-Eldin, A., Eds.; Academic Press: Cambridge, MA, USA; AOCS Press: Urbana, IL, USA, 2009; pp. 345–358.
3. Peričin, D.; Radulović-Popović, L.; Vaštag, Ž.; Mađarev-Popović, S.; Trivić, S. Enymatic hydrolysis of protein isolate from hull-less pumpkin oil cake: Application of response surface methodology. *Food Chem.* **2009**, *115*, 753–757. [CrossRef]
4. Vaštag, Ž.; Popović, L.; Popović, S.; Krimer, V.; Peričin, D. Production of enzymatic hydrolysates with antioxidant and angiotensin-I converting enzyme inhibitory activity from pumpkin oil cake protein isolate. *Food Chem.* **2011**, *124*, 1316–1321. [CrossRef]

5. Ozuna, C.; León-Galván, M.F. Cucurbitaceae seed protein hydrolysates as a potential source of bioactive peptides with functional properties. *Biomed. Res. Int.* **2017**, *2107*, 2121878. [CrossRef] [PubMed]
6. Chan, W.M.; Ma, C.Y. Acid modification of proteins from soymilk residue (okara). *Food Res. Int.* **1999**, *32*, 119–127. [CrossRef]
7. Majzoobi, M.; Abedi, E.; Farahnaky, A.; Aminlari, M. Functional properties of acetylated glutenin and gliadin at varying pH values. *Food Chem.* **2012**, *133*, 1402–1407. [CrossRef]
8. Franzen, K.J.; Kinsella, J.E. Functional properties of succinylated and acetylated leaf protein. *J. Agric. Food Chem.* **1976**, *24*, 914–919. [CrossRef]
9. Beuchat, L.R. Functional and electrophoretic characteristics of succinylated peanut flour protein. *J. Agric. Food Chem.* **1977**, *25*, 258–261. [CrossRef]
10. Kabirrulah, M.; Wills, R.B.H. Functional properties of acetylated and succinylated sunflower protein isolate. *J. Food Technol.* **1982**, *17*, 235–249. [CrossRef]
11. Gruener, L.; Ismond, A.H. Effects of acetylation and siccinylation on the pysicochemical properties of the canola 12S globulin. Part I. *Food Chem.* **1997**, *60*, 357–363. [CrossRef]
12. Melnychyn, P.; Stapley, R.B. Acylated Soybean Protein for Coffee Whiteners. U.S. Patent 3,764,711, 9 October 1973.
13. Mosher, A. Flavouring Agent and Process for Preparaing Same. U.S. Patent 3,840,674, 8 October 1974.
14. Creamer, L.K.; Roeper, J.; Lohrey, E.H. Preparation and evaluation of some acid soluble casein derivatives. *N. Z. J. Dairy Sci. Technol.* **1971**, *6*, 107–114.
15. Liu, L.L.; Wang, H.; Ren, G.Y.; Duan, X.; Li, D.; Yin, G.J. Effects of freeze-drying and spray drying processes on functional properties of phosphorylation of egg white protein. *Int. J. Agric. Biol. Eng.* **2015**, *8*, 116–123.
16. Zhao, C.-B.; Zhang, H.; Xu, X.-Y.; Cao, Y.; Zheng, M.-Z.; Liu, J.-S.; Wu, F. Effect of acetylation and succinylation on physicochemical properties and structural characteristics of oat protein isolate. *Process. Biochem.* **2017**, *57*, 117–123. [CrossRef]
17. Fan, M.; Huang, Q.; Zhong, S.; Li, X.; Xiong, S.; Xie, J.; Yin, T.; Zhang, B.; Zhao, S. Gel properties of myofibrillar protein as affected by gelatinization and retrogradation behaviors of modified starches with different crosslinking and acetylation degress. *Food Hydrocoll.* **2019**, *96*, 604–616. [CrossRef]
18. Acetic Anhydride. CAS No.: 108-24-7. Available online: http://www.eeaa.gov.eg/cmuic/cmuic_pdfs/generalpub/Acetic%20 Anhydride.PDF (accessed on 23 February 2021).
19. Rutkowski, A.; Kozłowska, H. *Preparaty Żywnościowe z Białka Roślinnego*, 1st ed.; Wydawnictwo Naukowo-Techniczne: Warsaw, Poland, 1981; p. 370. (In Polish)
20. Moure, A.; Sineiro, J.; Domínquez, H.; Parajó, J.C. Functionality of oilseed protein products: A review. *Food Res. Int.* **2006**, *39*, 945–963. [CrossRef]
21. El-Adawy, T.A. Functional properties and nutritional quality of acetylated and succinylated mung bean protein isolate. *Food Chem.* **2000**, *70*, 83–91. [CrossRef]
22. Lawal, O.S.; Adebowale, K.O. The acetylated protein derivatives of *Canavalia ensiformis* (jack bean): A study of functional characteristics. *Food Sci. Technol. LWT* **2006**, *39*, 918–929. [CrossRef]
23. Lawal, O.S.; Adebowale, K.O.; Adebowale, Y.A. Functional properties of native and chemically modified protein concentrates from bambarra groundnut. *Food Res. Int.* **2007**, *40*, 1003–1011. [CrossRef]
24. Miedzianka, A.; Pęksa, A.; Aniołowska, M. Properties of acetylated potato protein preparations. *Food Chem.* **2012**, *133*, 1283–1291. [CrossRef]
25. Khader, A.E.; Salem, O.M.; Zedan, M.A.; Mahmoud, S.F. Impact of substituting non-fat dry milk with acetylated whey protein concentrates the quality of chocolate ice milk. *Egypt. J. Dairy Sci.* **2001**, *29*, 299–312.
26. Zedan, M.A.; Zedan, A.N.; Mahmoud, S.F. Effect of fortification of cow milk with acetylated whey protein concentrates on the quality of set yoghurt. *Egypt. J. Dairy Sci.* **2001**, *29*, 285–297.
27. Zduńczyk, Z.; Minakowski, D.; Frejnagel, S.; Flis, M. Skład chemiczny i wartość pokarmowa makuchu z dyni. *Rośliny Oleiste* **1998**, *19*, 205–209. (In Polish)
28. Nakai, S. Structure-functiona relationships of food proteins: With an emphasis on the importance of protein hydrophobicity. *J. Agric. Food Chem.* **1983**, *31*, 676–683. [CrossRef]
29. Kotecka-Majchrzak, K.; Sumara, A.; Fornal, E.; Montowska, M. Proteomic analysis of oilseed cake: A comparative study of species-specific proteins and peptides extracted from ten seed species. *J. Sci. Food Agric.* **2021**, *101*, 297–306. [CrossRef] [PubMed]
30. Fernández-Quintela, A.; Macarulla, M.T.; Del Barrio, A.S.; Martínez, J.A. Composion and functional properties of protein isolates obtained from commercial legumes grown in northern Spain. *Plant Foods Hum. Nutr.* **1997**, *51*, 331–342. [CrossRef] [PubMed]
31. Ma, C.Y.; Wood, D.F. Functional properties of oat proteins modified by acylation, trypsyn hydrolysis or linoleate treatment. *J. Am. Oil Chem. Soc.* **1987**, *64*, 1726–1731. [CrossRef]
32. Lawal, O.S.; Dawodu, M.O. Maleic anhydride derivatives of a protein isolate: Preparation and functional evaluation. *Eur. Food Res. Technol.* **2007**, *226*, 187–198. [CrossRef]
33. Achouri, A.; Zhang, W. Effect of succinylation on the physicochemical properties of soy protein hydrolysate. *Food Res. Int.* **2001**, *34*, 507–514. [CrossRef]
34. Polzonetti, V.; Natalini, P.; Vincenzetti, S.; Vita, A.; Pucciarelli, S. Olives and Olive Oil in Health and Disease Prevention. In *Modulatory Effect of Oleuropeinon Digestive Enzymes*, 1st ed.; Preedy, V.R., Watson, R.R., Eds.; Academic Press: San Diego, CA, USA, 2010; pp. 1327–1333.

35. Shukla, T.P. Chemical modification of food protein. In *Food Protein Deterioration: Mechanism and Functionality*, 1st ed.; Chery, J.P., Ed.; American Chemical Society ACS: Washington, DC, USA, 1982; p. 275.
36. Salgó, A.; Ganzler, K.; Jécsai, J. Simple Enzymic Methods for Prediction of Plant Protein Digestibility. In *Amino Acid Composition and Biological Value of Cereal Proteins*, 1st ed.; Lásztity, R., Hidvégi, M., Eds.; Springer: Dordrecht, The Netherlands, 1985; pp. 311–321.
37. Bergner, H.; Seidler, W.; Simon, O.; Schmandke, H. Digestibility and dietary quality of nonacetylated and acetylated *Vicia faba* proteins in maintenance: Studies on ^{15}N-labelled adult rats. *Ann. Nutr. Metab.* **1984**, *28*, 156–163. [CrossRef]
38. Osborne, T.B. *The Vegetable Proteins*, 2nd ed.; Longmans, Green and Co.: London, UK, 1924.
39. Colman, P.; Suzuki, E.; Van Donkelaar, A. The structure of cucurbitin: Subunit symmetry and organization in situ. *Eur. J. Biochem.* **1980**, *103*, 585–588. [CrossRef] [PubMed]
40. Popović, L.M.; Peričin, M.; Vaštag, Ž.G.; Popović, S.Z. Optimalization of transglutaminase cross-linking of pumpkin oil cake globulin; Improvement of the solubility and gelation properties. *Food Bioprocess Technol.* **2013**, *6*, 1105–1111. [CrossRef]
41. Bučko, S.D.; Katona, J.M.; Popovič, L.M.; Vaštag, Ž.; Petrovič, L.B. Functional properties of pumpkin (*Cucurbita pepo*) seed protein isolate and hydrolysate. *J. Serb. Chem. Soc.* **2016**, *81*, 35–46. [CrossRef]
42. Marcone, F.M.; Kakuda, Y.; Yada, R.Y. Salt-soluble seed globulins of various dicotyledonous and monocotyledonous plants. I. Isolation—Purification and characterization. *Food Chem.* **1997**, *62*, 27–47. [CrossRef]
43. Vassiliou, A.G.; Neumann, G.M.; Condron, R.; Poly, G.M. Purification and mass spectrometry-assisted sequencing of basic antifungal proteins from seeds of pumpkin (*Cucurbita maxima*). *Plant Sci.* **1998**, *134*, 141–162. [CrossRef]
44. Fang, E.F.; Wong, J.H.; Lin, P.; Ng, T.B. Biochemical characterization of the RNA-hydrolytic activity of a pumpkin 2S albumin. *FEBS Lett.* **2010**, *584*, 4089–4096. [CrossRef] [PubMed]
45. Fruhwirth, G.O.; Hermetter, A. Seeds and oil of the Styrian oil pumpkin: Components and biological activities. *Eur. J. Lipid Sci. Technol.* **2007**, *109*, 1128–1140. [CrossRef]
46. Rezig, L.; Rianblanc, A.; Chonaibi, M.; Guegen, J.; Handi, S. Functional properties of protein fractions obtained from pumpkin (*Cucurbita Maxima*) seed. *Int. J. Food Prop.* **2015**, *19*, 172–186. [CrossRef]
47. Pham, T.T.; Tran, T.T.T.; Ton, N.M.N.; Le, V.V.M. Effects of pH and salt concentration on functional properties of pumpkin seed protein fractions. *J. Food Process. Preserv.* **2016**, *41*, 13–73. [CrossRef]
48. Zayas, J.F. *Functionality of Proteins in Food*, 1st ed.; Springer: Berlin, Germany, 1997; p. 390.
49. El-Adawy, T.E.; Taha, K.M. Characteristics and composition of different seed oils and flours. *Food Chem.* **2001**, *74*, 47–54. [CrossRef]
50. Rodrigues, I.M.; Coelho, J.F.J.; Carvalho, M.G. Isolation and valorisation of vegetable proteins from oilseed plants: Methods, limitations and potential. *J. Food Eng.* **2012**, *109*, 337–346. [CrossRef]
51. Liu, L.H.; Hung, T.V. Functional properties of acetylated chickpea proteins. *J. Food Sci.* **1998**, *63*, 331–337. [CrossRef]
52. Dua, S.; Mahajan, A.; Mahajan, A. Improvement of functional properties of rapeseed (*Brassica campestris* Var. Toria) preparations by chemical modification. *J. Agric. Food Chem.* **1996**, *44*, 706–710. [CrossRef]
53. Bora, P.S. Short communication. Effect of acetyltion on the functional properties of lentil (*Lens culinaris*) globulin. *J. Sci. Food Agric.* **2003**, *83*, 139–141. [CrossRef]
54. Kinsella, J.E. Functional properties of soy proteins. *J. Am. Oil Chem. Soc.* **1979**, *56*, 242–258. [CrossRef]
55. Sathe, S.K.; Desphande, S.S.; Salunkhe, D.K. Functional properties of winged bean (*Psophocarpus tetragonolobus* L.) proteins. *J. Food Sci.* **1982**, *47*, 503–509. [CrossRef]
56. Krause, J.-P. Comparison of the effect of acylation and phosphorylation on surface pressure, surface potaential and foaming properties of protein isolates from rapeseed (*Brasica napus*). *Ind. Crops Prod.* **2002**, *15*, 221–228. [CrossRef]
57. Wani, A.A.; Sogi, D.S.; Singh, P.; Wani, I.A. Characterisation and functional properties of watermelon (*Citrullus lanatus*) seeds proteins. *J. Sci. Food Agric.* **2011**, *91*, 113–121. [CrossRef]
58. Barac, M.; Cabrilo, S.; Pesic, M.; Stanojevic, S.; Ristic, N. Profile and functional properties of seed proteins from six pea (*Pisum sativum*) genotypes. *Int. J. Mol. Sci.* **2010**, *11*, 4973–4990. [CrossRef]
59. Atuonwu, A.C.; Akobundu, E.N.T. Nutritional and sensory quality of cookies supplemented with deffated pumpkin (*Cucurbita pepo*) seed flour. *Pak. J. Nutr.* **2010**, *9*, 672–677. [CrossRef]
60. Shilpashree, B.G.; Arora, S.; Chawla, P.; Tomar, S.K. Effect of succinylation on physicochemical and functional properties of milk protein concentrate. *Food Res. Int.* **2015**, *72*, 223–230. [CrossRef]
61. AOAC. *Official Methods of Analysis*, 16th ed.; Association of Official Analytical Chemists: Washington, DC, USA, 2005.
62. AOAC. *Official Methods of Analysis of AOAC International*, 16th ed.; AOAC: Gaithersburg, MD, USA, 1996.
63. Mariotti, F.; Tomé, D.; Mirand, P. Converting nitrogen into protein—Beyond 6.25 and Jones' factors. *Crit. Rev. Food Sci. Nutr.* **2008**, *48*, 177–184. [CrossRef]
64. AOAC. *Official Methods of Analysis of AOAC*, 16th ed.; AOAC: Washington, DC, USA, 1995.
65. Habeeb, A.F.S.A. Determination of free amino groups in proteins by trinitrobenzenesulfonic acid. *Anal. Biochem.* **1966**, *14*, 328–336. [CrossRef]
66. Kowalczewski, P.Ł.; Olejnik, A.; Białas, W.; Rybicka, I.; Zielińska-Dawidziak, M.; Siger, A.; Kubiak, P.; Lewandowicz, G. The nutritional value and biological activity of concentrated protein fraction of potato juice. *Nutrients* **2019**, *11*, 1523. [CrossRef]
67. Zielińska-Dawidziak, M.; Tomczak, A.; Burzyńska, M.; Rokosik, E.; Dwiecki, K.; Piasecka-Kwiatkowska, D. Comparison of *Lupinus angustifolius* protein digestibility in dependence on protein, amino acids, trypsin inhibitors and polyphenolic compounds content. *Int. J. Food Sci. Technol.* **2020**, *55*, 2029–2040. [CrossRef]

68. *ISO 1871:2009: Food and Feed Products—General Guidelines for the Determination of Nitrogen by the Kjeldahl Method*; ISO: Geneva, Switzerland, 2009.
69. Laemmli, U.K. Cleavage of structural proteins during the assembly of the head of bacteriophage T4. *Nature* **1970**, *227*, 680–685. [CrossRef]
70. Achouri, A.; Nail, V.; Boye, J.I. Sesame protein isolate: Fractionation, secondary structure and functional properties. *Food Res. Int.* **2012**, *46*, 360–369. [CrossRef]
71. Jeżowski, P.; Polcyn, K.; Tomkowiak, A.; Rybicka, I.; Radzikowska, D. Technological and antioxidant properties of proteins obtained from waste potato juice. *Open Life Sci.* **2020**, *15*, 379–388. [CrossRef]
72. Wu, H.; Wang, Q.; Ma, T.; Ren, J. Comparative studies on the functional properties of various protein concentrate preparations of peanut protein. *Food Res. Int.* **2009**, *42*, 343–348. [CrossRef]

Article

Membrane Filtration-Assisted Enzymatic Hydrolysis Affects the Biological Activity of Potato Juice

Przemysław Łukasz Kowalczewski [1],*, Anna Olejnik [2], Iga Rybicka [3], Magdalena Zielińska-Dawidziak [4], Wojciech Białas [2] and Grażyna Lewandowicz [2]

1   Department of Food Technology of Plant Origin, Faculty of Food Science and Nutrition, Poznań University of Life Sciences, 31 Wojska Polskiego St., 60-624 Poznań, Poland
2   Department of Biotechnology and Food Microbiology, Poznań University of Life Sciences, 48 Wojska Polskiego St., 60-627 Poznań, Poland; anna.olejnik@up.poznan.pl (A.O.); wojciech.bialas@up.poznan.pl (W.B.); grazyna.lewandowicz@up.poznan.pl (G.L.)
3   Department of Technology and Instrumental Analysis, Poznań University of Economics and Business, Al. Niepodległości 10, 61-875 Poznań, Poland; iga.rybicka@ue.poznan.pl
4   Department of Biochemistry and Food Analysis, Faculty of Food Science and Nutrition, 48 Mazowiecka St., Poznań University of Life Sciences, 60-623 Poznań, Poland; magdalena.zielinska-dawidziak@up.poznan.pl
*   Correspondence: przemyslaw.kowalczewski@up.poznan.pl

Abstract: The results of recently published studies indicate that potato juice is characterized by interesting biological activity that can be particularly useful in the case of gastrointestinal symptoms. Moreover, the studies also described the high nutritional value of its proteins. This article is a report on the impact of the enzymatic hydrolysis of proteins combined with membrane filtration. The obtained potato juice protein hydrolysate (PJPH) and its concentrate (cPJPH) were characterized in terms of their nutritional value and biological activity. The amino acid profile and scoring, the content of mineral compounds, and the antioxidant and in vitro cytotoxic activity were assessed. The study proved that the antioxidant activity of PJPH is higher than that of fresh potato juice, and the cytotoxicity against human gastric carcinoma cell line (Hs 746T), human colon cancer cell line (Caco-2), human colorectal adenocarcinoma cell line (HT-29), and human normal colon mucosa cell line (CCD 841 CoN) showed biological activity specifically targeted against cancer cells. Therefore, it can be concluded that the membrane filtration-assisted enzymatic hydrolysis of potato juice proteins may increase their biological activity and allow for potato juice to be used in the production of medicinal preparations.

Keywords: antiproliferative activity; antioxidant activity; cancer cells; cytotoxicity; in vitro study; nutritional value

Citation: Kowalczewski, P.Ł.; Olejnik, A.; Rybicka, I.; Zielińska-Dawidziak, M.; Białas, W.; Lewandowicz, G. Membrane Filtration-Assisted Enzymatic Hydrolysis Affects the Biological Activity of Potato Juice. *Molecules* **2021**, *26*, 852. https://doi.org/10.3390/molecules26040852

Academic Editor: Ricardo Calhelha
Received: 19 January 2021
Accepted: 3 February 2021
Published: 6 February 2021

Publisher's Note: MDPI stays neutral with regard to jurisdictional claims in published maps and institutional affiliations.

1. Introduction

In recent years, problems with the management and appropriate use of by-products in the food industry have become an increasing challenge. One of the most interesting side streams in the food industry is potato juice (PJ), arising during the production of potato starch [1]. According to the published data, up to 500 kg of PJ can be made from 1 ton of potatoes [2,3]. PJ consists of both mineral compounds and organic substances, primarily proteins. Non-protein organic substances mainly include vitamins (B1, B2, B6, PP, C, and E), as well as antinutritional substances (phytates) and even toxic substances (glycoalkaloids) [4–7]. Potatoes reveal huge intraspecific diversity, nevertheless, it should be noted that the most popular varieties of potatoes are an extraordinarily rich source of macro- and microelements. The iron content in 100 g of the dry matter of juice from the most popular potato varieties is over ten times the recommended daily intake (RDI) for this element. Potassium and calcium are present at 400% and 150% of the recommended daily values, respectively [8]. The difference in the content of toxic alkaloids in individual potato varieties varies within five orders of magnitude; however, the content of glycoalkaloids

(GAs) in potatoes intended for consumption is much lower than the permissible limits [8,9]. Currently, PJ is used as a feed component, a limited use which fails to recognize and exploit the full potential of this material. Some researchers attempted to use PJ as an ingredient in microbial media and to obtain valuable metabolites [10–14]. However, it seems more promising to use it for the production of health-promoting foods [15–17], as PJ stands out not only for its high nutritional value, but also for its distinctive biological activity.

Freshly squeezed PJ was used in European German-speaking countries as a medicine in traditional folk medicine. It was believed that it is effective in treating stomach ulcers. At the end of the 19th century, Swiss physician Maximilian Bircher-Benner initiated the use of PJ as a therapeutic agent. However, the scientific verification of the effectiveness and safety of this material as a medicine only began in the 21st century [18,19]. The first studies indicated the key role of the protein fraction in the therapeutic (mainly anti-inflammatory) effects of PJ. The protease inhibitor fraction was found to be particularly active [20,21]. Later studies demonstrated the broader biological activity of PJ that was not always associated with the protein fraction [5,6,22]. The cytotoxic activity of PJ towards intestinal cancer cells is particularly worthy of attention [22,23]. The individual substance responsible for this activity has not yet been identified, although the key role of GAs has not been excluded. At the end of the 20th century, it was shown that GAs reveal in vitro activity against neoplastic cells [24]. Subsequently, it was proven that solanine and chaconine have the ability to induce tumor cell apoptosis [25,26]. It should be emphasized that biological activity, including anti-inflammatory activity, did not decline as an effect of thermal treatment [22,27]. Moreover, the anti-inflammatory activity of PJ subjected to thermal treatment was demonstrated in in vivo studies. In particular, Kujawska et al. [28] showed that spray-dried PJ could be used for ameliorating inflammation-related diseases of the gastrointestinal tract.

The enzymatic hydrolysis of proteins has received increasing interest over the last year because this process makes it possible to achieve multiple, non-contradictory purposes. Primarily, it makes it possible to extract protein fractions from unconventional sources and make them more digestible [29]. Moreover, enzymatic hydrolysis enables the reduction of the allergenicity of nutritionally important proteins [30]. Most often, however, the possibility of producing bioactive peptides is exploited [31–34]. This process was extensively studied for whey proteins and can be especially efficient when it is performed using a membrane reactor [35–38]. Numerous studies regarding potato proteins indicated the potential of enzymatic hydrolysis for obtaining bioactive products; however, the possibility of using a membrane reactor for that process has not yet been studied [39–42]. Nevertheless, the use of a membrane reactor, which makes the precise separation of individual fractions possible, may be of key importance for the functional properties of the obtained hydrolysate [43]. In our previous work, we showed that different preparations derived from PJ could be used for the manufacturing of functional foods (pasta, frankfurters, breads, or pâtés); however, the attractiveness of the products to consumers strongly depends on the form of protein in the foods [15–17,44]. Moreover, the method used for the isolation of the protein fraction from PJ also influenced the biological activity of the products [22,23,28]. The application of a membrane reactor for the enzymatic hydrolysis of the protein fraction of PJ could result in a product with high nutritional value, attractive functional properties, and increased biological activity. Therefore, the aim of the study was to verify the hypothesis presented above.

2. Results and Discussion

2.1. Chemical Composition and Nutritional Value of Potato Juice Protein Hydrolysate

Table 1 presents the results of the protein and mineral compound content in the analyzed fresh potato juice (PJ), the potato juice protein hydrolysate (PJPH), and the concentrated potato juice protein (cPJPH). The use of an ultrafiltration module in the enzymatic hydrolysis process made it possible to obtain a product containing protein in a soluble form, at a concentration seven times higher than in the raw material. The additional use of a nanofiltration module resulted in a further concentration of the protein, with almost double the quantity. The mineral compounds present in the fresh juice were concentrated more than twenty times, mainly using the ultrafiltration module. If we compare the effect of the membrane-assisted enzymatic hydrolysis with the membrane separation of potato proteins described in our previous work [23], the ratio of the protein to mineral fraction content is different. This phenomenon is mainly related to the use of ultrafiltration modules with different cut-offs. The use of a 5 kDa cut-off ultrafiltration membrane to concentrate the fresh PJ, containing non-degraded protein macromolecules, resulted in a more effective separation of both fractions. The membrane-assisted enzymatic hydrolysis using an ultrafiltration module with a cut-off of 1 kDa and a nanofiltration module with a cut-off of 300–500 Da maintained the valuable minerals in the hydrolysis product.

Table 1. Chemical composition of the analyzed products.

Parameter	PJ	PJPH	cPJPH
Protein [%]	2.55 ± 0.11 [c]	16.85 ± 0.12 [b]	29.77 ± 0.23 [a]
Ash [%]	0.97 ± 0.05 [b]	23.02 ± 2.61 [a]	24.34 ± 1.13 [a]

PJ—fresh potato juice; PJPH—potato juice protein hydrolysate; cPJPH—concentrated potato juice protein hydrolysate. Mean values with different letters ([a–c]) in the rows are significantly different at $\alpha = 0.05$.

The observations presented above also reflect changes in the individual ion content (Table 2). Both the PJPH and the cPJPH had a very high and comparable content of K (18.8 g and 19.6 g/100 g, respectively). They also had a low content of Na, at 160 mg/100 g in the PJPH and 176 mg/100 g in the cPJPH, which is nutritionally relevant due to the excessive sodium intake in most of the population worldwide [45,46]. Moreover, the PJPH and the cPJPH had a high content of Mg, Mn, and Cu, but their contents significantly differed between the samples analyzed. The content of Mg in the PJPH was 513 mg/100 g, which corresponded to about 140% of the Nutrient Reference Value (NRV) for this mineral, while the content of Mn was 6.18 mg/100 g (above 300% of the NRV), and the content of Cu was 1.19 mg/100 g (about 120% of the NRV). The cPJPH contained about 180% of the NRV for Mg, 370% of the NRV for Mn, and 200% of the NRV for Cu from a 100 g sample. The content of Ca and Zn was high but did not exceed 100% of the NRV in either the PJPH or the cPJPH. The content of Ca was 20% (PJPH) and 29% (cPJPH) of the NRV, and the content of Zn was 60% (PJHP) and 74% (cPJPH) of the NRV. Only the content of Fe was found to be at the low level of 0.44 mg/100 g and 0.56 mg/100 g for both the PJPJ and the cPJPH, respectively, corresponding to less than 5% of the NRV for this mineral. Heavy metals were mainly concentrated at the ultrafiltration stage, which is related to the effective retention of other ions on the nanofiltration membrane. Specifically, the lead content in the cPJPH was below the limit for food supplements [47].

Table 2. Mineral composition of the analyzed products.

Mineral	NRV [mg]	PJPH	cPJPH
K [mg/100 g]	2000	18823 ± 590	19623 ± 526
Mg [mg/100 g]	375	513 ± 10	664 ± 10
Na [mg/100 g]	N/A	161 ± 4	176 ± 4
Ca [mg/100 g]	800	160 ± 6	228 ± 5
Zn [mg/100 g]	10	6.02 ± 0.13	7.45 ± 0.15
Mn [mg/100 g]	2	6.18 ± 0.13	7.38 ± 0.12
Cu [mg/100 g]	1	1.19 ± 0.02	1.98 ± 0.07
Fe [mg/100 g]	14	0.44 ± 0.04	0.56 ± 0.08
Cd [µg/g]	-	10 ± 1	8.75 ± 0.14
Pb [µg/g]	-	4.31 ± 0.11	2.01 ± 0.13

PJPH—potato juice protein hydrolysate; cPJPH—concentrated potato juice protein hydrolysate; NRV—nutrient reference value; N/A—not applicable.

Potato proteins have a high nutritional value due to their amino acid composition. Essential amino acids (EEA) accounted for 32% of the total amino acids (TAA) in the preparation analyzed in this study. These results were not much lower than the results presented by Gorissen et al. [48], which suggested that potato proteins meet the requirement for EAA, which constituted 38% of the TAA in the potato proteins studied. The PJPH unexpectedly had a lower amino acid score (AAS) than usually observed for potato proteins [23]. As shown in Table 3, the limiting amino acid was leucine, with a content that was merely 3.4% of the TAA (compared to 8.3% suggested by Gorissen et al. [48]). A decreased content of lysine was also noted (3.2% of the TAA). However, in the literature, it was found that the product of lysine degradation could be glutamate [49], and a high content of glutamate and glutamic acid was noted (10.1 g/16 g N) in the PJPH. The amino acid profile could also be influenced by the applied hydrolytic enzymes, the storage conditions [50], and the potato variety used to prepare the PJPH [51]. The PJPH was found to be an excellent source of tryptophan (~167% compared to the WHO/FAO standard [52]).

Table 3. The amino acid profile and amino acid score (AAS) for adults according to the standards reported by the FAO/WHO [52].

Amino Acid	FAO/WHO Standard [mg/g]	PJPH [g/16 g N]	AAS
Essential Amino Acids			
Histidine	16	1.92 ± 0.05	100
Isoleucine	30	2.44 ± 0.07	81.5
Leucine	61	2.35 ± 0.10	38.5
Lysine	48	2.28 ± 0.12	47.4
Methionine + Cystine	23	1.82 ± 0.21	79.0
Phenylalanine + Tyrosine	41	4.25 ± 0.33	100
Threonine	25	2.19 ± 0.08	87.7
Tryptophan	6.6	1.10 ± 0.08	100
Valine	40	4.44 ± 0.17	100
Dispensable Amino Acids			
Alanine	-	8.56 ± 0.29	-
Arginine	-	7.84 ± 0.31	-
Aspartic acid	-	15.06 ± 0.43	-
Glutamic acid	-	10.13 ± 0.50	-
Glycine	-	1.46 ± 0.22	-
Proline	-	1.69 ± 0.19	-
Serine	-	2.93 ± 0.13	-

2.2. Antioxidant Activity

The number of factors that pose a direct threat to human health are rapidly increasing as a result of development, industrialization, and urbanization. One such factor is oxidative stress, i.e., the excessive accumulation of free radicals in our bodies, which results in oxidative damage to the cells as well as damage to the DNA, lipids, and proteins, which can lead to a number of diseases, including neurodegenerative diseases [53–56]. Therefore, there is a constant search for antioxidant compounds, of which plants are precious sources [57–60]. The content of antioxidant compounds may increase significantly depending on the growth conditions and the action of stress factors [61,62]. A comparison of the published data on the antioxidant activity of fresh potato juice [22] and the hydrolysate obtained in this study (Table 4) showed that the enzymatic hydrolysis of the juice caused a significant, 10-fold increase in antioxidant activity. Moreover, the content of polyphenolic compounds in the analyzed hydrolysate was high, several times higher than the content in potatoes with colored flesh [63], and rich in anthocyanins and polyphenols. Data from the literature confirmed that enzymatic hydrolysis can significantly increase antioxidant activity, as well as release bound polyphenolic compounds [64–66].

Table 4. Antioxidant activity was expressed as Trolox equivalent antioxidant capacity (TEAC), and total phenolic compounds (TPC).

Parameter	PJPH	cPJPH
TEAC [mmol/g]	0.89 ± 0.05 [b]	0.96 ± 0.03 [a]
TPC [mg/g]	28.29 ± 1.88 [b]	31.11 ± 2.16 [a]

PJPH—potato juice protein hydrolysate; cPJPH—concentrated potato juice protein hydrolysate. Mean values with different letters ([a,b]) in the rows are significantly different at $\alpha = 0.05$.

2.3. In Vitro Cytotoxicity Assay

The most interesting, least described, and perhaps most important activity of PJ is probably its cytotoxicity against cancer cells [22,23]. On the basis of the obtained results presented in Table 5, it was found that the hydrolysate of PJ protein has a cytotoxic effect, which can be increased with concentration using membrane separation. Moreover, the use of a nanofiltration membrane to concentrate the obtained hydrolysate additionally increased the biological activity of cPJPH. By comparing the IC_{50} values, it was found that the concentration process caused a 1.4–3.3-fold increase in the cytotoxicity of PJPH depending on the cell line tested. The highest cytotoxicity of cPJPH was found in the Caco-2 colon cancer cells. In contrast, the weakest cytotoxic effects were observed in the culture of normal colon mucosa CCD 841 CoN cells. The first cytotoxic dose (IC_{10}) of cPJPH to Caco-2 cells was 7.3-fold lower than the IC_{10} to CCD 841 CoN cells. Moreover, the half-maximal inhibitory concentration (IC_{50}) and the lethal concentration (IC_{90}) of cPJPH determined for Caco-2 cells were significantly lower (3.4- and 1.6-fold, respectively) than those obtained for CCD 841 CoN cells. The high cytotoxic activity of both PJPH and cPJPH was also observed in the colon cancer HT-29 cell culture. As a result of PJPH concentration, a significant increase in cytotoxic potential was found in the stomach cancer Hs746T cell cultures (Table 5). Commonly used anti-cancer drugs cause significant damage to the body of patients because of their non-selective action [67–69]. Therefore, there is a need for substances that will act in a more targeted manner. The results of the cytotoxic activity of cPJPH indicated significantly lower IC_{50} doses for neoplastic cells compared to normal cells, which may be of particular interest in the context of further research into the use of PJ ingredients for the treatment of gastrointestinal cancer.

Table 5. Cytotoxic doses for stomach cancer cells (Hs746T line), colon cancer cells (HT-29 and Caco-2 lines), and colon normal cells (CCD 841 CoN line) [mg_{dm}/mL].

Cell Line	IC_{10}		IC_{50}		IC_{90}	
	PJPH	cPJPH	PJPH	cPJPH	PJPH	cPJPH
Hs 746T	4.43 ± 0.21 [b]	1.80 ± 0.18 [b]	6.30 ± 0.09 [b]	2.95 ± 0.13 [b]	8.96 ± 0.61 [b]	4.84 ± 0.07 [b]
Caco-2	2.58 ± 1.13 [c]	0.55 ± 0.17 [c]	5.26 ± 0.78 [b,c]	1.62 ± 0.24 [c]	11.35 ± 2.00 [a]	4.88 ± 0.46 [b]
HT-29	1.99 ± 0.19 [c]	2.16 ± 0.31 [b]	4.43 ± 0.21 [c]	3.11 ± 0.16 [b]	9.90 ± 0.61 [a,b]	4.48 ± 0.20 [b]
CCD 841 CoN	5.56 ± 0.30 [a]	4.04 ± 0.09 [a]	7.20 ± 0.10 [a]	5.47 ± 0.01 [a]	9.34 ± 0.39 [a,b]	7.40 ± 0.16 [a]

Mean values denoted by different letters ([a–c]) in columns differ statistically significantly ($p < 0.05$).

The comparison of the various methods used to process PJ suggests some conclusions regarding the question of which substances in PJ are actually responsible for its cytotoxic effect on cancer cells. While concentration by ultrafiltration using a 5 kDa cut-off membrane resulted in a product with a high content of nutritious protein [23], the membrane-assisted enzymatic hydrolysis presented in this study resulted in a product with significantly increased cytotoxic activity against cancer cells. Moreover, the thermal deproteination of PJ also resulted in products with higher cytotoxic activity against cancer cells compared to the raw materials [22]. These results prove that the anti-proliferative effect of PJ on cancer cells is not related to protein fraction but to other molecules of rather low molecular mass. Surely, this hypothesis requires confirmation with further analyses, but the phenomena observed so far provide important evidence that may direct further studies.

3. Materials and Methods

3.1. Enzymatic Hydrolysis of Potato Juice Proteins

The experimental material, potato juice (PJ), was collected during the starch production season from the production line of "Trzemeszno" Sp. z o.o. Potato Industry Company (Trzemeszno, Poland). The enzyme Savinase® (Sigma-Aldrich, Saint Louis, MO, USA), isolated from the *Bacillus* species, was used as the proteolytic preparation. On the basis of previous preliminary studies (data not shown), an enzyme dose of 4 µL/g of potato protein was adopted. The enzyme was added at the start of the hydrolysis process, according to the initial volume of PJ used in the experiment, and further portions were added every 60 min because of the continuous process, in which the finished hydrolysis product was removed and a new portion of PJ was added in its place. A polyethersulfone spiral-wound ultrafiltration membrane with a molecular weight of 1 kDa cut-off and an area of 3.5 m^2 (type 3838, SUEZ Water Technologies & Solutions, Budapest, Hungary) was used to perform the enzymatic hydrolysis and, consequently, to obtain a PJ protein hydrolysate (PJPH). The non-hydrolyzed PJ was returned to the initial tank of the system (recirculation). The PJPH was then concentrated on a polyamide thin film composite nanofiltration membrane, with a molecular weight of 300–500 Da cut-off and an area of 4.0 m^2 (type 3838, SUEZ Water Technologies & Solutions, Budapest, Hungary), to obtain a concentrated fraction of hydrolyzed, soluble potato proteins (retentate, denoted in the text as cPJPH) and a non-protein low molecular weight fraction (filtrate). A flowchart of the process used for these products is presented in Figure 1.

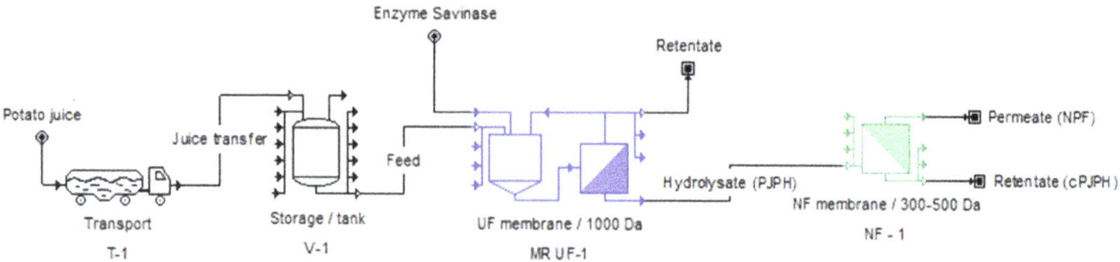

Figure 1. Schematic diagram of the applied membrane separation system for ultrafiltration.

3.2. Chemical Analysis

The Kjeldahl method was used to determine the total nitrogen content, which was then used to calculated the protein content using a nitrogen-to-protein conversion factor of 6.25 [70]. The international standard method ISO 763 [71] was used to measure the total ash content.

The concentrations of minerals Ca, Cu, Fe, K, Mg, Mn, Na, and Zn were determined using flame atomic absorption spectroscopy (FAAS) (SpectrAA-800, Varian, Palo Alto, CA, USA) that was preceded by microwave mineralization with nitric acid [72]. The recommendations for Ca, Cu, Fe, Mg, Mn, and Zn were established at the level of the Nutrient Reference Value (NRV) [73]. The contents of the minerals were expressed in g/100 g of the dry mass of the sample.

3.3. Amino Acid Composition and Scoring

Histidine (His), isoleucine (Ile), leucine (Leu), lysine (Lys), methionine (Met), phenylalanine (Phe), threonine (Thr), valine (Val), cysteine (Cys), tyrosine (Tyr), glycine (Gly), arginine (Arg), proline (Pro), aspartic acid (Asp), glutamic acid (Glu), alanine (Ala), and serine (Ser) were determined using ultra-performance liquid chromatography (Shimadzu Nexera 2.0, Kyoto, Japan), equipped with a PDA (at 260 nm, sampling rate of 20 points/s) and FL detector (Kyoto, Japan), that was preceded by acidic hydrolysis (110 °C, 23 h) [74]; meanwhile, sulfuric amino acids were prepared by oxidation (4 °C, 16 h) followed by acidic hydrolysis (110 °C, 2.5 h) [75]. The results were expressed in g/16 g N (which is equivalent to g/100 g of protein).

The FAO recommended method was used to calculate the amino acid score (AAS) for adults [52]:

$$AAS = \frac{\text{essential amino acids contents in PJPC } [\%]}{\text{recommended essential amino acids } [\%]}.$$

3.4. Total Phenolic and Antioxidant Activity of the Hydrolysates

The extraction of bioactive compounds was performed using lyophilized products. A 0.5 g sample was extracted with 40 mL of 80% ethanol for 2 h and then centrifuged ($4000 \times g$, 10 min) using a laboratory centrifuge (Rotofix 32 A, Hettich, Germany). The obtained supernatants were then decanted and filtered through a 0.22 μm filter. The samples were stored in a -20 °C freezer until use.

The Folin–Ciocalteu colorimetric method [76] was applied to measure the total phenolic compounds (TPC) using a spectrophotometer (Multiskan GO, Thermo Fisher Scientific, Vantaa, Finland). The results were expressed as a gallic acid equivalent (mg GAE/g).

The ABTS radical cation decolorization assay was determined by the method of Re et al. [77], with slight modifications that are described elsewhere [62]. A 2 mL sample of the ABTS solution was mixed with 0.98 mL of PBS and 0.02 mL of the PJPH extract. After 6 min of incubation at 30 °C, an absorbance at 734 nm was measured spectrophotometrically (Multiskan GO, Thermo Fisher Scientific, Vantaa, Finland). Trolox was used as a standard, and the results were presented as Trolox equivalents (mg/g of sample).

3.5. In Vitro Cytotoxicity Assay

The human colorectal adenocarcinoma cell line HT-29 (Cat. no: 85061109), human gastric carcinoma Hs 746T cell line (ATCC® HTB-135™), human colon cancer Caco-2 cell line (ATCC® HTB-37™), and human normal colon mucosa CCD 841 CoN cell line (ATCC® CRL-1790™) were used in this study and cultured according to the method that was previously described in detail in this report [23]. The cell viability and metabolic activity were determined using the 3-(4,5-dimethylthiazol-2-yl)-2,5-diphenyltetrazolium bromide (MTT) colorimetric assay [78]. The first cytotoxic dose (IC_{10}), the median effective concentration (IC_{50}), and the lethal dose (IC_{90}) were calculated on the basis of the MTT results.

3.6. Statistical Analysis

Statistica 13 software (Dell Software Inc., Round Rock, TX, USA) was used to perform a one-way analysis of variance (ANOVA). A post-hoc Tukey HSD multiple comparison test was used to identify statistically homogeneous subsets at $\alpha = 0.05$.

4. Conclusions

The enzymatic hydrolysis of potato juice in a membrane reactor influenced both the nutritional value as well as the biological activity of this raw material. The use of ultrafiltration systems resulted in a product containing soluble proteins in a concentration several times higher than that of a simple enzymatic hydrolysis. Mineral compounds were also concentrated in this process. The additional use of a nanofiltration module resulted in a further concentration of the solutes. The products of the enzymatic hydrolysis of potato juice were characterized by a significantly higher antioxidant activity and concentration of polyphenolic compounds than the raw materials.

It was also found that the enzymatic hydrolysis of potato juice in the reactor with the ultrafiltration membrane separation system increased the cytotoxic activity of the processed material. IC_{50} toxic doses of the hydrolysate for cancer cells were significantly lower than those of fresh potato juice. Moreover, IC_{50} toxic doses of the concentrate were lower for cancer cells than for normal cells. Therefore, the additional use of a nanofiltration system to concentrate the obtained hydrolysate further increased the cytotoxicity of the product against cancer cells.

Author Contributions: Conceptualization, P.Ł.K. and G.L.; Data curation, P.Ł.K., A.O. and M.Z.-D.; Formal analysis, P.Ł.K.; Funding acquisition, P.Ł.K.; Investigation, P.Ł.K., A.O. and I.R.; Methodology, P.Ł.K., A.O., M.Z.-D., W.B. and G.L.; Supervision, G.L.; Validation, W.B.; Writing—original draft, P.Ł.K., A.O., I.R., M.Z.-D. and G.L. All authors have read and agreed to the published version of the manuscript.

Funding: The work was financially supported by the grant of the National Science Centre (Poland) No. 2019/03/X/NZ9/00229.

Institutional Review Board Statement: Not applicable.

Informed Consent Statement: Not applicable.

Data Availability Statement: The datasets generated during and/or analysed during the current study are available from the corresponding author on reasonable request.

Conflicts of Interest: The authors declare no conflict of interest. The funders had no role in the design of the study; in the collection, analyses, or interpretation of the data; in the writing of the manuscript, or in the decision to publish the results.

Sample Availability: Samples of PJPH and cPJPH are available from the corresponding author.

References

1. Kot, A.M.; Pobiega, K.; Piwowarek, K.; Kieliszek, M.; Błażejak, S.; Gniewosz, M.; Lipińska, E. Biotechnological Methods of Management and Utilization of Potato Industry Waste—A Review. *Potato Res.* **2020**, *63*, 431–447. [CrossRef]
2. Lasik, M.; Nowak, J.; Kent, C.; Czarnecki, Z. Assessment of Metabolic activity of single and mixed microorganism population assigned for potato wastewater biodegradation. *Polish J. Environ. Stud.* **2002**, *11*, 719–726.
3. Miedzianka, J.; Pęksa, A.; Pokora, M.; Rytel, E.; Tajner-Czopek, A.; Kita, A. Improving the properties of fodder potato protein concentrate by enzymatic hydrolysis. *Food Chem.* **2014**, *159*, 512–518. [CrossRef] [PubMed]
4. Zwijnenberg, H.J.; Kemperman, A.J.B.; Boerrigter, M.E.; Lotz, M.; Dijksterhuis, J.F.; Poulsen, P.E.; Koops, G.-H. Native protein recovery from potato fruit juice by ultrafiltration. *Desalination* **2002**, *144*, 331–334. [CrossRef]
5. McGill, C.R.; Kurilich, A.C.; Davignon, J. The role of potatoes and potato components in cardiometabolic health: A review. *Ann. Med.* **2013**, *45*, 467–473. [CrossRef] [PubMed]
6. Camire, M.E.; Kubow, S.; Donnelly, D.J. Potatoes and Human Health. *Crit. Rev. Food Sci. Nutr.* **2009**, *49*, 823–840. [CrossRef] [PubMed]
7. Jeżowski, P.; Polcyn, K.; Tomkowiak, A.; Rybicka, I.; Radzikowska, D. Technological and antioxidant properties of proteins obtained from waste potato juice. *Open Life Sci.* **2020**, *15*, 379–388. [CrossRef]
8. Burlingame, B.; Mouillé, B.; Charrondière, R. Nutrients, bioactive non-nutrients and anti-nutrients in potatoes. *J. Food Compos. Anal.* **2009**, *22*, 494–502. [CrossRef]
9. Pęksa, A.; Gołubowska, G.; Aniołowski, K.; Lisińska, G.; Rytel, E. Changes of glycoalkaloids and nitrate contents in potatoes during chip processing. *Food Chem.* **2006**, *97*, 151–156. [CrossRef]
10. Bzducha-Wróbel, A.; Pobiega, K.; Błażejak, S.; Kieliszek, M. The scale-up cultivation of Candida utilis in waste potato juice water with glycerol affects biomass and $\beta(1,3)/(1,6)$-glucan characteristic and yield. *Appl. Microbiol. Biotechnol.* **2018**, *102*, 9131–9145. [CrossRef] [PubMed]
11. Bzducha-Wróbel, A.; Koczoń, P.; Błażejak, S.; Kozera, J.; Kieliszek, M. Valorization of Deproteinated Potato Juice Water into β-Glucan Preparation of *C. utilis* Origin: Comparative Study of Preparations Obtained by Two Isolation Methods. *Waste Biomass Valorization* **2019**. [CrossRef]
12. Kot, A.M.; Błażejak, S.; Kieliszek, M.; Gientka, I.; Bryś, J. Simultaneous Production of Lipids and Carotenoids by the Red Yeast Rhodotorula from Waste Glycerol Fraction and Potato Wastewater. *Appl. Biochem. Biotechnol.* **2019**, *189*, 589–607. [CrossRef]
13. Bzducha-Wróbel, A.; Błażejak, S.; Kieliszek, M.; Pobiega, K.; Falana, K.; Janowicz, M. Modification of the cell wall structure of Saccharomyces cerevisiae strains during cultivation on waste potato juice water and glycerol towards biosynthesis of functional polysaccharides. *J. Biotechnol.* **2018**, *281*, 1–10. [CrossRef] [PubMed]
14. Bzducha-Wróbel, A.; Błażejak, S.; Molenda, M.; Reczek, L. Biosynthesis of $\beta(1,3)/(1,6)$-glucans of cell wall of the yeast *Candida utilis* ATCC 9950 strains in the culture media supplemented with deproteinated potato juice water and glycerol. *Eur. Food Res. Technol.* **2015**, *240*, 1023–1034. [CrossRef]
15. Kowalczewski, P.; Lewandowicz, G.; Makowska, A.; Knoll, I.; Błaszczak, W.; Białas, W.; Kubiak, P. Pasta Fortified with Potato Juice: Structure, Quality, and Consumer Acceptance. *J. Food Sci.* **2015**, *80*, S1377–S1382. [CrossRef] [PubMed]
16. Kowalczewski, P.Ł.; Lewandowicz, G.; Krzywdzińska-Bartkowiak, M.; Piątek, M.; Baranowska, H.M.; Białas, W.; Jeziorna, M.; Kubiak, P. Finely comminuted frankfurters fortified with potato juice—Quality and structure. *J. Food Eng.* **2015**, *167*, 183–188. [CrossRef]
17. Kowalczewski, P.; Różańska, M.; Makowska, A.; Jeżowski, P.; Kubiak, P. Production of wheat bread with spray-dried potato juice: Influence on dough and bread characteristics. *Food Sci. Technol. Int.* **2019**, *25*, 223–232. [CrossRef]

18. Chrubasik, S.; Chrubasik, C.; Torda, T.; Madisch, A. Efficacy and tolerability of potato juice in dyspeptic patients: A pilot study. *Phytomedicine* **2006**, *13*, 11–15. [CrossRef] [PubMed]

19. Vlachojannis, J.E.; Cameron, M.; Chrubasik, S. Medicinal use of potato-derived products: A systematic review. *Phyther. Res.* **2010**, *24*, 159–162. [CrossRef] [PubMed]

20. Pouvreau, L.; Gruppen, H.; Piersma, S.R.; van den Broek, L.A.M.; van Koningsveld, G.A.; Voragen, A.G.J. Relative Abundance and Inhibitory Distribution of Protease Inhibitors in Potato Juice from cv. Elkana. *J. Agric. Food Chem.* **2001**, *49*, 2864–2874. [CrossRef] [PubMed]

21. Ruseler-van Embden, J.G.H.; van Lieshout, L.M.C.; Smits, S.A.; van Kessel, I.; Laman, J.D. Potato tuber proteins efficiently inhibit human faecal proteolytic activity: Implications for treatment of peri-anal dermatitis. *Eur. J. Clin. Invest.* **2004**, *34*, 303–311. [CrossRef] [PubMed]

22. Kowalczewski, P.Ł.; Olejnik, A.; Białas, W.; Kubiak, P.; Siger, A.; Nowicki, M.; Lewandowicz, G. Effect of Thermal Processing on Antioxidant Activity and Cytotoxicity of Waste Potato Juice. *Open Life Sci.* **2019**, *14*, 150–157. [CrossRef]

23. Kowalczewski, P.Ł.; Olejnik, A.; Białas, W.; Rybicka, I.; Zielińska-Dawidziak, M.; Siger, A.; Kubiak, P.; Lewandowicz, G. The Nutritional Value and Biological Activity of Concentrated Protein Fraction of Potato Juice. *Nutrients* **2019**, *11*, 1523. [CrossRef] [PubMed]

24. Kuo, K.-W.; Hsu, S.-H.; Li, Y.-P.; Lin, W.-L.; Liu, L.-F.; Chang, L.-C.; Lin, C.-C.; Lin, C.-N.; Sheu, H.-M. Anticancer activity evaluation of the *Solanum* glycoalkaloid solamargine: Triggering apoptosis in human hepatoma cells. *Biochem. Pharmacol.* **2000**, *60*, 1865–1873. [CrossRef]

25. Mohamed Saleem, T.S.; Chetty, C.M.; Ramkanth, S.; Alagusundaram, M.; Gnanaprakash, K.; Thiruvengada Rajan, V.S.; Angalaparameswari, S. *Solanum nigrum* Linn.—A review. *Phcog Rev.* **2009**, *3*, 342–345.

26. Ji, Y.B.; Gao, S.Y.; Ji, C.F.; Zou, X. Induction of apoptosis in HepG2 cells by solanine and Bcl-2 protein. *J. Ethnopharmacol.* **2008**, *115*, 194–202. [CrossRef] [PubMed]

27. Rytel, E.; Tajner-Czopek, A.; Kita, A.; Sokół-Łętowska, A.; Kucharska, A.Z.; Wojciechowski, W. Effect of temperature and pH value on the stability of bioactive compounds and antioxidative activity of juice from colour-fleshed potatoes. *Int. J. Food Sci. Technol.* **2019**. [CrossRef]

28. Kujawska, M.; Olejnik, A.; Lewandowicz, G.; Kowalczewski, P.; Forjasz, R.; Jodynis-Liebert, J. Spray-Dried Potato Juice as a Potential Functional Food Component with Gastrointestinal Protective Effects. *Nutrients* **2018**, *10*, 259. [CrossRef]

29. Nongonierma, A.B.; FitzGerald, R.J. Unlocking the biological potential of proteins from edible insects through enzymatic hydrolysis: A review. *Innov. Food Sci. Emerg. Technol.* **2017**, *43*, 239–252. [CrossRef]

30. Abd El-Salam, M.H.; El-Shibiny, S. Reduction of Milk Protein Antigenicity by Enzymatic Hydrolysis and Fermentation. A Review. *Food Rev. Int.* **2019**, 1–20. [CrossRef]

31. Liu, Y.-F.; Oey, I.; Bremer, P.; Carne, A.; Silcock, P. Bioactive peptides derived from egg proteins: A review. *Crit. Rev. Food Sci. Nutr.* **2018**, *58*, 2508–2530. [CrossRef] [PubMed]

32. Zhang, M.; Mu, T.-H. Identification and characterization of antioxidant peptides from sweet potato protein hydrolysates by Alcalase under high hydrostatic pressure. *Innov. Food Sci. Emerg. Technol.* **2017**, *43*, 92–101. [CrossRef]

33. Habinshuti, I.; Mu, T.-H.; Zhang, M. Ultrasound microwave-assisted enzymatic production and characterisation of antioxidant peptides from sweet potato protein. *Ultrason. Sonochem.* **2020**, *69*, 105262. [CrossRef]

34. Nazir, M.A.; Mu, T.; Zhang, M. Preparation and identification of angiotensin I-converting enzyme inhibitory peptides from sweet potato protein by enzymatic hydrolysis under high hydrostatic pressure. *Int. J. Food Sci. Technol.* **2020**, *55*, 482–489. [CrossRef]

35. O'Halloran, J.; O'Sullivan, M.; Casey, E. Production of Whey-Derived DPP-IV Inhibitory Peptides Using an Enzymatic Membrane Reactor. *Food Bioprocess Technol.* **2019**, *12*, 799–808. [CrossRef]

36. Cheison, S.C.; Wang, Z.; Xu, S.-Y. Multivariate strategy in screening of enzymes to be used for whey protein hydrolysis in an enzymatic membrane reactor. *Int. Dairy J.* **2007**, *17*, 393–402. [CrossRef]

37. Guadix, A.; Camacho, F.; Guadix, E.M. Production of whey protein hydrolysates with reduced allergenicity in a stable membrane reactor. *J. Food Eng.* **2006**, *72*, 398–405. [CrossRef]

38. Cheison, S.C.; Kulozik, U. Impact of the environmental conditions and substrate pre-treatment on whey protein hydrolysis: A review. *Crit. Rev. Food Sci. Nutr.* **2017**, *57*, 418–453. [CrossRef]

39. Yao, S.; Udenigwe, C.C. Peptidomics of potato protein hydrolysates: Implications of post-translational modifications in food peptide structure and behaviour. *R. Soc. Open Sci.* **2018**, *5*, 172425. [CrossRef]

40. Waglay, A.; Karboune, S. Enzymatic generation of peptides from potato proteins by selected proteases and characterization of their structural properties. *Biotechnol. Prog.* **2016**, *32*, 420–429. [CrossRef] [PubMed]

41. C K Rajendran, S.R.; Mason, B.; Udenigwe, C.C. Peptidomics of Peptic Digest of Selected Potato Tuber Proteins: Post-Translational Modifications and Limited Cleavage Specificity. *J. Agric. Food Chem.* **2016**, *64*, 2432–2437. [CrossRef] [PubMed]

42. Kamnerdpetch, C.; Weiss, M.; Kasper, C.; Scheper, T. An improvement of potato pulp protein hydrolyzation process by the combination of protease enzyme systems. *Enzyme Microb. Technol.* **2007**, *40*, 508–514. [CrossRef]

43. Eckert, E.; Han, J.; Swallow, K.; Tian, Z.; Jarpa-Parra, M.; Chen, L. Effects of enzymatic hydrolysis and ultrafiltration on physicochemical and functional properties of faba bean protein. *Cereal Chem.* **2019**, *96*, 725–741. [CrossRef]

44. Baranowska, H.M.; Masewicz, Ł.; Kowalczewski, P.Ł.; Lewandowicz, G.; Piątek, M.; Kubiak, P. Water properties in pâtés enriched with potato juice. *Eur. Food Res. Technol.* **2018**, *244*, 387–393. [CrossRef]

45. Grillo, A.; Salvi, L.; Coruzzi, P.; Salvi, P.; Parati, G. Sodium Intake and Hypertension. *Nutrients* **2019**, *11*, 1970. [CrossRef] [PubMed]

46. Gress, T.W.; Mansoor, K.; Rayyan, Y.M.; Khthir, R.A.; Tayyem, R.F.; Tzamaloukas, A.H.; Abraham, N.G.; Shapiro, J.I.; Khitan, Z.J. Relationship between dietary sodium and sugar intake: A cross-sectional study of the National Health and Nutrition Examination Survey 2001–2016. *J. Clin. Hypertens.* **2020**, *22*, 1694–1702. [CrossRef]

47. Commission Regulation (EC) No. 1881/2006 setting maximum levels for certain contaminants in foodstuffs. *Off. J. Eur. Union L* *364* **2006**, 5–24. Available online: https://eur-lex.europa.eu/legal-content/EN/ALL/?uri=CELEX%3A32006R1881 (accessed on 12 July 2020).

48. Gorissen, S.H.M.; Crombag, J.J.R.; Senden, J.M.G.; Waterval, W.A.H.; Bierau, J.; Verdijk, L.B.; van Loon, L.J.C. Protein content and amino acid composition of commercially available plant-based protein isolates. *Amino Acids* **2018**, *50*, 1685–1695. [CrossRef] [PubMed]

49. Møller, B.L. Lysine Catabolism in Barley (*Hordeum vulgare* L.). *Plant Physiol.* **1976**, *57*, 687–692. [CrossRef] [PubMed]

50. Davids, S.J.; Yaylayan, V.A.; Turcotte, G. Use of unusual storage temperatures to improve the amino acid profile of potatoes for novel flavoring applications. *LWT Food Sci. Technol.* **2004**, *37*, 619–626. [CrossRef]

51. Williams, J. Influence of variety and processing conditions on acrylamide levels in fried potato crisps. *Food Chem.* **2005**, *90*, 875–881. [CrossRef]

52. FAO. *FAO Dietary Protein Quality Evaluation in Human Nutrition*; Report of an FAO Expert Consultation; FAO Food and Nutrition Paper 92. FAO: Rome, Italy, 2013; ISBN 978-92-5-107417-6.

53. Blokhina, O. Antioxidants, Oxidative Damage and Oxygen Deprivation Stress: A Review. *Ann. Bot.* **2003**, *91*, 179–194. [CrossRef] [PubMed]

54. Gutteridge, J.M.C.; Halliwell, B. Mini-Review: Oxidative stress, redox stress or redox success? *Biochem. Biophys. Res. Commun.* **2018**, *502*, 183–186. [CrossRef]

55. Rao, A.V.; Balachandran, B. Role of Oxidative Stress and Antioxidants in Neurodegenerative Diseases. *Nutr. Neurosci.* **2002**, *5*, 291–309. [CrossRef] [PubMed]

56. Gilgun-Sherki, Y.; Melamed, E.; Offen, D. Oxidative stress induced-neurodegenerative diseases: The need for antioxidants that penetrate the blood brain barrier. *Neuropharmacology* **2001**, *40*, 959–975. [CrossRef]

57. Ražná, K.; Sawinska, Z.; Ivanišová, E.; Vukovic, N.; Terentjeva, M.; Stričík, M.; Kowalczewski, P.Ł.; Hlavačková, L.; Rovná, K.; Žiarovská, J.; et al. Properties of *Ginkgo biloba* L.: Antioxidant Characterization, Antimicrobial Activities, and Genomic MicroRNA Based Marker Fingerprints. *Int. J. Mol. Sci.* **2020**, *21*, 3087. [CrossRef] [PubMed]

58. Rovná, K.; Ivanišová, E.; Žiarovská, J.; Ferus, P.; Terentjeva, M.; Kowalczewski, P.Ł.; Kačániová, M. Characterization of R*osa canina* Fruits Collected in Urban Areas of Slovakia. Genome Size, iPBS Profiles and Antioxidant and Antimicrobial Activities. *Molecules* **2020**, *25*, 1888. [CrossRef] [PubMed]

59. Andre, C.; Larondelle, Y.; Evers, D. Dietary Antioxidants and Oxidative Stress from a Human and Plant Perspective: A Review. *Curr. Nutr. Food Sci.* **2010**, *6*, 2–12. [CrossRef]

60. Kujawska, M.; Jourdes, M.; Kurpik, M.; Szulc, M.; Szaefer, H.; Chmielarz, P.; Kreiner, G.; Krajka-Kuźniak, V.; Mikołajczak, P.Ł.; Teissedre, P.-L.; et al. Neuroprotective Effects of Pomegranate Juice against Parkinson's Disease and Presence of Ellagitannins-Derived Metabolite—Urolithin A—In the Brain. *Int. J. Mol. Sci.* **2019**, *21*, 202. [CrossRef]

61. Witkowska-Banaszczak, E.; Radzikowska, D.; Ratajczak, K. Chemical profile and antioxidant activity of *Trollius europaeus* under the influence of feeding aphids. *Open Life Sci.* **2018**, *13*, 312–318. [CrossRef]

62. Kowalczewski, P.Ł.; Radzikowska, D.; Ivanišová, E.; Szwengiel, A.; Kačániová, M.; Sawinska, Z. Influence of Abiotic Stress Factors on the Antioxidant Properties and Polyphenols Profile Composition of Green Barley (*Hordeum vulgare* L.). *Int. J. Mol. Sci.* **2020**, *21*, 397. [CrossRef] [PubMed]

63. Pęksa, A.; Miedzianka, J.; Nemś, A. Amino acid composition of flesh-coloured potatoes as affected by storage conditions. *Food Chem.* **2018**, *266*, 335–342. [CrossRef] [PubMed]

64. Bamdad, F.; Wu, J.; Chen, L. Effects of enzymatic hydrolysis on molecular structure and antioxidant activity of barley hordein. *J. Cereal Sci.* **2011**, *54*, 20–28. [CrossRef]

65. Jia, J.; Zhou, Y.; Lu, J.; Chen, A.; Li, Y.; Zheng, G. Enzymatic hydrolysis of Alaska pollack (*Theragra chalcogramma*) skin and antioxidant activity of the resulting hydrolysate. *J. Sci. Food Agric.* **2010**, *90*, 635–640. [CrossRef] [PubMed]

66. Lu, S.-Y.; Chu, Y.-L.; Sridhar, K.; Tsai, P.-J. Effect of ultrasound, high-pressure processing, and enzymatic hydrolysis on carbohydrate hydrolyzing enzymes and antioxidant activity of lemon (*Citrus limon*) flavedo. *LWT* **2020**, 110511. [CrossRef]

67. Albini, A.; Pennesi, G.; Donatelli, F.; Cammarota, R.; De Flora, S.; Noonan, D.M. Cardiotoxicity of Anticancer Drugs: The Need for Cardio-Oncology and Cardio-Oncological Prevention. *JNCI J. Natl. Cancer Inst.* **2010**, *102*, 14–25. [CrossRef]

68. Yeh, E.T.H.; Tong, A.T.; Lenihan, D.J.; Yusuf, S.W.; Swafford, J.; Champion, C.; Durand, J.-B.; Gibbs, H.; Zafarmand, A.A.; Ewer, M.S. Cardiovascular Complications of Cancer Therapy. *Circulation* **2004**, *109*, 3122–3131. [CrossRef] [PubMed]

69. Levi, M.; Sivapalaratnam, S. An overview of thrombotic complications of old and new anticancer drugs. *Thromb. Res.* **2020**, *191*, S17–S21. [CrossRef]

70. ISO. *ISO 1871: 2009 Food and Feed Products—General Guidelines for the Determination of Nitrogen by the Kjeldahl Method*; ISO: Geneva, Switzerland, 2009.

71. ISO. *ISO 763: 2003 Fruit and Vegetable Products—Determination of Ash Insoluble in Hydrochloric Acid*; ISO: Geneva, Switzerland, 2003.

72. Rybicka, I.; Gliszczyńska-Świgło, A. Minerals in grain gluten-free products. The content of calcium, potassium, magnesium, sodium, copper, iron, manganese, and zinc. *J. Food Compos. Anal.* **2017**, *59*, 61–67. [CrossRef]
73. European Food Safety Authority. Dietary Reference Values for nutrients Summary report. *EFSA Support. Publ.* **2017**, *14*, e15121. [CrossRef]
74. Tomczak, A.; Zielińska-Dawidziak, M.; Piasecka-Kwiatkowska, D.; Lampart-Szczapa, E. Blue lupine seeds protein content and amino acids composition. *Plant, Soil Environ.* **2018**, *64*, 147–155. [CrossRef]
75. AOAC International. *AOAC Official Method 994.12 Amino Acids in Feeds*; AOAC International: Rockville, MD, USA, 2000.
76. Singleton, V.L.; Orthofer, R.; Lamuela-Raventós, R.M. Analysis of total phenols and other oxidation substrates and antioxidants by means of folin-ciocalteu reagent. In *Methods in Enzymology*; Academic Press: Cambridge, MA, USA, 1999; Volume 299, pp. 152–178.
77. Re, R.; Pellegrini, N.; Proteggente, A.; Pannala, A.; Yang, M.; Rice-Evans, C. Antioxidant activity applying an improved ABTS radical cation decolorization assay. *Free Radic. Biol. Med.* **1999**, *26*, 1231–1237. [CrossRef]
78. Mosmann, T. Rapid colorimetric assay for cellular growth and survival: Application to proliferation and cytotoxicity assays. *J. Immunol. Methods* **1983**, *65*, 55–63. [CrossRef]

 *molecules*

Article

Whey Proteins as a Potential Co-Surfactant with *Aesculus hippocastanum* L. as a Stabilizer in Nanoemulsions Derived from Hempseed Oil

Wojciech Smułek [1], Przemysław Siejak [2], Farahnaz Fathordoobady [3], Łukasz Masewicz [2], Yigong Guo [3], Małgorzata Jarzębska [4], David D. Kitts [3], Przemysław Łukasz Kowalczewski [5], Hanna Maria Baranowska [2], Jerzy Stangierski [6], Anna Szwajca [7], Anubhav Pratap-Singh [3,*] and Maciej Jarzębski [2,*]

[1] Institute of Chemical Technology and Engineering, Poznan University of Technology, Berdychowo 4, 60-695 Poznań, Poland; wojciech.smulek@put.poznan.pl

[2] Department of Physics and Biophysics, Faculty of Food Science and Nutrition, Poznań University of Life Sciences, Wojska Polskiego 38/42, 60-637 Poznań, Poland; przemyslaw.siejak@up.poznan.pl (P.S.); lukasz.masewicz@up.poznan.pl (Ł.M.); hanna.baranowska@up.poznan.pl (H.M.B.)

[3] Food, Nutrition and Health Program, Faculty of Land & Food Systems, The University of British Columbia, 2205 East Mall, Vancouver, BC V6T 1Z4, Canada; farah.fathordoobady@ubc.ca (F.F.); yigong.guo@ubc.ca (Y.G.); david.kitts@ubc.ca (D.D.K.)

[4] Independent Researcher, 60-343 Poznań, Poland; malgorzata.jarzebska@o2.pl

[5] Department of Food Technology of Plant Origin, Poznań University of Life Sciences, 31 Wojska Polskiego St., 60-624 Poznań, Poland; przemyslaw.kowalczewski@up.poznan.pl

[6] Department of Food Quality and Safety Management, Faculty of Food Science and Nutrition, Poznań University of Life Sciences, Wojska Polskiego 31/33, 60-624 Poznań, Poland; jerzy.stangierski@up.poznan.pl

[7] Department of Synthesis and Structure of Organic Compounds, Faculty of Chemistry, Adam Mickiewicz University in Poznań, Uniwersytetu Poznańskiego 8, 61-614 Poznań, Poland; anna.szwajca@amu.edu.pl

* Correspondence: anubhav.singh@ubc.ca (A.P.-S.); maciej.jarzebski@up.poznan.pl (M.J.)

Citation: Smułek, W.; Siejak, P.; Fathordoobady, F.; Masewicz, Ł.; Guo, Y.; Jarzębska, M.; Kitts, D.D.; Kowalczewski, P.Ł.; Baranowska, H.M.; Stangierski, J.; et al. Whey Proteins as a Potential Co-Surfactant with *Aesculus hippocastanum* L. as a Stabilizer in Nanoemulsions Derived from Hempseed Oil. *Molecules* **2021**, *26*, 5856. https://doi.org/10.3390/molecules26195856

Academic Editor: Seyed Fakhreddin Hosseini

Received: 2 July 2021
Accepted: 20 September 2021
Published: 27 September 2021

Publisher's Note: MDPI stays neutral with regard to jurisdictional claims in published maps and institutional affiliations.

Abstract: The use of natural surfactants including plant extracts, plant hydrocolloids and proteins in nanoemulsion systems has received commercial interest due to demonstrated safety of use and potential health benefits of plant products. In this study, a whey protein isolate (WPI) from a byproduct of cheese production was used to stabilize a nanoemulsion formulation that contained hempseed oil and the *Aesculus hippocastanum* L. extract (AHE). A Box–Behnken experimental design was used to set the formulation criteria and the optimal nanoemulsion conditions, used subsequently in follow-up experiments that measured specifically emulsion droplet size distribution, stability tests and visual quality. Regression analysis showed that the concentration of HSO and the interaction between HSO and the WPI were the most significant factors affecting the emulsion polydispersity index and droplet size (nm) ($p < 0.05$). Rheological tests, Fourier transform infrared spectroscopy (FTIR) analysis and $L^*a^*b^*$ color parameters were also taken to characterize the physicochemical properties of the emulsions. Emulsion systems with a higher concentration of the AHE had a potential metabolic activity up to 84% in a microbiological assay. It can be concluded from our results that the nanoemulsion system described herein is a safe and stable formulation with potential biological activity and health benefits that complement its use in the food industry.

Keywords: nanoemulsion; whey protein; *Aesculus hippocastanum* L.; hempseed oil; emulsion stability; droplet size

1. Introduction

Emulsions are widely used in many industries and have an important role in the food industry to stabilize many different formulations that have either oil-in-water (O/W) or water-in-oil (W/O) dispersions [1]. Stability of emulsions is influenced by many factors, e.g., the concentration and ratio of individual phases, methods of preparation, storage condition, as well as the presence of compounds demonstrating the surface-active property (stabilizers) [2]. Therefore, new sources of natural, effective stabilizers, or surfactants, are in high demand to use plant-based ingredients that provide bioactive extracts [3–5], hydrocolloid activity [6], plant proteins [7–9]. Many biopolymers (proteins, polysaccharides or the combinations) that are used to assist in the formation of an emulsion have excellent effectiveness since they contain both hydrophilic and hydrophobic groups capable of lowering the surface tension between water and the oil phase [10]. Such proteins as collagen, whey protein, soy protein isolate and casein are the most commonly used due to a high emulsifying index and the water/oil holding capacity [11].

Hempseed oil (HSO), derived from the seeds of *Cannabis sativa* L., is recognized for both nutritional, health-promoting and bioactive properties [12,13]. It is a good source of both n-3 and n-6 essential fatty acids, specifically linoleic acid (55%) and alpha-linolenic acid (20%), with concentrations relatively higher than in other vegetable oils [14]. HSO is also a source of gamma linoleic acid (GLA) not found in many vegetable oils, but with noted bioactive properties. HSO's health advantages have been ascribed to the 3:1 n-3/n-6 fatty acid ratio, which is an established optimum for human nutrition uptake [14], and the presence of GLA [15]. Additionally, the presence of minor components, such as tocopherol, tocotrienols, carotenes, minerals, terpenoids and β-sitosterol also add to the nutritional value of HSO. Besides these health effects, HSO is often used in formulations that require high miscibility with water [16], which reduces the reliance of additional surfactants for use in nanoemulsions. Accordingly, HSO has been used in cosmetic, nutraceutical, food and functional food industries.

Dairy proteins have been widely used as emulsifiers in food industry. They adsorb to the oil droplet interface, producing a consistent protective film that can help prevent droplet aggregation [17]. They have also been used as emulsifiers in nanoemulsion formulations. Whey protein (WP), with a noted high nutritional value, is often used as an ingredient in many food products and for the production of sports nutrition [18,19]. Whey proteins are a good source of essential amino acids, and globular proteins isolated from whey are noted for having biological activities: antioxidant, immunostimulatory and anti-obesity [20]. They are also effective emulsifiers in food formulations because their amphiphilic properties enable relatively stable gels and microcapsules without additional chemical additives. For example, WP has been used to encapsulate probiotics [21–23]. Iqbal et al. [24] used the WI in the formation to generate a three-dimensional network in an oil/water phase. However, the resulting emulsions did not exhibit ideal plastic behavior. Since WP can be prepared as an isolate or a concentrate, the WPI has been used to stabilize nanoemulsion formulations containing peanut oil, corn oil, β-carotene in sunflower oil and α-Tocopherol in palm oil. It was found that the lowest degradation of β-carotene at 55 °C occurred when WPI was used as stabilizer in the nanoemulsion system. This was probably related to the substantial antioxidant property of WPI [25].

Aesculus hippocastanum L. is also known as horse chestnut [26] and contains many biologically active compounds including escin, a mixture of saponins with medicinal properties. The health beneficial effects of AH include anti-inflammatory and anti-edematous [27] properties, with potential benefits for kidney disease in diabetic nephropathy [28]. Due to the high content of polyphenols, such as quercetin and kaempferol [29], *Aesculus hippocastanum* L. bark also exhibits antioxidant and antitumor activities [30]. In addition, the extracts from *Aesculus hippocastanum* L. (AHE) with a high content of saponins [31] exhibited high interfacial activity due to hydrophobic aglycone and hydrophilic sugar residues [12]. The United States Department of Agriculture (USDA) considers the AHE to be generally regarded as safe, and the European Union has also approved it for use as a foam stabilizer in beverages [12]. AHE was found to be an effective stabilizer when used as a stabilizer in hempseed oil (HSO) nanoemulsions [12]. HSO–nanoemulsion formulations containing 2 g/L AHE also produce smaller droplets.

The purpose of this study was to prepare an O/W formulation of an HSO nanoemulsion system that was stabilized with an optimal concentration of the WPI and the AHE. The goal was to obtain both acceptable stability of the emulsion and also potential biological activity of its ingredients for both food and biomedical applications. We also focused on using the WPI as a potential cosurfactant.

2. Results and Discussion

2.1. Stability Tests

Three level Box–Behnken experiment design with response surface methodology was applied for the experiment setup (Table 1). To evaluate the efficiency of the emulsion preparation, a two-step homogenization process using both a high-speed homogenizer and an ultrasound homogenizer was used. To compare the behavior of the emulsion composition, we determined the emulsification index (EI) [31] which measured the emulsified layer relative to the total volume of the mixture. The EI results were compared with visual observation, which are presented in Figure 1 and show samples 24 h after homogenization. The visual measures confirmed that the two-step preparation process strongly increased the homogeneity of the emulsion in contrast to the one-step process [12,30]. One day after the preparation, no visible differences in samples were produced, and homogeneity was noted within the total volume. Visual observation confirmed a slight color difference in the samples, which were attributed to the different HSO and AHE concentrations, respectively.

Table 1. Box–Behnken design samples description; AHE—*Aesculus hippocastanum* L. extract; HSO—hempseed oil.

Sample Name	AHE Concentration (%)	Whey Concentration (%)	HSO Concentration (%)	Total Volume with Water (mL)
S07	1	1	3	
S14	1	2.5	1	
S10	1	2.5	5	
S02	1	4	3	
S04	2	1	1	
S09	2	1	5	
S01	2	2.5	3	
S05	2	2.5	3	
S06 *	2	2.5	3	25
S12 *	2	2.5	3	
S15 *	2	2.5	3	
S17	2	4	1	
S08	2	4	5	
S03	3	1	3	
S16	3	2.5	1	
S11	3	2.5	5	
S13	3	4	3	

* Excluded from droplet size distributions tests.

We employed additional centrifugation tests to accelerate evaluation of the sample emulsion. The ratio of the optical density (measured at 600 nm) of the samples before and after centrifugation was used as an indicator for emulsion stability. The results were collected and presented in Table 2. A simple correlation between the concentration of individual emulsion components on emulsion stability after centrifugation could not be established. Only samples with the lowest HSO content (i.e., 1%) were characterized by higher stability. The highest values for this parameter (>50%) were also observed for samples S16 and S17, respectively, with the same concentration of HSO (1%). These results suggest that greater whey concentrations promoted the formation of relatively more stable emulsions.

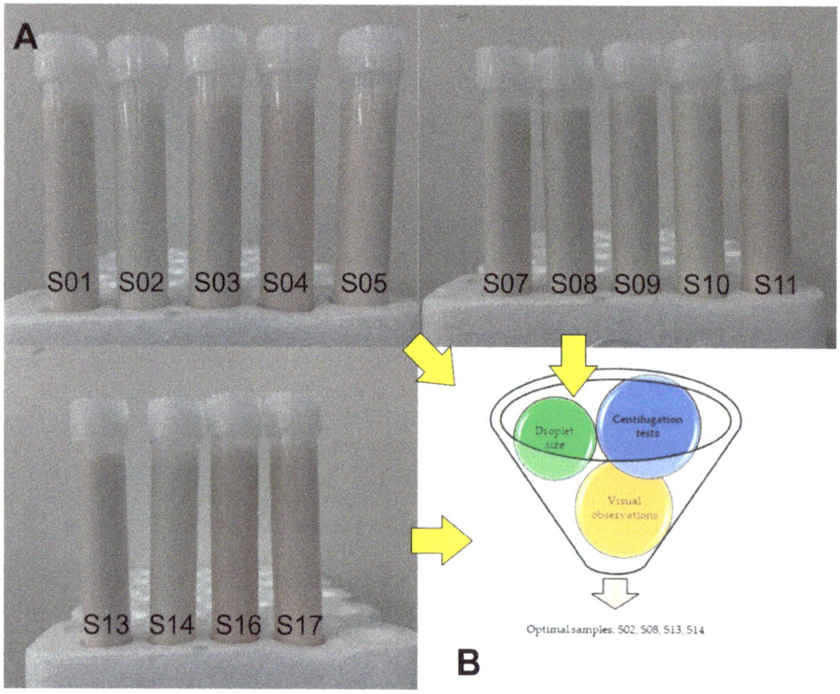

Figure 1. (**A**) The emulsion samples after 24 h. (**B**) Criteria of selecting the optimal samples composition for further studies.

Table 2. Preliminary stability tests after centrifugation. OD_C/OD_0—ratio of the optical density at 600 nm of samples after and before centrifugation.

Sample	AHE [g/L]	Whey [g/L]	HSO [%]	OD_C/OD_0 [-]
1	2	2.5	3	26%
2	1	4	3	44%
3	3	1	3	18%
4	2	1	1	28%
5	2	2.5	3	33%
6	2	2.5	3	30%
7	1	1	3	36%
8	2	4	5	17%
9	2	1	5	41%
10	1	2.5	5	35%
11	3	2.5	5	21%
12	2	2.5	3	33%
13	3	4	3	17%
14	1	2.5	1	31%
15	2	2.5	3	26%
16	3	2.5	1	54%
17	2	4	1	55%

2.2. Droplet Size Studies

Droplet size and PDI (polydisperisty index) are key factors affecting emulsion stability. Based on the experimental design used for preliminary studies (Table 3), different concentrations of HSO, AHE and whey protein were used to find the optimal droplet size distribution of nanoemulsion formulations. According to our previous studies [9,32], we decided to eliminate runs S06, S12 and S15 from the experimental design model (Table 1) for the evaluation of the droplet size. Analysis of the results presented in Table 3 for the center points of the Box–Behnken design were performed on samples S01, S05, S06, S12, S15, and similar results for Z-ave, PDI, average peak maximum by intensity were recorded for the S01 and S05 samples (see the standard deviation range of three replications).

In contrast to previous studies [33,34], we determined the droplet size of the emulsion systems as delivered after preparing the emulsion without additional dilution since dilution may affect the emulsion stability. Furthermore, based on the principles of the dynamic light scattering (DLS) method used herein and apparatus limitations, we chose to round droplet size numbers up to a full number. Moreover, based on previous experience from studies with different high-PDI systems [35], we considered droplet/particle size by intensity and number. Detailed studies of droplet size distributions clearly showed that even a small fraction of larger particles/droplets or accidental contaminations impacted the Z-ave and size distribution. Results presented in Table 2 exhibit that the major percentage of droplets were much smaller than average size distribution by intensity peak maximum. For example, the average maximum of the peak by number for sample S08 was 23 nm; however, its light scattering intensity was 318 nm.

Table 3. Preliminary tests, droplet size distribution results: Z-ave (nm); PDI—polydisperisty index; average size distribution by intensity peak maximum (nm) and average size distribution by number peak maximum (nm).

Sample Name	Z-ave (nm)	PDI	Average Size Distribution by Intensity Peak Maximum (nm)	Average Size Distribution by Number Peak Maximum (nm)
S01	229 ± 5	0.344 ± 0.078	369 ± 80	59 ± 31
S02	214 ± 1	0.303 ± 0.022	317 ± 8	30 ± 5
S03	239 ± 7	0.274 ± 0.063	333 ± 44	69 ± 48
S04	234 ± 4	0.182 ± 0.009	279 ± 31	121 ± 39
S05	221 ± 6	0.302 ± 0.010	316 ± 26	27 ± 8
S06			N/A	
S07	215 ± 3	0.291 ± 0.026	286 ± 22	33 ± 5
S08	202 ± 4	0.412 ± 0.016	318 ± 20	23 ± 7
S09	246 ± 2	0.471 ± 0.035	456 ± 46	38 ± 19
S10	195 ± 5	0.439 ± 0.044	306 ± 33	31 ± 19
S11	212 ± 3	0.441 ± 0.029	326 ± 84	74 ± 44
S12			N/A	
S13	218 ± 7	0.322 ± 0.040	322 ± 33	71 ± 46
S14	215 ± 2	0.190 ± 0.018	253 ± 6	129 ± 12
S15			N/A	
S16	232 ± 4	0.183 ± 0.019	288 ± 17	103 ± 42
S17	243 ± 3	0.218 ± 0.038	291 ± 22	98 ± 50

N/A: not applicable; ±SD (standard deviation of three replicates).

According to the regression analysis and the ANOVA, we were able to predict Z-Ave for samples using a quadratic model ($p < 0.05$). The relative significance in order of individual factors according to the greatest effect was determined to be as follows: HSO (%) > whey (%) > AHE. The interaction of HSO and whey was a significant factor affecting Z-ave of samples ($p < 0.05$). The significant interaction effect of HSO and a poloxamer used as a surfactant to produce an HSO-based nanoemulsion was previously reported [36]. The ANOVA and the regression analysis also showed that the PDI of nanoemulsion samples could be fit to a linear model with HSO (%) as the only significant factor ($p < 0.05$). To find an optimal and stable formulation and further possible applications of the emulsion system, other characterizations such as droplet size distribution, visual, centrifugation and microscopic properties of the emulsion need to be considered in addition to Z-ave and the PDI. Additional stability studies, such as identification of destabilization mechanisms including creaming, sedimentation, flocculation, coalescence can also confirm the stability potential of an emulsion system. Based on the preliminary stability tests as well as droplet size distribution, we selected four samples for further analysis (Figure 1B).

Figure 2 presents detailed analysis of droplet size distribution from four selected samples. All the selected samples contained a whey concentration of 4, except for S14, which had a whey concentration of 2.5%. According to the DLS results of S14 (Figure 2D), the scattered light intensity and droplet size by number were not different. This result was confirmed by a low PDI value (0.190). Microscopic observation of sample S14 (Figure 2D) also confirmed homogeneity of the samples. In addition, in comparison with other samples, single droplets were easy to distinguish at low magnification (e.g., $\times 40$).

Based on our previous observations for the HSO emulsion system stabilized by the AHE [12], we decided to evaluate emulsion droplet distribution and sample homogeneity using light microscopy with an inverted microscope. To prevent fast evaporation of samples, the emulsions were inserted in a corvette equipped with slide channels. The images presented in Figure 3 were taken a few minutes after the emulsion injection. This method prevents the flow of injected liquid artefacts. For unstable systems, a short break in time enabled the observed coalescence of the droplets. With lower magnification, high homogeneity of samples S02, S08 and S13 was observed (Figure 3A–C). Only a small fraction of larger droplets could be distinguished. This observation corresponded with the DLS results in case of droplet size distribution by intensity, where a more intensive signal was detected for the samples with larger hydrodynamic diameters (Figure 2A–C). The presence of the smaller droplets fraction determined by DLS was probably caused by two-step processing using ultrasound treatment for preparing this emulsion. Some of the studies suggested that ultrasound treatment as well as ultrahigh-pressure homogenization caused an increase in the α-helix structure content of whey protein and a decrease in the β-sheet component of whey protein [37,38].

Microscopic imaging for nanoemulsion systems with proteins presented by Zhu et al. [39] corresponded to our findings. In their studies, optical microscopic imaging was compared with confocal laser scanning microscopy (CLSM). It was suggested that using CLSM allowed for the determination of a core–shell structure in the emulsion systems where proteins were located at the surface of the emulsion droplets. More detailed studies are needed for our nanoemulsion systems. First of all, signals from all the fluorescent structures should be identified and separated (i.e., from HSO, AHE) [12,31]. Then, additional studies on the impact of whey protein on the fluorescent behavior of the components should be investigated. Ren and Giusti [40] showed that anthocyanin-rich extracts decreased the fluorescence intensity of whey protein while increasing λ_{max}. The study concluded that thermally induced whey protein was effective in protecting anthocyanin from color degradation. Using an optical microscope, we focused only on the verification of the homogeneity of the samples as well as possible coalescence (which was not observed in the optimal nanoemulsions). Nevertheless, we strongly recommend using more than one technique for the analysis of the emulsion systems droplets size.

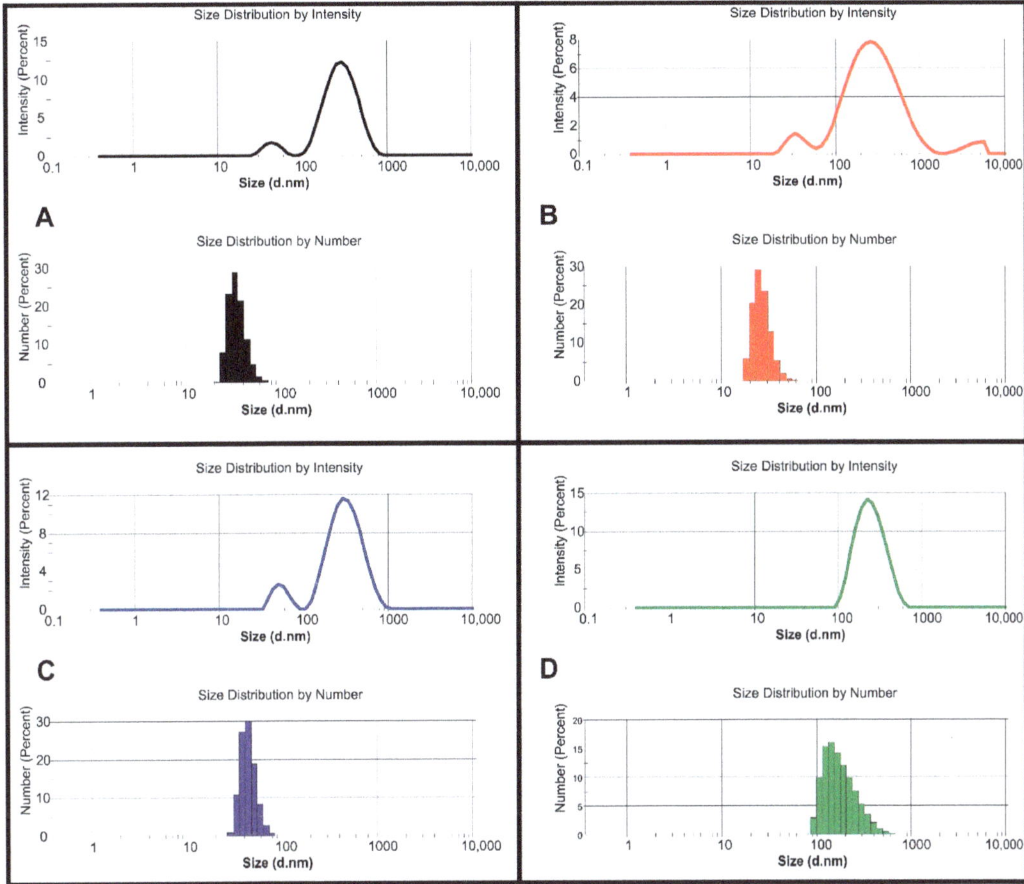

Figure 2. Droplet size distribution by intensity and by number: (**A**) S02, (**B**) S08, (**C**) S13, (**D**) S14.

2.3. Rheological Tests

Figure 4a shows the flow curves of the various samples, which illustrate the samples' rheological properties. The flow curves showed a non-Newtonian behavior, with a decreasing slope (viscosity) up to a certain cut-off, suggesting typical pseudoplastic behavior. However, the viscosity increased after a certain shear rate, which is a typical property of a dilatant fluid. The changes in the dynamic viscosity of the samples with increasing shear rate are shown in Figure 4b. It was apparent that the S08 sample had the highest viscosity, followed by S13 and S02, with S14 being least viscous at various shear rates. All the samples exhibited two types of behavior: first, shear thinning, and then shear thickening, after a particular shear rate cut-off. This became more apparent in Figure 4c when the power law was used to model these curves to evaluate the coefficient that depicted clearly the two zones, with a cut-off close to 130 Hz. As a result, a broken power law model (Equation (1)) was used to describe the flow behavior of the samples.

$$\eta = K_1(\dot{\gamma})^{n_1-1} \text{ for } \dot{\gamma} < 130 = K_2(\dot{\gamma})^{n_2-1} \text{ for } \dot{\gamma} \geq 130 \qquad (1)$$

where η is the dynamic viscosity, K_1 and K_2 are the consistency coefficients, $\dot{\gamma}$ is the shear rate and n_1 and n_2 are the flow behavior indices.

Table 4 shows the broken power law model parameters. The consistency coefficient K_1 at shear rates less than 130 Hz was found to be highest for S08, while those of other samples were not significantly different. This measure is an indicator of the initial system viscosity [41], suggesting that sample S08 had higher viscosity to begin with, which was sustained even at changing shear rates.

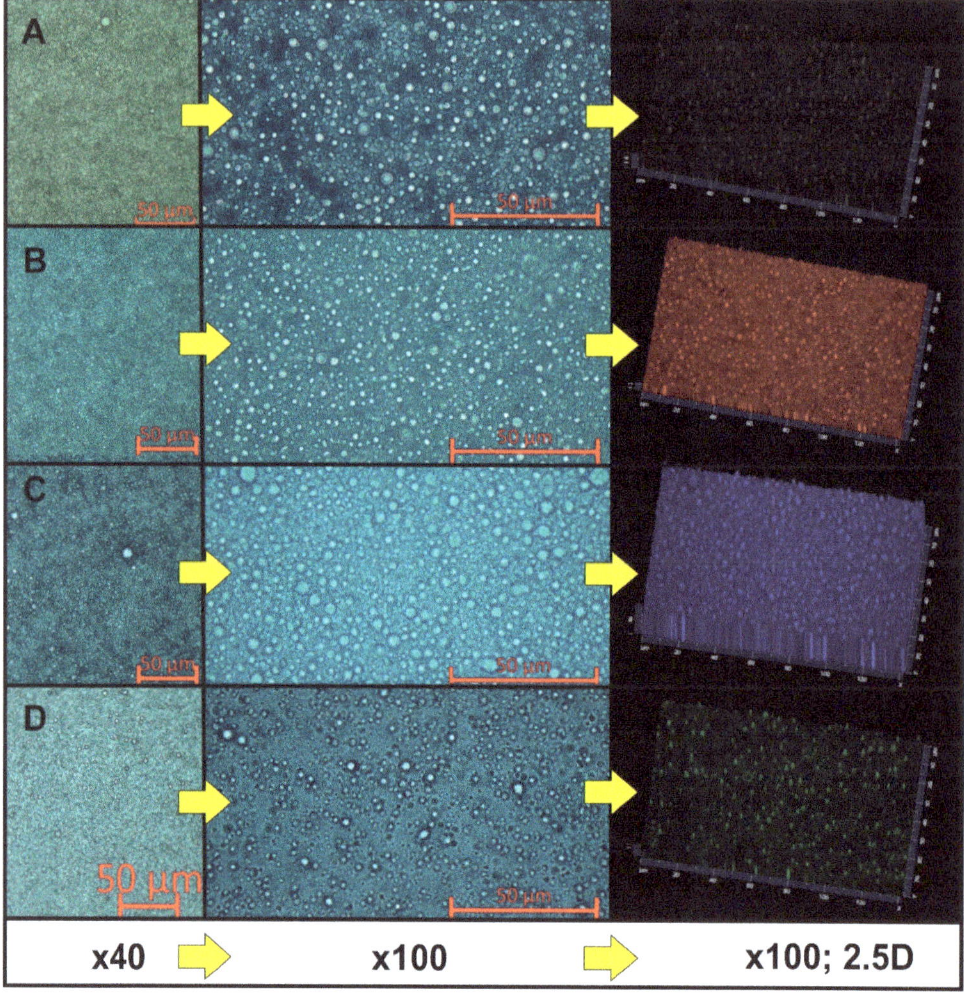

Figure 3. Microscopic images of the emulsion droplets (magnification, ×40, ×100, and ×100; 2.5D image): (**A**) S02, (**B**) S08, (**C**) S13, (**D**) S14 (the images were adjusted as the best fit and colored with software).

Table 4. Parameters of the broken power model describing the viscosity curves of the samples: consistency coefficient (K_1 and K_2), flow index (n_1 and n_2).

Sample Name	Consistency Coefficient K_1	Flow Behavior Index n_1	Consistency Coefficient K_2	Flow Behavior Index n_2
S02	1.85 ± 0.03 [a]	0.931 ± 0.002 [c]	0.093 ± 0.009 [a]	1.54 ± 0.02 [b]
S08	2.01 ± 0.04 [b]	0.923 ± 0.003 [b]	0.131 ± 0.008 [c]	1.48 ± 0.02 [a]
S13	1.86 ± 0.02 [a]	0.934 ± 0.004 [c]	0.118 ± 0.006 [b]	1.50 ± 0.01 [a]
S14	1.85 ± 0.04 [a]	0.916 ± 0.003 [a]	0.115 ± 0.007 [b]	1.50 ± 0.02 [a]

[a–c] Values in the same column with the same superscript alphabet letters are not significantly different from each other according to Duncan's grouping of means.

The higher viscosity of S08 as compared to the other samples could be attributed to the highest HSO (5%) content amongst the four tested samples. Furthermore, S14 was found to possess the lowest viscosity on account of having the lowest HSO content (1%). This showed that a higher oil loading was associated with higher viscosity, which is consistent with observations of other researchers [42–44]. Rha [45] and Jarzebski et al. [12] attributed this phenomenon of increasing viscosity with increasing oil loading to the greater formation of interphase layers, creating a larger barrier between the emulsion components.

At a shear rate of 130 Hz, all the samples exhibited a transition from shear thinning behavior to shear thickening behavior. Again, samples S08, having the highest oil loading, and sample S14, with the lowest oil and whey loading, demonstrated the highest and the lowest viscosity, respectively. This transition from shear thinning to the shear thickening behavior could be attributed to the phenomenon that at very high shear rates, tremendous turbulence occurs. When such turbulence occurs after a certain shear rate cut-off (which was 130 Hz in our case), any increase in the shear rate will result in increased turbulence, resulting in increased viscous dissipation and higher resistance to flow, which in turn makes the flow appear as shear-thickening. This effect is a typical example of the Taylor vortex flow in shear-thinning fluids [46,47]. The interpretation of Chhabra and Richardson [48] could be used to explain such transitionary behavior of our shear-thinning samples. At rest, the emulsion has sufficient interfacial tension to be stable. At low shear rates, lubrication for particle motion of the continuous phase between the plates is provided by the dispersed oil phase resulting in decreased stress with increasing shear (shear-thinning behavior). However, at high shear rates, the emulsion breaks, and the dispersed phase is completely separated from the continuous phase due to centrifugal forces. Furthermore, continuous and dispersed phases expand or dilate slightly under increasing shear strain, resulting in increased friction and shear stress, causing the dynamic viscosity to increase rapidly with shear rate.

2.4. FTIR

The FTIR spectra were recorded with air or water as the background. Both series of results are presented in Figure 5.

The HSO spectrum shows clear signals originating from the C–H bonds (around 2950 cm^{-1}) and C = O (at 1700 cm^{-1}), respectively. The absence of signals that are characteristic of hydroxyl group bonds proves that there were no significant amounts of free fatty acids and hydroxylated acids in the oil. A relatively weak but distinctive signal slightly above 3000 cm^{-1} confirms the existence of double bonds between carbon atoms, which most likely originated from unsaturated fatty acid residues in HSO, as confirmed by others [49].

Among the signals common to all emulsion samples, those arising from the O–H bonds (at about 3250 cm^{-1}), aliphatic C–H (at 2800–2950 cm^{-1}) and C = O (at 1700 cm^{-1}) should be distinguished. Although the differences in the intensities of the individual bands can be explained by the different composition of the emulsions, it is important to note that these differences were not only in intensity, but also in the position of the signals in the 1050–1150 cm^{-1} range. They could originate from vibrations of the C–O bonds (to a lesser extent, of C–N, which occurs in the protein structure). A similar shift, visible as a change in shape, is present for vibrations from the O–H group. This shift may indicate a variation in the strength and configuration of the hydrogen bonds between polar components of the tested emulsions, suggesting the complex interactions between the emulsion components, which may then have impact on their physicochemical properties [50,51] and plausible biological activity of the samples [52,53].

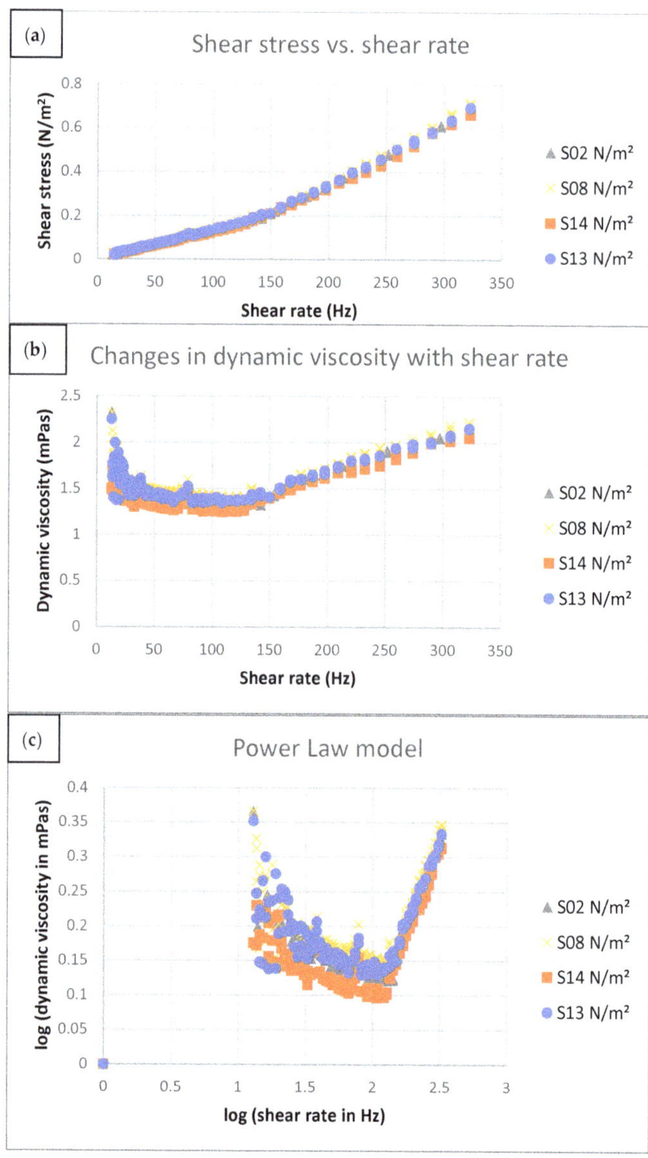

Figure 4. Rheological test results of the S02, S08, S13, and S14 samples: (**a**) flow curve, (**b**) dynamic viscosity changes, (**c**) power law model.

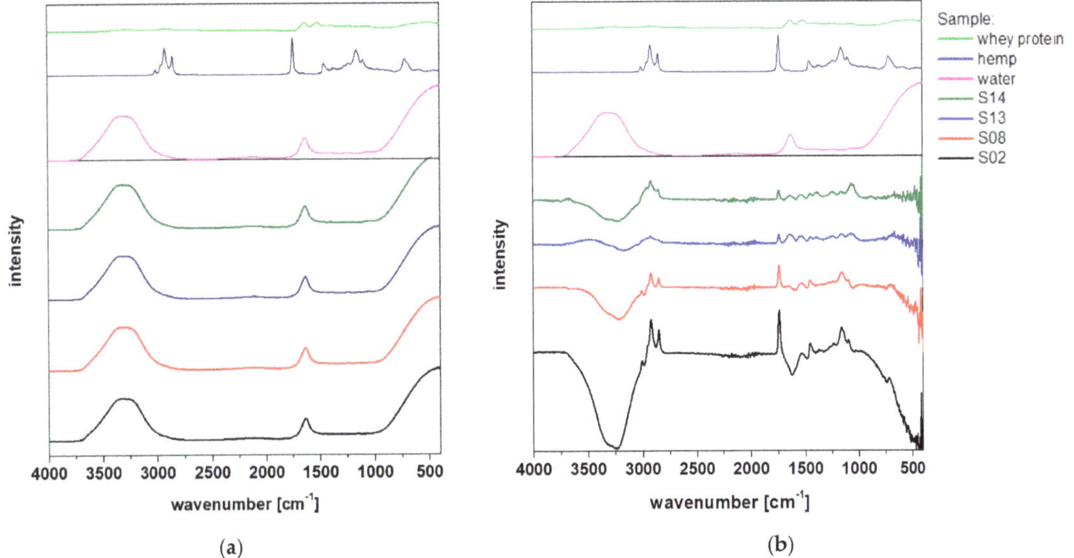

Figure 5. FTIR spectra of investigated samples with (**a**) air or (**b**) water as the background.

2.5. Color and Refractive Index Analysis

The results of the $L^*a^*b^*$ color analysis of samples in Figure 6 showed that the S02 sample represented more lightness, followed by S13, S08 and S14. Considering the ratio of ingredients in samples S02, S13 and S08, it was determined that the L^* value decreased with higher concentrations of HSO or the AHE, respectively. However, sample S14 having a low content of HSO (1%) and the AHE (1%) featured a low L^* value. We hypothesize that other factors might affect the lightness of samples in addition to the concentration of ingredients. Based on droplet size distribution studies, the PDI and Z-ave of S14 were lower compared to other samples (Table 1). In addition, according to McClements and Demetriades (1998), interactions between the ingredients within the emulsion and some other factors such as transmission, reflection, scattering and absorption can also affect the color and appearance of the emulsion [54].

It was also found that the difference between the a^* values was negligible. In this analysis, sample S14 showed lower b^* values. This observation may also relate to the higher ratio of whey to HSO in this formulation, which caused a shift in favor of a yellow shade. To monitor the color of the emulsions with a unique index, the WI values of the samples were calculated based on the L^*, a^* and b^* values. The WI values were found to be 42.68, 35.86, 40.00 and 29.87 for S02, S08, S13 and S14, respectively, indicating marked difference between sample appearance.

The refractive index recorded for the dispersed phase represented the average refractive indices of individual droplets (Table 5). The theory that the thickness of the phase interface is small enough comparative to the wavelength of the related light would indicate that the interface of droplets plus the oil phase act as an individual dispersed phase [55].

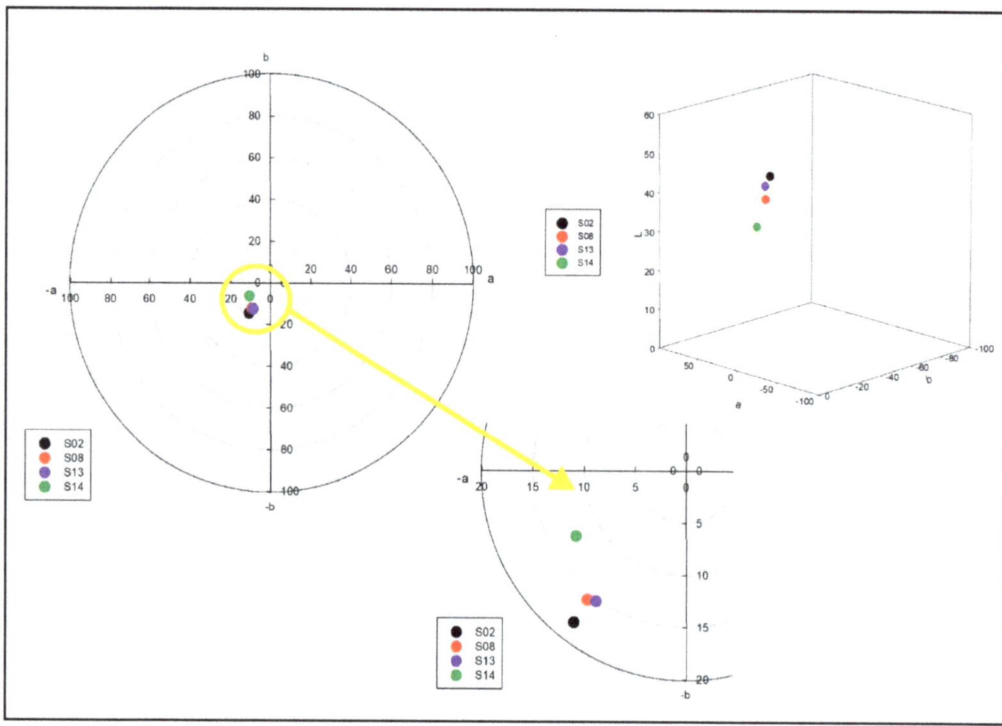

Figure 6. *L*a*b** color analysis results of the samples S02, S08, S13, S14.

Table 5. Refractive index of the investigated samples.

Sample Name	S02	S08	S13	S14
Refractive index	1.3368	1.3400	1.3384	1.3349

Refractive index accuracy was estimated to be 0.0001.

The total appearance of an emulsion can be determined by both light scattering and absorption. Scattering is mainly accountable by recording the turbidity or lightness of an emulsion, while absorption determines the chromatic properties (redness, blueness, greenness, etc.) [56]. The relationship between the color of an emulsion and the refractive index ratio has an important role in foods containing high proportions of the aqueous phase. The refractive index of the samples (Table 4) was similar to the RI of the dispersed phase (water). This property allowed us to use an optical transmission method to analyze emulsions. The level of lipid oxidation in an emulsion, for example, can be determined by adding some chemicals (such as glycerol) to adjust the RI to 1.0, then measuring the spectra of the absorption [57].

2.6. Biological Activity

To assess the microbiological safety of optimized emulsions, we determined the effect of emulsions on the *Lactobacillus* sp. 2675 strain, a common probiotic (Figure 7). As a reference sample, cultures without an emulsion were considered (with 0% changes in cell metabolic activity). The highest increase in the bacterial cell metabolic activity (84%) was noticed for sample S08. The observed increase for sample S13 was relatively lower (66%). The increase in metabolic activities were lowest for samples S14 and S02 (37% and 31%, respectively). It can be assumed that higher concentrations of the AHE promoted higher probiotic bacteria growth; however, the small number of experiments needs further confirmation of results.

Nevertheless, the critical observations were that all samples showed a positive effect on the metabolic activity of *Lactobaccilus* sp. 2675, indication non-toxic properties of the emulsions. Our results correspond directly with results presented by Gharehcheshmeh et al. [58], who reported no effect of a sweet almond and sesame oil nanoemulsions on the growth of *L. delbrueckii* subsp. *bulgaricus*. There are however, few studies showing an effect of AHE on probiotic bacteria. However, former studies showing an impact on probiotic bacteria of polyphenol-rich extracts from blueberry [57,58] or apples [59] agree with our results.

Nevertheless, the crucial observations are that all the samples exhibited a positive effect on the metabolic activity, which proves the nontoxic properties of the emulsions. Our results correspond directly with the results presented by Gharehcheshmeh et al. who studied the impact of sweet almond and sesame oil nanoemulsions on the growth of *L. delbrueckii* subsp. *bulgaricus*. They did not observe any inhibitory effect of nanoemulsions.

It is worth adding that the blueberry (*Vaccinium corymbosum*) extract [59,60] also showed beneficial effects on probiotic microorganisms due to the presence of polyphenols, which is also a distinguishing feature of the AHE we used [31]. It should also be noted that plant-derived substances can become an additional nutrient for probiotic microorganisms while providing a protective substance against the adverse effects of the external environment. Ahmad et al. [61] obtained promising results using polyphenols extracted from apples to protect the *Bifidobacterium lactis* bacteria during freezing. In the light of these studies, we can see the great potential of our obtained emulsions to stimulate growth and protect probiotic bacteria.

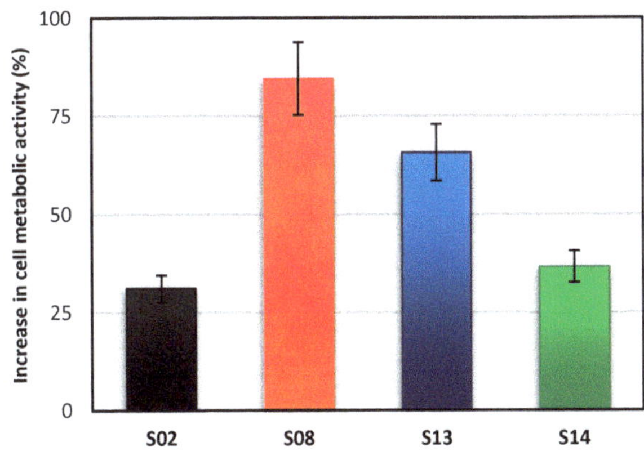

Figure 7. Changes in the metabolic activity of the *Lactobacillus* sp. 2675 cells exposed for 24 h to emulsion samples.

2.7. Water Activity

Finally, water activity (aw) measurements were taken for the optimized samples (see Figure 8). The lowest water activity was determined in sample S02, where the amount of HSO and the WPI were the highest, but the concentration of the AHE surfactant was the lowest. On the other hand, for the emulsion with the lowest concentration of HSO, the WPI and the AHE (S14) produced the highest water activity. This effect on water activity could have been the result of better uniformity of emulsion droplets and the low concentration of whey. Very similar values of water activity were obtained for the samples with the same concentration of whey, but varied with HSO and the AHE (S08 and S13). Nevertheless, based on the microbiological activity results and visual observation of the samples after two-week storage at room temperature (Figure 1), we concluded that water activity of the samples should be monitored regularly.

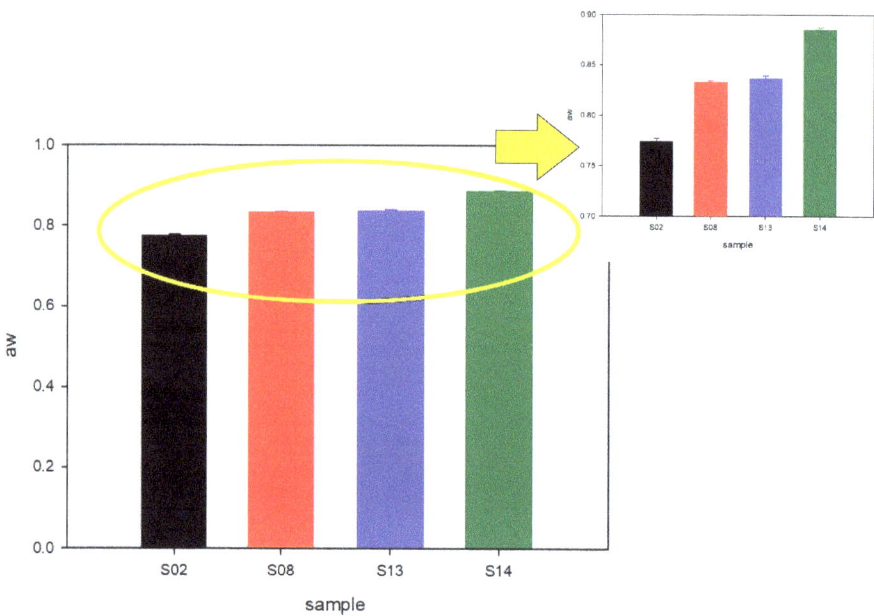

Figure 8. Water activity (aw) results.

3. Materials and Methods

3.1. Materials

For all the experiments performed in this study, the chemicals used were of analytical grade. The solvents and reagents were purchased from Sigma-Aldrich (Poznań, Poland). Hempseed oil (HSO) was purchased from Złoto Polskie (Kalisz, Poland). Whey protein (Isolac® Instant 125H) was purchased from Carbery Group (Ballineen, Ireland). The plant material, *Aesculus hippocastanum* L. bark, was obtained from Flos (Mokrsko, Poland), and was used to prepare a saponin-rich extract as described [12,31]. The total saponin content in the extract, which was equal to $4 \pm 1\%$, was determined using the method described by Hiai et al. [62]. The other main components of the extract included flavonoids, sugars and phenolic acids [31].

3.2. Emulsion Preparation

Emulsion samples with a volume of 20 mL were prepared in sterile 50 mL plastic laboratory tubes. The two-step process used [9] had small modifications, such as in the first step where the components were homogenized using a hand homogenizer CAT X120 equipped with a T10 shaft. The samples were mixed at 10,000 rpm for 600 s. In the second step, the samples were homogenized (sonicator Sonoplus, Bandelin, Berlin, Germany) in the following conditions: 10 min, in 10 s/10 s action/break cycles, amplitude of 16%, and cooled with tap water.

3.3. Experimental Design

In this study, a three level Box–Behnken experiment design with response surface methodology (RSM) (Design Expert software version 13.0) was used for determining the nanoemulsion criteria. The variables included the AHE, WPI and HSO concentrations ($\% \ w/v$) at three coded levels, i.e., $-1, 0, +1$. The specific ranges of the variables were selected according to prior knowledge and preliminary studies performed on the formulation of the hempseed oil nanoemulsion. The Box-Behnken design provided 17 experimental trials with five replications for the center point. The optimal formulations for further stability tests and visual observations were chosen based on the droplet size distribution results of the experimental trails.

3.4. Methods

3.4.1. Stability Tests: EI Index, UV-Vis, Centrifugation

The emulsion stability was evaluated after 24 h by measuring the proportion of the emulsified phase content to the total volume of the homogenized mixture (EI index). Measurements details were described in [31]. In addition, 5 mL of the samples were taken immediately after homogenization, their optical density at 600 nm (OD_0) was measured, and then the samples were centrifuged (10,000 RCF, 10 min) and the optical density (OD_C) was measured again. Evaluation of emulsion stability after centrifugation was determined by the ratio of OD_C to OD_0.

3.4.2. Droplet Size Distribution

Zetasizer Nano-ZS (Malvern, Malvern, UK) was used for measurements of the hydrodynamic diameters (d_H) of droplets of the emulsion samples. The DLS autocorrelation functions were registered from the scattered light, which was recorded at an angle of 173°. The droplet size distribution measurements were performed with the automatic settings mode at 23.5 °C. Immediately prior to the examinations, the samples were kept at 23.5 °C for 5 min inside the measurement cell (the temperature was adjusted to the average laboratory temperature, where the samples were prepared and stored). Measurement series values and their respective standard deviations were obtained from the average of three measurements.

3.4.3. Microscopic Investigations

The microscopic studies were performed using an inverted microscope ZEISS Axio Vert.A1 (Zeiss, Shanghai, China) equipped with a color camera Axiocam 208 (Zeiss, China). Imaging was performed using two kinds of objectives with different magnifications: LD A-Plan ($\times 40/0.55$; phase 1 (air)) and A-Plan ($\times 100/1.25$; phase 2 (oil)). For the imaging, the emulsions were inserted into a 1 μ-Slide VI 0.1 cuvette (ibidi GmbH, Gräfelfing, Germany). For the presentation, the resolution of the images was automatically adjusted by the best fit with the ZEN2.5 software (Zeiss, Jena, Germany).

3.4.4. Rheological Tests

A ViscoQC 300 viscometer (Anton Paar GmbH, Graz, Austria) was used to determine the rheological properties of the emulsions. The tests were conducted at room temperature using the "double-gap" DG26 system with an L1 spindle. The speed of the spindle ranged from 7 to 250 rpm, with an increasing trend, which corresponds to a shear rate of 10–322 1/s. Each shear rate was imposed for 1 min to stabilize the viscosity. Each test was performed in triplicate, with fresh samples.

3.4.5. FTIR

The FTIR spectra were obtained using a Spectrum Two FT-IR spectrometer equipped with a Universal ATR with a diamond crystal (PerkinElmer, Waltham, MA, USA). The data were collected over the 4000–500 cm^{-1} spectral range. Typically, a few microliters were used on the diamond, and the measurements were repeated three times for each sample.

3.4.6. Color Analysis

For color evaluation, an NH310 portable spectrophotometer (Shenzhen Threenh Technology Co., Ltd., Shenzhen, China) equipped with internal software was applied. Before the examinations, 2 mL of the sample were inserted into a transparent plastic cuvette. The measurements were carried out in a cuvette inserted into the dedicated measurement chamber. The color tests were repeated 10 times, and the average values with the SD were recorded. The whiteness index (*WI*) of the emulsions was calculated using the following equation [63]:

$$WI = 100 - \left[\left(100 - L* \right)^2 + a*^2 + b*^2 \right]^{1/2}$$

3.4.7. Refractive Index

The refractive indices (RI) were determined using a PAL-RI optical electronic refractometer (ATANGO CO., LTD., Tokyo, Japan). The measurements were repeated 10 times, and the mean value with the SD is given as a final result.

3.4.8. Biological Activity

The bioactivity of the selected emulsions was determined using an environmental cell metabolic activity test of the *Lactobacillus* sp. 2675 strain. The test was performed according to the methodology described by Pacholak et al. (2019). In these assays, 0.25 mL of the emulsion, 0.25 mL of the cell suspension in a nutrient broth (OD_{600nm} ca. 1.0 after 24 h of incubation at 30 °C) and 0.05 mL of the MTT indicator solution (5 mg/mL) were used. After 24 h of incubation, the reduction of yellow MTT into its purple formazan, which is catalyzed by cellular respiratory pathway enzymes, was measured. The samples in which the emulsion was replaced with deionized water were used as the reference sample.

3.4.9. Water Activity

Measurements of water activity of the emulsions were conducted using water diffusion and activity analyzer ADA-7 (COBRABID, Poznan, Poland). The system is equipped with automatic time recording of water evacuation runs from individual samples. Detailed characteristics of the experimental method are specifically described in [64]. The emulsion samples were placed in the measuring vessels. The volume of the sample was 2 mL. The temperature during measurements was stabilized at 20.0 °C ± 0.1 °C using Peltier modules. The chamber was dried to the water activity of 0.05. The duration of one measurement was set to 1400 s.

Author Contributions: Conceptualization, W.S., M.J. (Maciej Jarzębski) and A.P.-S.; methodology, W.S., F.F., A.P.-S. and M.J. (Maciej Jarzębski); software, W.S., P.S., F.F., Ł.M., A.P.-S. and M.J. (Maciej Jarzębski); validation, W.S., P.S., F.F., Ł.M., Y.G., M.J. (Małgorzata Jarzębska), D.D.K., P.Ł.K., H.M.B., J.S., A.S., A.P.-S. and M.J. (Maciej Jarzębski); formal analysis, W.S., P.S., F.F., Ł.M., Y.G., M.J. (Małgorzata Jarzębska), D.D.K., P.Ł.K., H.M.B., J.S., A.S., A.P.-S. and M.J.(Maciej Jarzębski); investigation, W.S., P.S., F.F., Ł.M., H.M.B., J.S., A.S., M.J. (Maciej Jarzębski) and A.P.-S.; resources, W.S., P.S. and P.Ł.K.; data curation, W.S., P.S., F.F., Ł.M., Y.G., M.J. (Małgorzata Jarzębska), D.D.K., P.Ł.K., H.M.B., J.S., A.S., A.P.-S. and M.J. (Maciej Jarzębski); writing, W.S., P.S., F.F., Ł.M., Y.G., M.J. (Małgorzata Jarzębska), D.D.K., P.Ł.K., H.M.B., J.S., A.S., A.P.-S. and M.J. (Maciej Jarzębski); visualization, W.S., P.S., Ł.M., A.P.-S. and M.J. (Maciej Jarzębski); supervision, M.J. (Maciej Jarzębski), W.S. and A.P.-S.; funding acquisition, P.S., P.Ł.K. and A.P.-S. All authors have read and agreed to the published version of the manuscript.

Funding: The authors would like to acknowledge funding support from the Natural Sciences and Engineering Research Council of Canada's Discovery Grant Programme (Grant No. RGPIN-2018-04735 to Anubhav Pratap-Singh).

Institutional Review Board Statement: Not applicable.

Informed Consent Statement: Not applicable.

Data Availability Statement: All collected data were presented in the manuscript.

Conflicts of Interest: The authors declare no conflict of interest.

Sample Availability: Samples of HSO are commercially available; samples of AHE and the HSO nanoemulsion might be prepared upon request to the corresponding author.

References

1. Hasenhuettl, G.L. Overview of Food Emulsifiers. In *Food Emulsifiers and Their Applications*; Springer: New York, NY, USA, 2008; pp. 1–9.
2. Dickinson, E. Emulsion Stability. In *Food Hydrocolloids*; Springer US: Boston, MA, USA, 1994; pp. 387–398.
3. Jarzębski, M.; Smułek, W.; Kościński, M.; Białopiotrowicz, T.; Kaczorek, E. *Verbascum nigrum* L. (mullein) extract as a natural emulsifier. *Food Hydrocoll.* **2018**, *81*, 341–350. [CrossRef]
4. Ralla, T.; Salminen, H.; Edelmann, M.; Dawid, C.; Hofmann, T.; Weiss, J. Sugar Beet Extract (*Beta vulgaris* L.) as a New Natural Emulsifier: Emulsion Formation. *J. Agric. Food Chem.* **2017**, *65*, 4153–4160. [CrossRef] [PubMed]
5. Drozłowska, E.; Bartkowiak, A.; Trocer, P.; Kostek, M.; Tarnowiecka-Kuca, A.; Łopusiewicz, Ł. Formulation and Evaluation of Spray-Dried Reconstituted Flaxseed Oil-In-Water Emulsions Based on Flaxseed Oil Cake Extract as Emulsifying and Stabilizing Agent. *Foods* **2021**, *10*, 256. [CrossRef] [PubMed]
6. Dickinson, E. Hydrocolloids acting as emulsifying agents—How do they do it? *Food Hydrocoll.* **2018**, *78*, 2–14. [CrossRef]
7. Dammak, I.; Sobral, P.J.D.A.; Aquino, A.; Das Neves, M.A.; Conte-Junior, C.A. Nanoemulsions: Using emulsifiers from natural sources replacing synthetic ones—A review. *Compr. Rev. Food Sci. Food Saf.* **2020**, *19*, 2721–2746. [CrossRef] [PubMed]
8. Pojić, M.; Misan, A.; Tiwari, B.K. Eco-innovative technologies for extraction of proteins for human consumption from renewable protein sources of plant origin. *Trends Food Sci. Technol.* **2018**, *75*, 93–104. [CrossRef]
9. Jarzębski, M.; Fathordoobady, F.; Guo, Y.; Xu, M.; Singh, A.; Kitts, D.D.; Kowalczewski, P.; Jeżowski, P.; Singh, A.P. Pea Protein for Hempseed Oil Nanoemulsion Stabilization. *Molecules* **2019**, *24*, 4288. [CrossRef] [PubMed]
10. Zembyla, M.; Murray, B.S.; Sarkar, A. Water-in-oil emulsions stabilized by surfactants, biopolymers and/or particles: A review. *Trends Food Sci. Technol.* **2020**, *104*, 49–59. [CrossRef]
11. Domínguez, R.; Munekata, P.E.; Pateiro, M.; López-Fernández, O.; Lorenzo, J.M. Immobilization of oils using hydrogels as strategy to replace animal fats and improve the healthiness of meat products. *Curr. Opin. Food Sci.* **2020**, *37*, 135–144. [CrossRef]
12. Jarzębski, M.; Smułek, W.; Siejak, P.; Rezler, R.; Pawlicz, J.; Trzeciak, T.; Jarzębska, M.; Majchrzak, O.; Kaczorek, E.; Kazemian, P.; et al. *Aesculus hippocastanum* L. as a Stabilizer in Hemp Seed Oil Nanoemulsions for Potential Biomedical and Food Applications. *Int. J. Mol. Sci.* **2021**, *22*, 887. [CrossRef]
13. Mikulcová, V.; Kašpárková, V.; Humpolíček, P.; Buňková, L. Formulation, Characterization and Properties of Hemp Seed Oil and Its Emulsions. *Molecules* **2017**, *22*, 700. [CrossRef] [PubMed]
14. Fathordoobady, F.; Singh, A.; Kitts, D.D.; Singh, A.P. Hemp (Cannabis Sativa L.) Extract: Anti-Microbial Properties, Methods of Extraction, and Potential Oral Delivery. *Food Rev. Int.* **2019**, *35*, 664–684. [CrossRef]
15. Kitts, D.D.; Singh, A.; Fathordoobady, F.; Doi, B.; Singh, A.P. Plant Extracts Inhibit the Formation of Hydroperoxides and Help Maintain Vitamin E Levels and Omega-3 Fatty Acids during High Temperature Processing and Storage of Hempseed and Soybean Oils. *J. Food Sci.* **2019**, *84*, 3147–3155. [CrossRef] [PubMed]
16. Yin, J.; Xiang, C.; Wang, P.; Yin, Y.; Hou, Y. Biocompatible nanoemulsions based on hemp oil and less surfactants for oral delivery of baicalein with enhanced bioavailability. *Int. J. Nanomed.* **2017**, *12*, 2923–2931. [CrossRef] [PubMed]
17. Adjonu, R.; Doran, G.; Torley, P.; Agboola, S. Whey protein peptides as components of nanoemulsions: A review of emulsifying and biological functionalities. *J. Food Eng.* **2014**, *122*, 15–27. [CrossRef]
18. Smithers, G.W. Whey and whey proteins—From 'gutter-to-gold'. *Int. Dairy J.* **2008**, *18*, 695–704. [CrossRef]
19. Prazeres, A.R.; Carvalho, F.; Rivas, J. Cheese whey management: A review. *J. Environ. Manag.* **2012**, *110*, 48–68. [CrossRef]
20. Patel, S. Functional food relevance of whey protein: A review of recent findings and scopes ahead. *J. Funct. Foods* **2015**, *19*, 308–319. [CrossRef]
21. Krunic, T.; Rakin, M.; Bulatovic, M.; Zaric, D. The Contribution of Bioactive Peptides of Whey to Quality of Food Products. In *Food Processing for Increased Quality and Consumption*; Elsevier: Amsterdam, The Netherlands, 2018; pp. 251–285. [CrossRef]
22. Guzey, D.; McClements, D.J. Formation, stability and properties of multilayer emulsions for application in the food industry. *Adv. Colloid Interface Sci.* **2006**, *128–130*, 227–248. [CrossRef]
23. Mantovani, R.A.; Furtado, G.D.F.; Netto, F.M.; Cunha, R.L. Assessing the potential of whey protein fibril as emulsifier. *J. Food Eng.* **2018**, *223*, 99–108. [CrossRef]
24. Iqbal, S.; Hameed, G.; Baloch, M.K.; McClements, D.J. Formation of semi-solid lipid phases by aggregation of protein microspheres in water-in-oil emulsions. *Food Res. Int.* **2012**, *48*, 544–550. [CrossRef]
25. Mao, L.; Yang, J.; Xu, D.; Yuan, F.; Gao, Y. Effects of Homogenization Models and Emulsifiers on the Physicochemical Properties of β-Carotene Nanoemulsions. *J. Dispers. Sci. Technol.* **2010**, *31*, 986–993. [CrossRef]
26. Allen, D.J.; Khela, S. Aesculus hippocastanum. In *IUCN Red List Threat. Species*; International Union for Conservation of Nature and Natural Resources: Gland, Switzerland, 2017.
27. Dudek-Makuch, M.; Studzińska-Sroka, E. Horse chestnut—Efficacy and safety in chronic venous insufficiency: An overview. *Rev. Bras. Farm.* **2015**, *25*, 533–541. [CrossRef]
28. Elmas, O.; Erbas, O.; Yigitturk, G. The efficacy of Aesculus hippocastanum seeds on diabetic nephropathy in a streptozotocin-induced diabetic rat model. *Biomed. Pharmacother.* **2016**, *83*, 392–396. [CrossRef] [PubMed]
29. Sirtori, C.R. Aescin: Pharmacology, pharmacokinetics and therapeutic profile. *Pharmacol. Res.* **2001**, *44*, 183–193. [CrossRef]

30. Makino, M.; Katsube, T.; Ohta, Y.; Schmidt, W.; Yoshino, K. Preliminary study on antioxidant properties, phenolic contents, and effects of Aesculus hippocastanum (horse chestnut) seed shell extract on in vitro cyclobutane pyrimidine dimer repair. *J. Intercult. Ethnopharmacol.* **2017**, *6*, 414–419. [CrossRef]

31. Jarzębski, M.; Smułek, W.; Siejak, P.; Kobus-Cisowska, J.; Pieczyrak, D.; Baranowska, H.M.; Jakubowicz, J.; Sopata, M.; Białopiotrowicz, T.; Kaczorek, E. Aesculus hippocastanum L. extract as a potential emulsion stabilizer. *Food Hydrocoll.* **2019**, *97*, 105237. [CrossRef]

32. Pratap-Singh, A.; Guo, Y.; Ochoa, S.L.; Fathordoobady, F.; Singh, A. Optimal ultrasonication process time remains constant for a specific nanoemulsion size reduction system. *Sci. Rep.* **2021**, *11*, 9241. [CrossRef] [PubMed]

33. Farshi, P.; Tabibiazar, M.; Ghorbani, M.; Mohammadifar, M.A.; Amirkhiz, M.B.; Hamishehkar, H. Whey protein isolate-guar gum stabilized cumin seed oil nanoemulsion. *Food Biosci.* **2019**, *28*, 49–56. [CrossRef]

34. Szumała, P.; Pacyna-Kuchta, A.; Wasik, A. Proteolysis of whey protein isolates in nanoemulsion systems: Impact of nanoemulsification and additional synthetic emulsifiers. *Food Chem.* **2021**, *351*, 129356. [CrossRef] [PubMed]

35. Jarzębski, M.; Siejak, P.; Sawerski, A.; Stasiak, M.; Ratajczak, K.; Masewicz, L.; Polewski, K.; Fathordoobady, F.; Guo, Y.; Pratap-Singh, A. Nanoparticles Size Determination by Dynamic Light Scattering in Real (Non-standard) Conditions Regulators— Design, Tests and Applications. In *Practical Aspects of Chemical Engineering*; Ochowiak, M., Woziwodzki, S., Mitkowski, P.T., Eds.; Springer Nature: Cham, Switzerland, 2020.

36. Fathordoobady, F.; Sannikova, N.; Guo, Y.; Singh, A.; Kitts, D.D.; Pratap-Singh, A. Comparing microfluidics and ultrasonication as formulation methods for developing hempseed oil nanoemulsions for oral delivery applications. *Sci. Rep.* **2021**, *11*, 72. [CrossRef] [PubMed]

37. Vivian, J.T.; Callis, P. Mechanisms of Tryptophan Fluorescence Shifts in Proteins. *Biophys. J.* **2001**, *80*, 2093–2109. [CrossRef]

38. Xu, J.; Mukherjee, D.; Chang, S.K. Physicochemical properties and storage stability of soybean protein nanoemulsions prepared by ultra-high pressure homogenization. *Food Chem.* **2018**, *240*, 1005–1013. [CrossRef]

39. Zhu, Y.; Li, Y.; Wu, C.; Teng, F.; Qi, B.; Zhang, X.; Zhou, L.; Yu, G.; Wang, H.; Zhang, S.; et al. Stability Mechanism of Two Soybean Protein-Phosphatidylcholine Nanoemulsion Preparation Methods from a Structural Perspective: A Raman Spectroscopy Analysis. *Sci. Rep.* **2019**, *9*, 6985. [CrossRef]

40. Ren, S.; Giusti, M. Monitoring the Interaction between Thermally Induced Whey Protein and Anthocyanin by Fluorescence Quenching Spectroscopy. *Foods* **2021**, *10*, 310. [CrossRef] [PubMed]

41. Kurek, M.A.; Pratap-Singh, A. Plant-Based (Hemp, Pea and Rice) Protein–Maltodextrin Combinations as Wall Material for Spray-Drying Microencapsulation of Hempseed (*Cannabis sativa*) Oil. *Foods* **2020**, *9*, 1707. [CrossRef]

42. Rave, M.C.; Echeverri, J.D.; Salamanca, C.H. Improvement of the physical stability of oil-in-water nanoemulsions elaborated with Sacha inchi oil employing ultra-high-pressure homogenization. *J. Food Eng.* **2019**, *273*, 109801. [CrossRef]

43. Zheng, H. *Introduction: Measuring Rheological Properties of Foods*; Springer: Berlin/Heidelberg, Germany, 2019; pp. 3–30. [CrossRef]

44. Zhu, Y.; Gao, H.; Liu, W.; Zou, L.; McClements, D.J. A review of the rheological properties of dilute and concentrated food emulsions. *J. Texture Stud.* **2019**, *51*, 45–55. [CrossRef]

45. Rha, C. Rheology of fluid foods. *Food Technol.* **1978**, *32*, 77–81.

46. Topayev, S.; Nouar, C.; Bernardin, D.; Neveu, A.; Bahrani, S.A. Taylor-vortex flow in shear-thinning fluids. *Phys. Rev. E* **2019**, *100*, 023117. [CrossRef]

47. Taylor, G.I. VIII. Stability of a viscous liquid contained between two rotating cylinders. *Philos. Trans. R. Soc. A Math. Phys. Eng. Sci.* **1923**, *223*, 289–343. [CrossRef]

48. Chhabra, R.P.; Richardson, J.F. Non-Newtonian fluid behaviour. In *Non-Newtonian Flow in the Process Industries*; Elsevier: Amsterdam, The Netherlands, 1999; pp. 1–36.

49. Callaway, J.C. Hempseed as a nutritional resource: An overview. *Euphytica* **2004**, *140*, 65–72. [CrossRef]

50. Wu, N.; Li, X.; Liu, S.; Zhang, M.; Ouyang, S. Effect of Hydrogen Bonding on the Surface Tension Properties of Binary Mixture (Acetone-Water) by Raman Spectroscopy. *Appl. Sci.* **2019**, *9*, 1235. [CrossRef]

51. David, R.; Neumann, A.W. A Theory for the Surface Tensions and Contact Angles of Hydrogen-Bonding Liquids. *Langmuir* **2014**, *30*, 11634–11639. [CrossRef]

52. Echeverría, J.; Urzúa, A.; Sanhueza, L.; Wilkens, M. Enhanced Antibacterial Activity of Ent-Labdane Derivatives of Salvic Acid (7α-Hydroxy-8(17)-ent-Labden-15-Oic Acid): Effect of Lipophilicity and the Hydrogen Bonding Role in Bacterial Membrane Interaction. *Molecules* **2017**, *22*, 1039. [CrossRef]

53. Bondar, A.-N.; White, S.H. Hydrogen bond dynamics in membrane protein function. *Biochim. Biophys. Acta (BBA)—Biomembr.* **2012**, *1818*, 942–950. [CrossRef]

54. McClements, D.; Demetriades, K. An Integrated Approach to the Development of Reduced-Fat Food Emulsions. *Crit. Rev. Food Sci. Nutr.* **1998**, *38*, 511–536. [CrossRef] [PubMed]

55. Ray, A.K.; Johnson, J.K.; Sullivan, R.J. Refractive Index of the Dispersed Phase in Oil-in-Water Emulsions: Its Dependence on Droplet Size and Aging. *J. Food Sci.* **1983**, *48*, 513–516. [CrossRef]

56. McClements, D.J. Theoretical prediction of emulsion color. *Adv. Colloid Interface Sci.* **2002**, *97*, 63–89. [CrossRef]

57. Chantrapornchai, W.; Clydesdale, F.; McClements, D. Influence of relative refractive index on optical properties of emulsions. *Food Res. Int.* **2001**, *34*, 827–835. [CrossRef]

58. Gharehcheshmeh, M.H.; Arianfar, A.; Mahdian, E.; Naji-Tabasi, S. Production and evaluation of sweet almond and sesame oil nanoemulsion and their effects on physico-chemical, rheological and microbial characteristics of enriched yogurt. *J. Food Meas. Charact.* **2020**, *15*, 1270–1280. [CrossRef]
59. Liu, D.; Lv, X.X. Effect of blueberry flower pulp on sensory, physicochemical properties, lactic acid bacteria, and antioxidant activity of set-type yogurt during refrigeration. *J. Food Process. Preserv.* **2018**, *43*, e13856. [CrossRef]
60. Cai, D.; Harrison, N.; Kling, D.; Gonzalez, C.; Lorca, G. Blueberries as an additive to increase the survival of Lactobacillus johnsonii N6.2 to lyophilisation. *Benef. Microbes* **2019**, *10*, 473–482. [CrossRef]
61. Ahmad, I.; Khalique, A.; Junaid, M.; Shahid, M.Q.; Imran, M.; Rashid, A.A. Effect of polyphenol from apple peel extract on the survival of probiotics in yoghurt ice cream. *Int. J. Food Sci. Technol.* **2020**, *55*, 2580–2588. [CrossRef]
62. Hiai, S.; Oura, H.; Nakajima, T. Color reaction of some sapogenins and saponins with vanillin and sulfurlc acid. *Planta Med.* **1976**, *29*, 116–122. [CrossRef] [PubMed]
63. Javidi, F.; Razavi, S.M.; Amini, A.M. Cornstarch nanocrystals as a potential fat replacer in reduced fat O/W emulsions: A rheological and physical study. *Food Hydrocoll.* **2018**, *90*, 172–181. [CrossRef]
64. Stangierski, J.; Rezler, R.; Baranowska, H.; Poliszko, S. Effect of enzymatic modification on frozen chicken surimi. *Czech J. Food Sci.* **2013**, *31*, 203–210. [CrossRef]

 *molecules*

Article

Functionality of Cricket and Mealworm Hydrolysates Generated after Pretreatment of Meals with High Hydrostatic Pressures

Alexandra Dion-Poulin, Myriam Laroche, Alain Doyen and Sylvie L. Turgeon *

Department of Food Sciences, Institute of Nutrition and Functional Foods (INAF), Université Laval, Quebec City, QC G1V 0A6, Canada; alexandra.dion-poulin.1@ulaval.ca (A.D.-P.); myriam.laroche.6@ulaval.ca (M.L.); alain.doyen@fsaa.ulaval.ca (A.D.)
* Correspondence: Sylvie.Turgeon@fsaa.ulaval.ca; Tel.: +1-418-656-2131 (ext. 404970)

Academic Editors: Przemyslaw Lukasz Kowalczewski, Anubhav Pratap Singh and David Kitts
Received: 15 October 2020; Accepted: 6 November 2020; Published: 17 November 2020

Abstract: The low consumer acceptance to entomophagy in Western society remains the strongest barrier of this practice, despite these numerous advantages. More positively, it was demonstrated that the attractiveness of edible insects can be enhanced by the use of insect ingredients. Currently, insect ingredients are mainly used as filler agents due to their poor functional properties. Nevertheless, new research on insect ingredient functionalities is emerging to overcome these issues. Recently, high hydrostatic pressure processing has been used to improve the functional properties of proteins. The study described here evaluates the functional properties of two commercial insect meals (*Gryllodes sigillatus* and *Tenebrio molitor*) and their respective hydrolysates generated by Alcalase®, conventionally and after pressurization pretreatment of the insect meals. Regardless of the insect species and treatments, water binding capacity, foaming and gelation properties did not improve after enzymatic hydrolysis. The low emulsion properties after enzymatic hydrolysis were due to rapid instability of emulsion. The pretreatment of mealworm meal with pressurization probably induced protein denaturation and aggregation phenomena which lowered the degree of hydrolysis. As expected, enzymatic digestion (with and without pressurization) increased the solubility, reaching values close to 100%. The pretreatment of mealworm meal with pressure further improved its solubility compared to control hydrolysate, while pressurization pretreatment decreased the solubility of cricket meal. These results may be related to the impact of pressurization on protein structure and therefore to the generation of different peptide compositions and profiles. The oil binding capacity also improved after enzymatic hydrolysis, but further for pressure-treated mealworm hydrolysate. Despite the moderate effect of pretreatment by high hydrostatic pressures, insect protein hydrolysates demonstrated interesting functional properties which could potentially facilitate their use in the food industry.

Keywords: entomophagy; *Gryllodes sigillatus*; *Tenebrio molitor*; edible insect meals; protein hydrolysate; high hydrostatic pressures; functional properties

---

1. Introduction

Recently, the interest in entomophagy, defined as the practice of eating insects, is growing due to its environmental and nutritional benefits compared to other livestock [1–9]. Edible insects have been targeted as potential alternative protein sources to resolve the problem of a global food crisis since, overall, their protein content is over 50% [4,10]. Despite the nutritional and sustainable advantages of insect consumption, low acceptability and negative consumer perception (insects as pests, disgusting and unsafe) of this unconventional food matrix remains the main issue in Western societies for the development of this food sector [11]. However, several studies have shown that acceptability is improved when incorporation of insects in food is unrecognizable [12–16]. Consequently, the use of insect ingredients, such as insect meals, concentrate or isolate, may be a promising strategy to improve consumer acceptance [11,15].

Insect meal is obtained simply by grinding whole dried insects. Oven-drying is the most widely used method for production of edible insect meal at a commercial scale [17,18]. However, studying other conventional and emerging drying technologies demonstrated that processing parameters largely influence the protein functionality [18]. More specifically, Kröncke et al. studied the effect of different drying methods on the solubility of mealworm (*Tenebrio molitor*) proteins [19]. Solubility is a very important functional property because it influences other functional properties such as emulsifying properties [3,20]. Thus, the solubility of mealworm (*T. molitor*) proteins decreased significantly, from 53% in the fresh state to only 14% after oven-drying [19]. This decrease in solubility was caused mainly by protein denaturation during heat treatment, which unfolds and exposes previously hidden hydrophobic groups [21].

Enzymatic hydrolysis is widely used to improve and modify protein functionality from a wide range of protein sources [22,23]. Currently, few studies are available on the functional properties of insect hydrolysates generated after enzymatic digestion. As examples, Wang et al. showed that the solubility of housefly (*Musca domestica*) protein was greater than 90% after enzymatic hydrolysis, and Hall et al. showed that enzymatic hydrolysis not only improved the solubility of cricket (*Gryllodes sigillatus*) proteins but also their emulsifying and foaming properties, which depend on hydrolysis parameters [3,24]. Purschke et al. improved the protein solubility, foaming properties and oil binding capacity of a migratory cricket (*Locusta migratoria* L.) through enzymatic hydrolysis using several proteases (alone or in combination), enzyme/substrate ratios and hydrolysis times [20]. Several protein pretreatment methods, such as microwaves [25,26], ultrasound [26–28] and high-voltage pulsed electric field [29–31], have been shown to improve hydrolysis rate and enhance bioactive peptide production. However, recently, interest in the use of high hydrostatic pressure for protein pretreatment is growing [32].

High hydrostatic pressures influence protein functionality by modulating the structures and conformation of the protein [33,34]. The denaturation of proteins caused by disruption of non-covalent bonds (hydrogen, hydrophobic and ionic bonds) exposes reaction sites and thus improves the efficiency of enzymatic hydrolysis [32,35]. Among the advantages of enzymatic hydrolysis assisted by high hydrostatic pressures are the hydrolysis of proteins normally resistant to enzymatic hydrolysis, the reduction of hydrolysis duration and an increased concentration of peptides, including bioactive peptides [32,36]. Hemker et al. showed that the solubility and emulsifying properties of fish (*Orechromis niloticus*) hydrolysate were increased after enzymatic hydrolysis assisted by high hydrostatic pressures while the water binding capacity was reduced [37]. To the best of our knowledge, no literature is available regarding the effect of high hydrostatic pressure-assisted enzymatic hydrolysis on the functionality of insect hydrolysates. Consequently, the objective of this study is to determine the functional properties of insect meals and insect peptide hydrolysates generated with or without pretreatment of the insect meal with high hydrostatic pressure. This work focuses on crickets (*G. sigillatus*) and mealworms (*T. molitor*), as these insect meals have different nutritional composition and are already produced and sold in Canada.

2. Materials and Methods

2.1. Materials

2.1.1. Insects

Commercial mealworm (*T. molitor*) and cricket meals (*G. sigillatus*), were purchased from Entomo Farms (Norwood, ON, Canada) and stored at 4 °C. Their proximate compositions are shown in Table 1.

Control hydrolysates and protein hydrolysates were produced from each meal by coupling enzymatic hydrolysis and high hydrostatic pressures.

2.1.2. Chemicals

Unless specified, all chemicals used for analytical purposes were analytical grade. Alcalase® (protease from Bacillus licheniformis), β-mercaptoethanol, D-L-Leucine, sodium tetraborate and o-phtaldialdehyde (OPA reagent) were purchased from Sigma Aldrich (St Louis, Missouri, MO, USA). Hexane, hydrochloric acid, methanol and sodium hydroxide were purchased from Fisher Scientific (Ottawa, ON, Canada). Citric Acid monohydrate, hydrochloric acid (36.5–38% *v/v*), NaCl, sodium hydroxide pellets and sodium phosphate dibasic were purchased from VWR International (Mississauga, ON, Canada). Sodium dodecyl sulfate (SDS) was purchased from Bio Basic (Markham, ON, Canada). Food grade canola oil was purchased from a local grocery store in Quebec City.

2.2. Preparation of Insect Protein Hydrolysates

Control hydrolysate was prepared following the procedure described by Liceaga-Gesualdo and Li-Chan, with some modifications [38]. A mass of 500 g of insect meal was dispersed in deionized water at 5% (*w/w*) and magnetically stirred at 4 °C for 12 h. The temperature of dispersion was adjusted to 55 °C and the pH was adjusted at 8.5 with 0.66 N NaOH [39,40]. Alcalase® was then added to the dispersion at 3% (E/S). During the 2 h enzymatic hydrolysis, the dispersion was constantly stirred, and the pH was constantly adjusted with alkaline solution (NaOH, 0.66 N). After hydrolysis, Alcalase® was inactivated by heat treatment at 80 °C for 15 min. The hydrolysate was cooled to 20 °C and centrifuged at 4500× *g* for 45 min at 20 °C. Finally, the supernatant was filtered using a strainer to remove some lipids [41] and freeze-dried. For the high-pressure conditions, both insect dispersions were prepared as described for the control. Pressurization parameters (200 and 380 MPa for 1 min for cricket and mealworm meals, respectively) were based on results obtained by Boukil et al. [42]. All hydrolysates were stored at 4 °C between experiments.

2.3. Proximate Composition of Protein Insect Ingredients

Total crude protein content was determined by using the Kjeldahl method according to Association of Official Analytical Chemists (AOAC) 928.08 procedures [43]. Two different conversion factors of 4.76 and 5.60 were used for insect meals and hydrolysates respectively, due to the presence of chitin, a nonprotein nitrogen component in insects [44]. More specifically, the conversion factor of 4.76 was applied for insect meals whereas a conversion factor of 5.60 was used for hydrolysates as suggested by Janssen et al. [44]. Total crude fat content was determined using the Soxhlet extraction method described by Tzompa-Sosa et al. except that hexane was used as the extraction solvent [45]. Chitin content was determined using a gravimetric method based on the method of Spinelli et al. [46]. Chitin is considered the residue after the extraction with 2% (*w/v*) sodium hydroxide and demineralization with 5% (*w/v*) hydrochloric acid [46]. The standard AOAC methods 950.46A and 920.153 were used to determine dry matter and ash content, respectively [43]. The proximate composition analyses were performed in triplicate except for total crude protein and fat content, which were performed in duplicate.

2.4. Degree of Hydrolysis

Freeze-dried insect hydrolysates were rehydrated in deionized water at 5% (*w/w*) to determine the degree of hydrolysis. The degree of hydrolysis was determined by the o-phthaldialdehyde (OPA) method according to Church et al. [47]: Different concentrations (0.75–3 mM) of D-L-Leucine were used to obtain the standard curve [48]. The degree of hydrolysis was calculated according to the Equation (1) proposed by Hall et al. [3]:

$$DH\,(\%) = \left(\frac{h}{h_{tot}}\right) \times 100 \tag{1}$$

where h is the total peptide bonds cleaved and h_{tot} is the total peptide bonds. The h_{tot} of the equation was 8.64 milliequivalents (meq)/g for all samples [3]. Measurements of all samples were performed in triplicate.

2.5. Measurement of Particle Size

The distribution of particle sizes for hydrolysates and meals was determined using a laser diffraction system (Mastersizer 3000, Malvern Instrument Ltd., Worcestershire, UK) on hydrolysate and meal dispersions prepared by mixing 0.5 g sample (insect meals and hydrolysates) with 40 mL of McIlvaine buffer (pH 4.0, 5.5 and 7.0) for 12 h at 4 °C. The refraction index of sample dispersion corresponding to the protein refraction index was set to 1.45. The dispersant phase was deionized water and this refraction index was set to 1.33. The dispersion sample was added until an obscuration of approximately 10% was obtained. The particle size distribution was expressed as D3,2 (μm), the sautermean diameter. All conditions were conducted in duplicate.

2.6. Protein Solubility

Protein solubility was determined at different concentrations (3.0, 1.0 and 0.5% (*w/v*)) of hydrolysates and meals, and at different pHs (4.0, 5.5 and 7.0) by using the methodology described by Morr et al. [49]. Briefly, samples of meals and hydrolysates were dispersed in 50 mL McIlvaine buffer and stirred at 20 °C for 2 h. The dispersion was then centrifuged at 2000× g for 30 min at 20 °C. The Kjeldahl method was used to determine the nitrogen content of the supernatants. Protein solubility was calculated using the Equation (2) proposed by Hall et al. [3] and expressed as a percentage:

$$\text{Solubility } (\%) = \left(\frac{\text{protein content in supernatant}}{\text{protein content in sample}} \right) \times 100 \tag{2}$$

2.7. Rheological Behavior

The viscosity of dispersion of the samples was determined using an ARES-G2 rheometer (TA instrument, New Castle, DE, USA) with DIN geometry (diameter 27.77 mm, gap 5.849 mm) with a cup (diameter 29.9 mm). Briefly, 0.75 g of sample (insect meals and hydrolysates) was dispersed at 3% (*w/v*) in McIlvaine buffer (pH 4.0, 5.5 and 7.0) in a centrifuge tube and stirred for 30 s. The dispersion was stored at 4 °C for 12 h and stirred again for 30 s. The dispersion was then conditioned at 23 °C and a pre-shear at 10 s^{-1} for 60 s was applied. The flow experiment was performed by increasing the shear rate from 10 to 750 s^{-1}. Data were analyzed using TRIOS software (TA Instrument). The flow properties were determined by fitting the data to the Power Law Model on the flow sweep for a shear rate from 10 to 100 s^{-1}. Equation (3) describes the Power Law Model:

$$\sigma = k\dot{y}^{n} \tag{3}$$

where σ is the stress (Pa), k is the viscosity (Pa·s), y is the shear rate (s^{-1}) and n is the flow behavior, allowing fluids classification: pseudoplastic ($n < 1$), dilatant ($n > 1$) and Newtonian ($n = 1$).

2.8. Gelation Properties

Visual observation of gel formation was performed according to the method described by Yi et al. [41]. Specifically, 3% and 10% (*w/v*) dispersion samples (hydrolysates and insect meals) in McIlvaine buffer (pH 4.0, 5.5 and 7.0) were stirred and stored at 4 °C for 12 h. The dispersion was heated at 86 ± 1 °C in a water bath for 10 min, cooled for 1 min in an ice-water bath and stored at 4 °C for 12 h. Gelation was confirmed if the gel was not deformed when the tube was overturned [41] and the gel formation was determined using dynamic oscillatory measurements. Since no gelation properties were visually observed for hydrolysates, only edible insect meals were tested at 30%, 20% and 10% (*w/v*) dispersion in McIlvaine buffer (pH 4.0, 5.5. and 7.0) at two ionic strengths (0 and 1 M

NaCl). Measurements were made with an ARES-G2 rheometer (TA Instrument, New Castle, DE, USA) used with DIN geometry (diameter 27.70 mm, gap 5.849 mm) with a cup (diameter 29.9 mm). Two temperature ramps were used, 25 to 85 °C and 85 to 5 °C, at a rate of 5 °C/min. The final temperature of each ramp was kept constant for 10 min. The oscillatory parameters for both temperature ramps were 1.0 Hz for angular frequency and 0.05% for strain. An amplitude step was used with a strain of 0.05% to 500% and angular frequency of 1.0 Hz at 5 °C. Data were analyzed using TRIOS software (TA Instrument). It was considered gel if the elastic behavior (G′) was greater than viscoelastic behavior (G″) [50].

2.9. Foaming Properties

The foam capacity was determined by the protocol described by Guo et al.'s [51] method with some modifications: 0.75 g of sample (hydrolysates and insect meals) was dispersed at 3% (*w/v*) in McIlvaine buffer (pH 4.0, 5.5 and 7.0) and magnetically stirred at 4 °C for 12 h. The volume of the dispersion was measured in a 100 mL graduated cylinder. The dispersion was transferred to a mixing bowl and aired with an electric hand mixer (KitchenAid) at maximum power for 2 min. Finally, the foam was carefully transferred to the same 100 mL graduated cylinder. The foam capacity was calculated from the Equation (4) of Guo et al. [51]:

$$FC\ (\%) = \left(\frac{V_0 - V_i}{V_i} \right) \times 100 \tag{4}$$

where FC is the foam capacity, V_0 is the volume of the foam that has been formed and V_i is the volume of the dispersion prior to aeration.

2.10. Water and Oil Binding Capacities

Water binding capacity (WBC) was assayed according to Quinn and Paton with some modifications [52]. Briefly, 1 g of each sample (hydrolysates and insect meals) was mixed with 10 g McIlvaine buffer (pH 4.0, 5.5 and 7.0) and vortexed for 30 s. After 10 min at room temperature, samples were centrifuged at 2000× *g* for 30 min at 20 °C. Supernatants were decanted and the residual non-bound water drained by placing the centrifugation tube upside-down on a filter paper for 10 min. WBC was calculated using the Equation (5) of Bußler et al. [11]:

$$WBC \left[\frac{g_{water}}{g_{sample,\ DM}} \right] = \frac{m_0 - m_1}{m_{0,DM}} \tag{5}$$

where m_0 is the initial mass of the sample, m_1 is the final mass of the sample and $m_{0,DM}$ is the initial mass of the sample on dry basis. The oil binding capacity (OBC) was assayed according to the method of Haque and Mozaffar with some modifications [53]. Canola oil (5 g) was added to 1 g of sample (hydrolysates or insect meals). The experimental procedure and calculation of OBC were similar to WBC, except for the addition of a stirring step of 3 × 30 s with 4 min breaks between repetitions. The pellet was weighed immediately after decanting the supernatants. Determination of WBC and OBC were performed in triplicate.

2.11. Emulsifying Properties

The spectroturbidimetric procedure of Pearce and Kinsella with the modifications by Liceaga-Gesualdo and Li-Chan was used to determine the emulsifying activity index (EAI) and the emulsion stability index (ESI) [38,54]. Briefly, samples (hydrolysates and insect meals) were dispersed at 3.0%, 1.0% and 0.5% (*w/v*) in 100 mL McIlvaine buffer (pH 4.0, 5.5 and 7.0) and stored at 4 °C for 12 h. Then, sample dispersions were mixed with 25 mL of canola oil using an Ultra-Turrax at 13,500 rpm for 1 min and diluted into tubes containing 0.3% SDS solution to reach an absorbance greater than 0.1. Homogeneity of dilution was ensured by inverting the tube six times. Absorbance was read at 500 nm with a UV-visible spectrometer. EAI was calculated from the Equation (6) used by Hall et al. [3]:

$$\text{EAI} = \frac{2 \cdot T \cdot A \cdot df}{\varnothing \cdot c \cdot 100} \tag{6}$$

where T is the turbidity, A is the absorbance measured, df is the dilution factor, L is the light path (in meters), $\varnothing$ is the volume of the oil phase (0.25) and c is the concentration of the aqueous phase. The turbidity was calculated according to Pearce and Kinsella [54]. For ESI, the emulsion was read every 30 min for 90 min and was calculated from the equation used 7 by Hall et al. [3]:

$$\text{ESI} = 100 - \left(\frac{\text{EAI}_0 - \text{EAI}_t}{\text{EAI}_0}\right) \times 100 \tag{7}$$

where EAI_0 is the initial EAI (time zero) and EAI_t is the EAI at 30, 60 and 90 min.

2.12. Statistical Analysis

All analyses were performed in triplicate except for gelling properties (duplicate). A fully random plan was used except for water and oil binding capacities where repetitions were blocked. The factors were insect, treatment, pH, concentration and time. Tukey tests were used as multiple comparison tests with a significance level (α) of 5%. The results were reported as mean ± standard deviation (M ± SD).

3. Results and Discussion

3.1. Proximate Composition and Degree of Hydrolysis

Table 1 shows the proximate composition of cricket and mealworm meals, as well as their protein hydrolysates generated at atmospheric pressure (control—0.1 MPa) or using high hydrostatic pressures (HHP) treatment prior to enzymatic hydrolysis by Alcalase®. Pretreatment of insect meals with HHP did not change the proximate composition of either insect hydrolysate ($p > 0.05$). The dry matter values for all insect species and ingredients (meals and hydrolysates) were close to 100% (96.5 to 98.6%) and slightly higher for mealworms compared to crickets. This difference directly correlates with the differences observed for crude protein, lipid, chitin and ash contents. The chitin content was similar in both insect meals (4.2 to 4.8%) but was significantly lower in hydrolysates (0.02 to 0.07%) because this water-insoluble polysaccharide [40] was removed by the centrifugation step performed to recover the soluble protein fraction. Similarly, lipid content was decreased by about 50% in cricket hydrolysates and 21% in mealworm hydrolysates, compared to their respective initial meals, due to the centrifugation step [41]. The higher lipid concentrations recovered in mealworm (20.6 and 23.3%) vs. cricket hydrolysates (7.5 and 8.3%) were related to differences in fatty acid composition since the unsaturated fatty acid fraction of the mealworm matrix is higher than that in the cricket [45]. The filtration method after the centrifugation step did not successfully remove the entire unsaturated lipid fraction, which was in a liquid form at room temperature. As expected, ash content was higher in control and pressure-treated hydrolysates compared to initial meals for both insect species. This is explained by the addition of Na+ from the NaOH required to control pH during enzymatic hydrolysis [38]. Contrary to previous studies [4,55], the crude protein content was higher for cricket meal (55.5%) than for mealworm meals (39.6%). Insect diet and rearing techniques could account for this difference [56]. After enzymatic hydrolysis, the protein content was increased in both hydrolysates. This result is consistent with the findings of Hall et al. after enzymatic hydrolysis of cricket protein by Alcalase®, due to the increase in protein solubility [20,22,23,57]. Table 1 also presents the degree of hydrolysis (DH) after in vitro digestion of cricket and mealworm proteins by Alcalase®. The DH was similar for both insect species with values ranging from 28.1 to 33.8%. These values were consistent with values obtained by Boukil et al. [42]: lower than those obtained by Hall et al. [3] after enzymatic hydrolysis of cricket (G. sigillatus) proteins by Alcalase® (ranged from 42.1 to 52.4%; E/S = 3.0%; hydrolysis time ranged from 30 to 90 min) and higher than the published study of Purschke et al. [20], calculated after in vitro digestion of migratory locust proteins with Alcalase® (ranging from 11.6 to 15.2%; E/S = 1.0%; hydrolysis time ranging from 30 to 120 min). Different parameters known to influence the degree of hydrolysis (E/S ratio, temperature, pH and reaction duration) can explain these differences [57,58]. The protein quality is also important since production of edible insect meals at laboratory or commercial scale might impact protein denaturation and aggregation, and consequently, their solubility which could affect the efficiency of enzymatic hydrolysis [19]. As shown in Table 1, HHP pretreatment of mealworm meal decreased the degree of hydrolysis (25.6%) compared to control (33.8%). This result is probably due to protein denaturation and aggregation phenomena caused by HHP which may decrease the efficiency of enzymatic hydrolysis [59].

3.2. Particle Size Distribution

Table 2 shows particle size indexes of cricket and mealworm meals at pH 4.0, 5.5 and 7.0, as well as hydrolysates generated with or without pretreatment of insect meals with HHP. As expected, and whatever the pH, higher particle sizes were obtained for both edible insect meals compared to their respective hydrolysates. At pH 4.0, cricket meal particles ($D3,2 = 52.7$ μm) were bigger than those at pH 5.5 ($D3,2 = 47.0$ μm) and pH 7.0 ($D3,2 = 37.3$ μm). The same tendency was observed for cricket hydrolysates. Since pH 4.0 is close to the isoelectric point (pI) of commercial cricket meal (pI close to 3.85) [60], the repulsive forces between the proteins decrease, inducing their aggregation [61]. Surprisingly, the particle size ($D3,2$) of mealworm meal increased with increasing pH (4.0–7.0) (values ranging from 17.0 to 41.4 μm), whereas the pI of the proteins was 3.95 [60]. This tendency was the same for mealworm hydrolysates and could be explained by the formation of disulfide bonds between mealworm proteins or even by the formation of protein–lipid complexes [62]. For both insects, the pretreatment of meals with HHP did not modify the particle sizes.

Table 1. Proximate composition and degree of hydrolysis of cricket and mealworm meals and hydrolysates generated at atmospheric pressure or with the pretreatment of insect meals with high hydrostatic pressures (HHP) prior to enzymatic hydrolysis.

Insects	Treatments [1]	Crude Protein	Lipid	Chitin [2]	Dry Matter [2]	Ash	Degree of Hydrolysis
				% *w/w* (dry basis)			
G. sigillatus	M	55.5 ± 0.3 [b]	16.7 ± 0.1 [c]	4.8 ± 0.2 [A]	98.0 ± 0.1 [a, A]	4.8 ± 0.1 [d]	0
	HT	70.0 ± 0.4 [a]	7.5 ± 0.1 [d]	0.03 ± 0.01 [B]	96.5 ± 0.4 [a, B]	13.4 ± 0.1 [a]	28.1 ± 4.2 [a, b]
	HP	68.2 ± 1.1 [a, b]	8.3 ± 0.6 [d]	0.02 ± 0.02 [B]	97.5 ± 0.2 [a, C]	13.1 ± 0.1 [a]	29.6 ± 1.3 [a, b]
T. molitor	M	39.6 ± 4.4 [c]	36.8 ± 0.3 [a]	4.1 ± 0.6 [A]	98.6 ± 0.1 [b, A]	3.3 ± 0.0 [e]	0
	HT	59.6 ± 4.3 [a, b]	23.3 ± 0.2 [b]	0.06 ± 0.02 [B]	97.0 ± 0.1 [b, B]	9.3 ± 0.1 [c]	33.8 ± 1.5 [a]
	HP	67.5 ± 5.1 [a, b]	20.6 ± 2.1 [b, c]	0.07 ± 0.07 [B]	97.7 ± 0.2 [b, C]	9.8 ± 0.1 [b]	25.5 ± 2.8 [b]

[1] M: insect meal, HT: control hydrolysate, HP: Hydrolysate whose insect meal has been pretreated at high hydrostatic pressures before enzymatic digestion. Values, except crude protein, represent the mean of three replicates ± standard deviation. Results with different letters are significantly different ($p < 0.05$). [2] Interaction was not significant, so lower-case letters ([a, b, c, d, e]) represent insect effect on results and upper-case letters ([A, B, C]) represent treatment effect on results.

Table 2. Particle size distribution indexes of cricket and mealworm meals and hydrolysates generated with or without HHP pretreatment of meals.

Insects	Treatments	pH	D(3,2) [1,2] μm
G. sigillatus	M	4.0	52.7 ± 15.1 [a, A]
		5.5	47.0 ± 1.6 [a, A, B]
		7.0	37.3 ± 3.3 [a, B]
	HT	4.0	7.6 ± 5.3 [b, A]
		5.5	0.21 ± 0.03 [b, A, B]
		7.0	0.09 ± 0.00 [b, B]
	HP	4.0	6.4 ± 7.5 [b, A]
		5.5	0.69 ± 0.72 [b, A, B]
		7.0	0.11 ± 0.01 [b, B]
T. molitor	M	4.0	17.0 ± 3.1 [a, A]
		5.5	37.3 ± 8.8 [a, A, B]
		7.0	41.2 ± 17.6 [a, B]
	HT	4.0	5.8 ± 1.9 [b, A]
		5.5	4.9 ± 0.3 [b, A, B]
		7.0	6.2 ± 1.1 [b, B]
	HP	4.0	5.0 ± 1.7 [b, A]
		5.5	5.5 ± 0.8 [b, A, B]
		7.0	6.4 ± 1.4 [b, B]

M: insect meal, HT: control hydrolysate, HP: Hydrolysate generated by the treatment of insect meal with high hydrostatic pressures prior to enzymatic hydrolysis, D(3,2): area-based mean particle diameter. Values represent the mean of three replicates ± standard deviation. [1] Results with different letters within an insect are significantly different ($p < 0.05$). [2] Lower-case letters ([a], [b]) represent insect–treatment interaction and upper-case letters ([A], [B]) represent insect–pH interaction.

3.3. Solubility of Insect Meals and Hydrolysates

Table 3 shows the protein solubility of cricket and mealworm meals and their hydrolysates (control and generated from pressure-treated insect meals) at different pH values (4.0, 5.5 and 7.0) and concentrations (0.5–3.0% *w/v*). The sample concentration (0.5–3.0% *w/v*) and pH values (4.0, 5.5 and 7.0) modified the protein solubility of insect meals. Regardless of pH and concentration, cricket and mealworm meals had low solubility with values ranging from 17.1 to 18.7% and from 15.8 to 20.2%, respectively. These values are consistent with those available in the literature [3,11,61,63]. More specifically, Stone [60] obtained protein solubility ranging from 29.0 to 23.4% between pH 3.0 and 7.0 for commercial cricket and mealworm meals. A study published by Kröncke et al. confirmed that oven drying of *T. molitor* larvae decreased the quality of proteins and reduced their solubility by 74% [19]. This low solubility was mainly related to the drying method (oven-drying) applied at commercial scale before the larvae grinding step. The heat treatment denatured the protein, exposing hydrophobic groups and causing protein aggregation [21,64].

Table 3. Solubility of cricket and mealworm meals and hydrolysates at different concentrations and pH values, generated with or without HHP pretreatment before enzymatic digestion.

Insects	Treatment	Concentration (% w/v)	pH	Solubility (%)
G. sigillatus	M	0.5	4.0	17.9 ± 0.6 [a]
			5.5	18.5 ± 0.6 [a]
			7.0	18.7 ± 0.6 [a]
		1.0	4.0	17.1 ± 0.6 [a]
			5.5	18.7 ± 0.6 [a]
			7.0	18.6 ± 0.6 [a]
		3.0	4.0	17.1 ± 0.6 [a]
			5.5	18.5 ± 0.6 [a]
			7.0	17.9 ± 0.6 [a]
	HT	0.5	4.0	98.1 ± 0.9 [a, b, c]
			5.5	96.7 ± 0.9 [a, b, c]
			7.0	98.7 ± 0.9 [a, b]
		1.0	4.0	100.2 ± 0.1 [a]
			5.5	94.4 ± 0.9 [c, d]
			7.0	92.5 ± 0.9 [d]
		3.0	4.0	95.8 ± 0.9 [b, c, d]
			5.5	95.5 ± 0.9 [b, c, d]
			7.0	95.6 ± 0.9 [b, c, d]
	HP	0.5	4.0	92.9 ± 0.9 [a]
			5.5	92.4 ± 0.9 [a]
			7.0	90.9 ± 0.9 [a]
		1.0	4.0	94.7 ± 0.9 [a]
			5.5	91.4 ± 0.9 [a]
			7.0	92.9 ± 0.9 [a]
		3.0	4.0	92.5 ± 0.9 [a]
			5.5	92.1 ± 0.9 [a]
			7.0	92.1 ± 0.9 [a]
T. molitor	M	0.5	4.0	16.2 ± 1.0 [b]
			5.5	17.2 ± 1.0 [a, b]
			7.0	17.1 ± 1.0 [a, b]
		1.0	4.0	15.8 ± 1.0 [b]
			5.5	17.5 ± 1.0 [a, b]
			7.0	18.1 ± 1.0 [a, b]
		3.0	4.0	16.8 ± 1.0 [a, b]
			5.5	16.1 ± 1.0 [b]
			7.0	20.2 ± 1.0 [a]
	HT	0.5	4.0	75.7 ± 1.9 [a]
			5.5	73.1 ± 1.9 [a]
			7.0	72.9 ± 1.9 [a]
		1.0	4.0	77.0 ± 1.9 [a]
			5.5	79.0 ± 1.9 [a]
			7.0	74.8 ± 1.9 [a]
		3.0	4.0	73.4 ± 1.9 [a]
			5.5	73.4 ± 1.9 [a]
			7.0	76.7 ± 1.9 [a]
	HP	0.5	4.0	108.7 ± 1.5 [a]
			5.5	88.6 ± 1.5 [b, c]
			7.0	87.9 ± 1.5 [b, c]
		1.0	4.0	89.3 ± 1.5 [b, c]
			5.5	94.2 ± 1.5 [a, b]
			7.0	88.9 ± 1.5 [b, c]
		3.0	4.0	88.2 ± 1.5 [b, c]
			5.5	89.1 ± 1.5 [b, c]
			7.0	86.7 ± 1.5 [c]

M: insect meal, HT: control hydrolysate, HP: Hydrolysate generated by HHP treatment of insect meal prior to enzymatic hydrolysis. Values represent the mean of three replicates ± standard deviation. Results in the same insects-treatment with different letters ([a], [b], [c], [d]) are significantly different ($p < 0.05$).

The protein solubility of hydrolysates, which was similar for all pHs (4.0–7.0) and sample concentrations (0.5–3.0% *w/v*), was drastically improved compared to insect meals, with values ranging from 92.5 to 100.0% and 72.9 to 78.9%, for cricket and mealworm hydrolysates, respectively. This improvement in solubility is consistent with results obtained by Hall et al. and Wang et al. for cricket and house fly larvae after enzymatic hydrolysis by Alcalase® [3,24]. The digestion of protein into peptides increases ionizable groups, such as amino and carboxyl groups, improving hydration [65] and solubility. Furthermore, the higher solubility was also related to the fact that only the soluble fraction was freeze-dried during hydrolysate preparation [22].

Compared to the control, pressurization pretreatment of cricket meal slightly decreased the hydrolysate solubility, mainly at 1.0 and 3.0% for pH 4.0, 5.5 and 7.0 (Table 3). Protein solubility is influenced by the type of protein, protein concentration, pH and the presence of salts [66]. However, the similar compositions (Table 1) and particle sizes (Table 2) could not explain the differences in solubility between control and pressure-treated hydrolysates for both insect species. Gbogouri et al. [57], studying protein hydrolysate from salmon, showed that hydrolysates with a higher degree of hydrolysis generally had higher solubilities than those possessing lower DH. Indeed, smaller peptides from hydrolysates with high DH were considered to have more polar residues, which could enhance the quantity of hydrogen bonds with water, resulting in an increase in protein solubility in solution [37,67]. However, the similar DH values calculated for cricket hydrolysate (control and pressure treatments) (Table 1) could not explain their differences in solubility. HHP parameters, such as pressure, duration and temperature, also impact protein solubility. Under pressure, the loss of solubility is mainly related to the formation of insoluble high molecular weight protein aggregates due to exposure of hydrophobic residues and/or disulfide bond formation [68,69]. Consequently, during the enzymatic hydrolysis of pressure-treated protein, enzymes break the protein in different ways due to modifications of protein structure since some bonds became inaccessible to the enzymes and, on the contrary, others may be exposed due to conformational changes [36,70,71]. Therefore, Alcalase® hydrolysis of cricket meal pretreated by pressure could generate a different peptide profile that contains more hydrophobic peptides, which could negatively impact hydrolysate solubility. On the contrary, the solubility of mealworm hydrolysate generated from pressure-treated meals increased for all concentrations and pH values (Table 3). Kim et al., studying hemoglobin hydrolysate, mentioned that the hydrolysate solubility could be improved at low DH due to the generation of hydrophobic peptides in lower amounts [72]. Consequently, and as proposed for cricket hydrolysate, the protein unfolding and aggregation due to pressurization could generate a different peptide profile with a larger amount of hydrophilic peptides, improving hydrolysate solubility.

3.4. Rheological Behavior

Table 4 shows the viscosity and the flow behavior rate of cricket and mealworm meals and their hydrolysates (control and generated from pressure-treated insect meals) at different pHs (4.0, 5.5 and 7.0). Whatever the pH and the treatment (meals and hydrolysates), the viscosity of insect meals and hydrolysates for all conditions analyzed was very near 0 Pa·s. The viscosity of cricket and mealworm meals (0.7 mPa·s) was slightly lower than hydrolysates (1.2 mPa·s) due to the insoluble particles of insect meals which settle quickly, resulting in the rheological behavior of only the soluble phase being measured. Generally, the viscosity of hydrolysates is lower than the initial proteins due to the small sizes of the peptides [73–75]. For both insects, treating insect meals with HHP prior to enzymatic digestion did not change the viscosity compared to the control hydrolysate. Whatever the pH and treatment (meals and hydrolysates), the flow behavior rate was around 1, which indicated that all samples behaved as Newtonian fluids. Thus, the insect, treatment or pH did not modify the flow behavior rate ($p > 0.05$). Jung et al. have shown that the flow behavior index values change from pseudoplastic ($n < 1$) to Newtonian behavior ($n = 1$) after the enzymatic digestion of soy protein [73].

Table 4. Viscosity and flow behavior rate of insect meals and hydrolysates at different pHs, generated with or without HHP treatment prior to enzymatic digestion.

Insects	Treatments	pH	Viscosity mPa·s	Flow Behavior Rate
G. sigillatus	M	4.0	0.52 ± 0.53 [a]	1.26 ± 0.23 [a]
		5.5	0.84 ± 0.50 [a]	1.13 ± 0.15 [a]
		7.0	0.90 ± 0.83 [a]	0.83 ± 0.58 [a]
	HT	4.0	1.14 ± 0.18 [b]	1.07 ± 0.04 [a]
		5.5	1.23 ± 0.04 [b]	1.08 ± 0.03 [a]
		7.0	1.33 ± 0.10 [b]	1.05 ± 0.04 [a]
	HP	4.0	1.08 ± 0.21 [b]	1.06 ± 0.03 [a]
		5.5	1.04 ± 0.16 [b]	1.05 ± 0.01 [a]
		7.0	1.29 ± 0.22 [b]	1.05 ± 0.02 [a]
T. molitor	M	4.0	0.66 ± 0.55 [a]	1.31 ± 0.44 [a]
		5.5	0.68 ± 0.53 [a]	1.23 ± 0.28 [a]
		7.0	0.43 ± 0.66 [a]	1.44 ± 0.36 [a]
	HT	4.0	1.20 ± 0.05 [b]	1.04 ± 0.02 [a]
		5.5	1.10 ± 0.14 [b]	1.07 ± 0.04 [a]
		7.0	1.22 ± 0.00 [b]	1.03 ± 0.01 [a]
	HP	4.0	1.21 ± 0.09 [b]	1.04 ± 0.01 [a]
		5.5	1.11 ± 0.22 [b]	1.07 ± 0.03 [a]
		7.0	1.40 ± 0.03 [b]	1.04 ± 0.00 [a]

M: insect meal, HT: control hydrolysate, HP: Hydrolysate generated by HHP treatment of insect meal prior to enzymatic hydrolysis. Values represent the mean of three replicates ± standard deviation. Results with different letters ([a] and [b]) are significantly different ($p < 0.05$).

3.5. Gelation Properties

Gel formation was not observed with insect meals and hydrolysates (control and generated from pressure-treated insect meals) at 3% (*w/v*) or 10% (*w/v*), which is probably caused by the denaturation of proteins during the drying method used for commercial production of insect meals. Hydrolysates are also generally known to have poor gelling properties [58,74]. Protein concentration is a key factor influencing gelation properties and the gelation threshold depends on structural characteristics and gelling conditions such as pH and ionic strength [76–78]. Therefore, the concentrations used (3 and 10% *w/v*) in our study were probably insufficient to reach the gelation threshold.

The gelation abilities of higher concentrations of insect meals were evaluated using dynamic rheology. The results were similar for both insect meals regardless of the experimental conditions (10, 20 and 30% *w/v* at pH 7.0, with or without 1M NaCl). The rheological behavior of cricket meal is given as an example (Figure 1A). For cricket and mealworm meals, the elasticity modulus (G′) was higher than the loss modulus (G″), which indicates that gelation occurred, regardless of pH and ionic strength. However, the higher G′ values that should indicate gelation were observed before heat treatment, suggesting that a protein gel may not be responsible for this rheological behavior. The insect meals had very low solubility (15.84–20.17%, Table 3) and the insoluble particles settled quickly after agitation. In addition, certain ionic strengths could not be analyzed because the maximum axial force of the device was reached before the analysis could be performed, also suggesting that a packed precipitate formed at the bottom of the geometry. To confirm that no gelation occurred, the experimental conditions described at the beginning of this section were reproduced in glass tubes. Two distinct phases (soluble and insoluble phases) were observed (Figure 2A). The texture of the insoluble phase was similar to wet sand while the upper phase was liquid. After heat treatment, the two phases had a similar texture, but the proportion of insoluble phase seemed to increase compared to the soluble fraction (Figure 2A). From these observations, the changes in rheological behavior observed (in Figure 1A) could be explained by temperature-induced modification of the physico-chemical characteristics of the precipitated phase. Consequently, G′ and G″ moduli decreased during the heating phase and increased during the cooling phase, which is representative of protein interactions where the hydrophobic interactions increase with increasing temperature and hydrogen bonds are favored with cooling [79,80]. Zhao et al. [61] also observed this phenomenon for a mealworm protein concentrate. Moreover, the G′ and G″ moduli were slightly lower until the cooling step and similar after this step. These authors demonstrated that the effect of the viscoelastic moduli of the salt concentration was pH-dependent [81]. In the literature, only few studies mention gelation properties of insect proteins [41,61]. More specifically, Yi et al. [41] obtained gelation of an *A. domesticus* soluble fraction (3% *w/v*) at pH 7.0 and gelation of different insect meal soluble fractions (*T. molitor, A. diaperinus, Z. morio* and *B. dubia*), but at high concentration (30% *w/v*) for pH 7.0 and 5.0. Otherwise, the authors generally obtained only an aggregation induced by the heat treatment [41]. Zhao et al. [81] obtained a weak gel after adding 2% NaCl to a mealworm protein concentrate since NaCl can improve the gelation properties of proteins by reducing the repulsive forces between proteins.

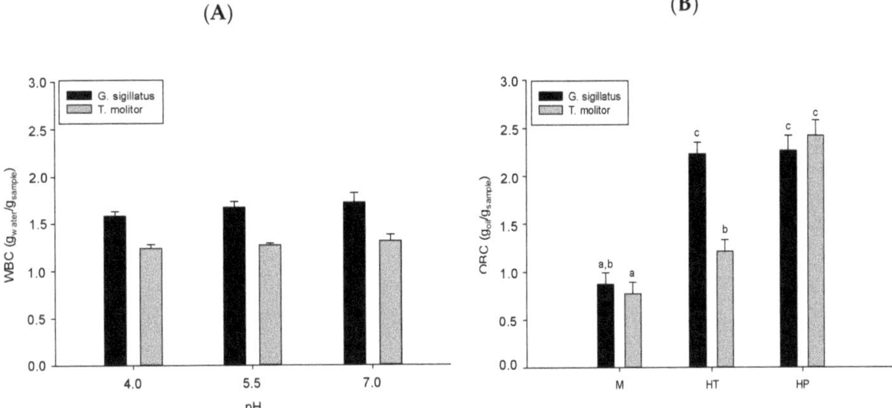

Figure 1. Water binding capacity (**A**) and oil binding capacity (**B**) of insect meals (M) and hydrolysates (HT: control hydrolysate, HP: hydrolysate generated by HHP treatment of meal prior to enzymatic hydrolysis). Different letters indicate significant difference ($p < 0.05$).

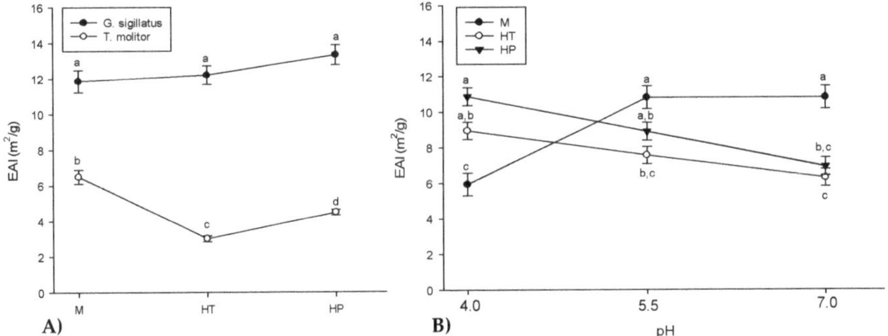

Figure 2. Emulsion activity index (EAI) of insect meals (M) and hydrolysates (HT: control hydrolysate, HP: hydrolysate generated by HHP treatment of meal prior to enzymatic hydrolysis) according to the insect source (cricket and mealworms) (**A**) and according to the pH (4.0–7.0) (**B**). Different letters indicate significant difference ($p < 0.05$).

3.6. Foaming Properties

No foaming properties could be measured for cricket and mealworm meals and hydrolysates (control and generated after HHP), mainly due to foam destabilization during aeration. Several studies have shown that insect proteins have poor foaming properties [41,63,82,83]. Stone obtained a foaming capacity of around 82% for a commercial cricket meal but no foaming properties were observed for commercial mealworm meal [60]. According to Hall et al., enzymatic digestion enhanced the foam capacity of cricket proteins due to structural and conformational modifications [3]. Enzymatic hydrolysis generates low molecular weight peptides and exposes surface-stabilizing residues at the air–water surface which can allow rapid migration, better flexibility and rearrangement at the interface and, therefore, improve foaming properties [38,58]. In our study, the poor foaming properties of the different insect ingredients is mainly due to their lipid content. Indeed, it is well-known that just 0.5% lipid can reduce the volume of foam and cause destabilization during aeration of egg whites [84]. Consequently, efficient defatting of insect meals is crucial to generate a food ingredient with good foaming properties.

3.7. Water and Oil Binding Capacities

Figure 1A shows the WBC of cricket and mealworm meals at pH 4.0, 5.5 and 7.0. The simple effect of insect (crickets and mealworms) and pH (4.0, 5.5 and 7.0) variation was significant ($p < 0.0001$), but their interactions were not ($p = 0.08$). WBC was calculated for cricket and mealworm hydrolysates (control and generated from pressure-treated insect meals) since their composition of soluble peptides did not retain water. However, Purschke et al. [20] obtained WBC values for commercially migratory locust hydrolysates (close to 1.50 g_{water}/g) after enzymatic hydrolysis with Neutrase or Flavourzyme. This can be explained by differences in the composition of edible insects, experimental conditions and hydrolysis parameters. As observed in Figure 2A, the WBC of cricket meal (1.58–1.72 g_{water}/g) was higher than mealworm meal (1.24–1.31 g_{water}/g), probably because mealworms have a lower hydrophilic amino acid content than crickets [55]. Our values for commercial cricket and mealworm meals (1.76 g_{water}/g and 1.62 g_{water}/g, respectively) were consistent with those published by Stone [60]. The WBC of insects is generally lower than several vegetable proteins such as soy protein isolate (4.47 g_{water}/g), red kidney beans (2.25–2.65 g_{water}/g) and Indian kidney beans (2.60 g_{water}/g) [85]. Many factors influence the WBC, such as the amino acid profile, conformation, hydrophobicity, pH, ionic strength, temperature and protein concentration [86]. The water binding capacity at pH 4.0 was significantly ($p < 0.05$) lower than at pH 5.5 or 7.0. As previously mentioned, the pI of commercial cricket and mealworm meals is close to 3.85 and 3.95, respectively. The pH affects the charge on proteins and, consequently, close to the pI, the WBC is minimal since protein–protein interactions are favored over protein–water interactions [86].

Figure 1B also shows the OBC of insect meals and hydrolysates (control and generated from pressure-treated insect meals). The OBC was similar for both insect meals with values ranging from 0.77 to 0.87 g_{oil}/g, but lower than the results obtained by Stone [60] for commercial cricket and mealworm meals (1.42 g_{oil}/g and 1.58 g_{oil}/g, respectively) and by Purschke et al. [20] for commercial migratory locust meals using rapeseed oil (1.10 g_{oil}/g). These differences may be due to the use of different oils, but many other factors such as the amino acid composition and degree of denaturation of the proteins may also have had an effect [85]. Similar to WBC, the insect OBC is generally lower than for other proteins, such as soy protein isolate (1.54 g_{oil}/g), red kidney beans (1.23–1.52 g_{oil}/g) and Indian kidney beans (2.40 g_{oil}/g) [85]. The OBC of control hydrolysates increased from 0.87 to 2.23 g_{oil}/g and from 0.77 to 1.21 g_{oil}/g for cricket and mealworms, respectively. This tendency, also observed by Purschke et al. [20] for migratory locust protein following enzymatic hydrolysis by Neutrase and Flavourzyme (alone or in combination), is due to the exposure of hydrophobic groups that were previously hidden in the edible insect protein structure [20,87]. While HHP did not impact the OBC of cricket hydrolysate, it doubled OBC for mealworm hydrolysate compared to the control (1.21 to 2.42 g_{oil}/g). The decrease in the degree of hydrolysis between control and HHP mealworm hydrolysates possibly explains this tendency, as Chalamaiah et al. [22] showed that a higher degree of hydrolysis reduces oil binding capacity. The HHP treatment probably modified the protein structures and peptide profile due to different enzyme breaks.

3.8. Emulsifying Properties

3.8.1. Emulsion Activity Index (EAI)

Figure 2A shows the EAI values of cricket and mealworm meals and their hydrolysates (control and generated from pressure-treated insect meals). Higher EAI values mean that the dispersed fat droplets are smaller in size and that proteins (or peptides) have more ability to absorb at the oil–water interface [88]. Globally, cricket ingredients (11.86–13.32 m^2/g) had higher EAI values compared to mealworms (3.01–6.50 m^2/g) which could be due to the difference in the hydrophobicity of the proteins and the smaller size of cricket proteins, which would allow faster diffusion at the oil–water interface [23,57,89]. Chatsuwan et al. [90] obtained EAI values ranging from 29.23 to 36.69 m^2/g for *P. succinta* and *C. rosea*, respectively. Regardless of the insect species, the lower EAI values in our study can be attributed to the drying method used to produce commercial insect meals, which modifies protein solubility [19] and emulsifying properties [91]. The EAI ($p > 0.05$) of the cricket hydrolysate control was not significantly different from the meal. This tendency was also observed by Hall et al. for some enzymatic hydrolysis conditions of cricket using Alcalase® (E/S of 0.5%; hydrolysis time of 90 min and E/S of 1.5%; hydrolysis time of 30 min) compared to unhydrolyzed proteins [3]. Conversely, the EAI value of the mealworm control hydrolysate was reduced compared to the meal. Hall et al. also obtained a decreased EAI after enzymatic hydrolysis of tropical banded cricket with similar enzymatic digestion conditions (E/S of 3%; hydrolysis time of 90 min) [3]. This reduction of EAI was possibly caused by the enzymatic hydrolysis parameters which would have reduced the interfacial activity of proteins due to excessive protein degradation [57,92]. However, several studies have shown that the enzymatic hydrolysis of several proteins using Alcalase® improved their EAI values [23,38,88]. This has been explained by the enzyme's specificity in cutting aromatic residues to reveal hydrophobic peptides, which facilitates the formation of emulsions [73,93]. In addition, the increased peptide solubility after enzymatic hydrolysis promotes peptide absorption at the oil–water interface [23]. Applying HHP treatment prior to enzymatic hydrolysis had no impact on EAI values for cricket, since the degree of hydrolysis was similar. However, HPP treatment of the mealworm meal increased its EAI compared to the control hydrolysate, possibly due to modification of the peptide hydrophobicity [89] caused by different enzyme cut sites linked to the structural modifications of proteins after HHP. Despite this improvement of EAI, the value is still lower than unhydrolyzed mealworm meal.

Figure 2B shows the EAI of insect ingredients (meals and hydrolysates) at different pHs (4.0, 5.5 and 7.0), regardless of insect and concentration of the aqueous phase. The insect meals had significantly ($p < 0.05$) lower EAIs at pH 4.0 than at pH 5.5 and 7.0, near the pI of commercial insect meals (3.85 and 3.95 respectively, for cricket and mealworm meals) [60]. Measuring at the pI produced the lowest EAI (Table 3), consequently, diffusion and adsorption of protein at the oil–water interface were delayed [94]. This reduction of EAI at the pI was also observed for chickpea and whey protein isolate [95,96]. For hydrolysates (control and HHP-treated), the EAI decreased slightly with increasing pH (4.0–7.0) due, once again, to solubility modification. Pacheco-Aguilar et al. [88] obtained similar EAI values for fish (*Mercuccius productus*) protein hydrolysate at pHs 4.0 and 7.0, but higher values at pH 10.0 and a different degree of hydrolysis (DH 10–20%).

3.8.2. Emulsion Stability Index (ESI)

Table 5 shows the ESI of cricket and mealworm meals and hydrolysates (control and from pressure-treated insect meals), and different concentrations of aqueous phase (0.5–3.0% *w/v*) and pH (4.0–7.0). Significant triple interactions were obtained between the different treatments (meals and hydrolysates), concentrations (0.5, 1.0 and 3.0% *w/v*) and times (30, 60 and 90 min) ($p < 0.0001$) as well as insects, treatments and time ($p < 0.01$) (Table A1). Specifically, the variable time is an influencing factor. Indeed, the emulsions were very unstable since most of the destabilization process (28.8–59.5% and 39.1–76.6% for meals and hydrolysates, respectively) occurred within 30 min. A creaming phenomenon has been observed only a few minutes after the homogenization was stopped. The ESI values obtained by Hall et al. [3] after 30 min for cricket hydrolysate (E/S of 0.5–3%; hydrolysis time of 30–90 min) were generally below 60%. Several studies have shown that the emulsion stability of insect proteins was generally low (21.2–45.4%) due to their very low solubility [63,85,90]. The size of proteins (or peptides) and the amino acid composition and distribution also influence the strength of the interfacial layer and therefore the emulsion stability [3,88]. Enzymatic hydrolysis slightly enhanced the ESI values for cricket but did not seem to affect the mealworms (Table 5). For example, using a cricket concentration of 3.0% (*w/v*), pH 7.0 and 30 min treatment resulted in ESI values ranging from 51.5 to 72.1% after enzymatic digestion (with or without pressure-treated meals), while the ESI ranged from 59.5 to 53.0–55.4 for mealworms. Hydrolysis using Alcalase® has been shown to improve emulsion stability in different studies [3,38]. This might be caused by stronger and more cohesive interfacial layer films with peptides compared to initial proteins [38]. For both insects, the treatment of insect meals by HHP prior to the enzymatic digestion did not change the ESI values. For example, the ESI values at a concentration of 3.0% (*w/v*) and pH 7.0 ranged from 67.23 to 72.1% and 53.0 to 55.4 for crickets and mealworms, respectively. For both insects, a concentration of 3% (*w/v*) produced better stability than other concentrations analyzed (0.5–1.0%). The effect of pH and concentration was similar for hydrolysates generated at atmospheric pressures or by pressure-treated insect meals. Bußler et al. [11] observed that insect emulsion stability decreased near the isoelectric point due to weak electrostatic repulsion between the oil droplets, but this phenomenon was not observed in this study.

Table 5. ESI of cricket and mealworm meals and their hydrolysates (control and generated from pressure-treated insect meals) according to the aqueous phase concentration and pH.

Insects	Treatments	Concentration (% *w/v*)	pH	ESI (%) 30 min	60 min	90 min
G. sigillatus	M	0.5	4.0	34.5 ± 1.6	33.1 ± 1.7	31.6 ± 1.4
			5.5	35.0 ± 3.6	32.2 ± 2.3	31.4 ± 3.3
			7.0	33.7 ± 1.8	32.0 ± 2.1	30.4 ± 1.8
		1.0	4.0	35.4 ± 4.4	33.7 ± 3.2	30.1 ± 2.4
			5.5	28.8 ± 7.3	26.4 ± 6.6	24.6 ± 6.7
			7.0	36.6 ± 3.6	33.2 ± 2.9	31.4 ± 3.2
		3.0	4.0	51.5 ± 2.6	43.7 ± 3.5	42.2 ± 1.7
			5.5	50.1 ± 4.0	43.8 ± 1.5	40.7 ± 3.1
			7.0	51.4 ± 0.8	44.7 ± 3.1	41.1 ± 1.1
	HT	0.5	4.0	49.6 ± 3.4	47.3 ± 2.3	45.3 ± 0.8
			5.5	46.6 ± 2.8	45.1 ± 1.3	43.3 ± 1.6
			7.0	48.8 ± 2.2	47.0 ± 1.8	45.0 ± 1.1
		1.0	4.0	59.1 ± 0.6	54.7 ± 3.1	52.8 ± 1.7
			5.5	57.4 ± 2.3	52.5 ± 2.0	49.9 ± 1.1
			7.0	56.2 ± 2.8	53.8 ± 1.9	51.6 ± 3.3
		3.0	4.0	74.8 ± 2.1	71.8 ± 2.7	68.0 ± 3.0
			5.5	75.1 ± 0.9	69.1 ± 2.8	65.7 ± 2.8
			7.0	67.3 ± 1.7	63.8 ± 4.0	61.0 ± 2.1
	HP	0.5	4.0	42.1 ± 4.5	39.4 ± 5.9	38.0 ± 6.9
			5.5	39.5 ± 5.8	36.3 ± 6.6	35.6 ± 7.3
			7.0	40.3 ± 6.4	37.1 ± 7.3	36.0 ± 7.9
		1.0	4.0	60.5 ± 2.0	55.9 ± 2.2	49.9 ± 1.2
			5.5	59.0 ± 7.9	54.1 ± 4.2	49.3 ± 2.0
			7.0	50.9 ± 2.2	47.5 ± 1.7	45.1 ± 2.9
		3.0	4.0	76.6 ± 1.2	70.4 ± 4.9	67.8 ± 5.3
			5.5	74.6 ± 1.5	69.1 ± 2.2	65.5 ± 2.7
			7.0	72.1 ± 1.7	64.3 ± 2.5	64.0 ± 1.2

Table 5. *Cont.*

Insects	Treatments	Concentration (% *w/v*)	pH	ESI (%) 30 min	ESI (%) 60 min	ESI (%) 90 min
T. molitor	M	0.5	4.0	53.1 ± 2.4	50.9 ± 3.2	50.0 ± 3.2
			5.5	7.0 ± 3.8	43.8 ± 3.2	41.3 ± 3.3
			7.0	46.01 ± 1.0	41.9 ± 0.9	39.7 ± 0.5
		1.0	4.0	46.8 ± 10.9	44.0 ± 9.3	41.7 ± 9.5
			5.5	44.2 ± 5.7	40.4 ± 4.1	36.0 ± 4.3
			7.0	48.1 ± 1.5	42.8 ± 1.4	38.5 ± 2.0
		3.0	4.0	54.6 ± 6.4	48.8 ± 6.5	45.4 ± 6.1
			5.5	55.9 ± 2.1	49.2 ± 4.0	43.7 ± 4.5
			7.0	59.5 ± 3.7	51.2 ± 3.1	46.2 ± 4.3
	HT	0.5	4.0	48.9 ± 0.3	47.3 ± 1.0	46.7 ± 0.9
			5.5	40.5 ± 4.7	38.5 ± 6.7	38.2 ± 6.7
			7.0	39.1 ± 5.4	38.6 ± 5.4	37.5 ± 5.6
		1.0	4.0	44.9 ± 8.7	42.7 ± 10.0	42.9 ± 8.8
			5.5	46.9 ± 4.8	44.9 ± 5.4	44.0 ± 5.3
			7.0	47.9 ± 1.6	45.1 ± 2.5	44.3 ± 2.8
		3.0	4.0	58.9 ± 6.4	53.1 ± 5.7	51.1 ± 6.6
			5.5	51.2 ± 4.5	48.3 ± 5.9	46.9 ± 4.89
			7.0	53.0 ± 6.9	49.4 ± 7.4	46.6 ± 6.3
	HP	0.5	4.0	39.1 ± 4.8	38.5 ± 5.1	37.3 ± 4.7
			5.5	41.1 ± 0.9	39.8 ± 1.6	38.8 ± 0.8
			7.0	40.5 ± 0.7	39.2 ± 1.2	37.8 ± 0.8
		1.0	4.0	46.5 ± 1.4	42.4 ± 2.7	40.6 ± 3.4
			5.5	46.4 ± 5.4	43.2 ± 6.6	41.9 ± 7.4
			7.0	43.5 ± 8.6	41.1 ± 7.9	38.9 ± 7.6
		3.0	4.0	58.8 ± 2.7	54.4 ± 2.1	52.0 ± 3.8
			5.5	56.2 ± 3.4	50.4 ± 3.0	47.6 ± 3.9
			7.0	55.4 ± 4.8	47.5 ± 5.7	47.4 ± 4.3

M: insect meal, HT: control hydrolysate, HP: Hydrolysate whose insect flour has been pretreated at high hydrostatic pressures. Values represent mean observations of three replicates ± standard deviation. The interaction between treatment, concentration, pH and time was not significant.

4. Conclusions

This study demonstrated that some functional properties of commercial insect meals, mainly solubility and the OBC, can be enhanced by enzymatic hydrolysis by Alcalase®. More specifically, the WBC and the foaming and gelation properties were not improved by enzymatic digestion (control and pressure-treated insect meals). Moreover, meals and hydrolysates had low emulsifying properties and viscosity. These results corroborate their current use as filler agents to produce protein-enriched bread and pasta. The DH was further improved by the use of HHP, but only for the mealworms. Despite only a moderate effect of HHP, these results showed the potential of insect protein hydrolysate as a new food ingredient for food formulations. However, to produce insect-based ingredients with better functionalities, it is crucial to control the heating step during insect drying to minimize its negative impact on protein functionality while ensuring microbiological safety of the insect ingredients. The use of extrusion-cooking technology on insect meals and hydrolysates also represents an advantage for improving their functional properties.

Author Contributions: Conceptualization, A.D.-P., A.D. and S.L.T.; Methodology, A.D.-P., M.L., A.D. and S.L.T.; Software, A.D.-P.; Validation, A.D.-P., A.D. and S.L.T.; Formal Analysis, A.D.-P.; Investigation, A.D.-P., A.D. and S.L.T.; Resources, A.D.-P., M.L., A.D. and S.L.T.; Data Curation, A.D.-P.; Writing—Original Draft Preparation, A.DL-P.; Writing—Review and Editing, A.D.-P., A.D. and S.L.T.; Visualization, A.D.-P., A.D. and S.L.T.; Supervision, A.D. and S.L.T.; Project Administration, A.D. and S.L.T.; Funding Acquisition, A.D. and S.L.T. All authors have read and agreed to the published version of the manuscript.

Funding: This research was funded by Le Fonds de recherche du Québec—Nature et technologies (Grant # 2018-PR-208090).

Acknowledgments: The authors thank Le Fonds de recherche du Québec-Nature et technologies (Grant # 2018-PR-208090) for financial support. The authors also thank Diane Gagnon and Véronique Perreault for their technical support.

Conflicts of Interest: The authors declare no conflict of interest.

Appendix A

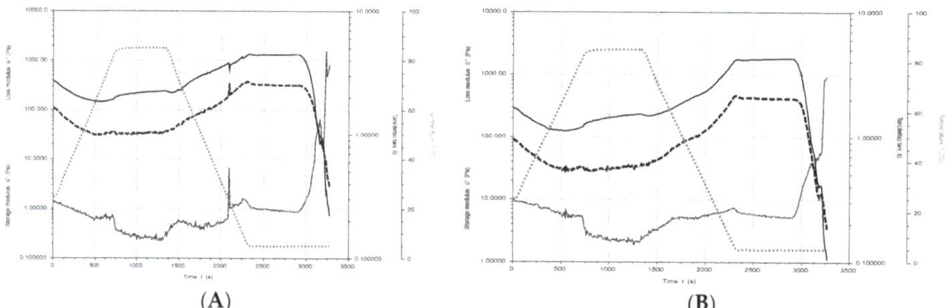

Figure A1. Dynamic moduli G′ and G″ of cricket meals at 30% (*w/v*) and pH 7.0 during heating and cooling. (**A**) without NaCl addition and (**B**) with 1 M NaCl.

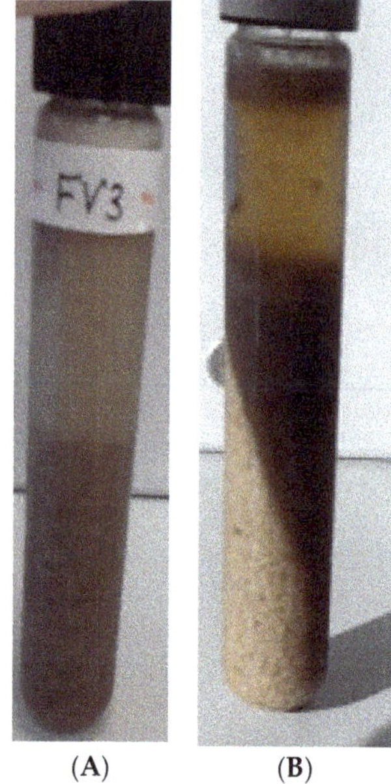

Figure A2. Mealworm meal at 3% (*w/v*) and pH 7.0 (**A**) before and (**B**) after heat treatment.

Table A1. Probability values for the effects of insects (crickets and mealworms), treatments (meals, control hydrolysate and hydrolysate generated after HHP treatment of meals), concentration (0.5, 1.0 and 3.0% *w/v*), pH (4.0, 5.5 and 7.0) and time (30, 60 and 90 min) on the emulsion stability index.

	Probability Values
Main effects	
Insects (I)	<0.0001
Treatments (T)	<0.0001
Concentration (C)	<0.0001
pH	0.0015
Time	<0.0001
Interactions	
I × T	<0.0001
I × pH	0.9160
T × pH	0.3082
I × C	<0.0001
T × C	<0.0001
C × pH	0.8537
I × Time	0.1246
T × Time	<0.0001
pH × Time	0.7627
C × Time	<0.0001
I × T × pH	0.7393
I × T × C	0.1191
I × C × pH	0.2663
T × C × pH	0.1481
I × T × Time	<0.0001
I × pH × Time	0.1367
I × C × Time	0.4107
T × pH × Time	0.0971
T × C × Time	0.0129
C × pH × Time	0.8910
I × T × C × pH	0.6985
I × T × pH × Time	0.6341
I × T × C × Time	0.4343
I × C × pH × Time	0.8402
T × C × pH × Time	0.7098
I × T × C × pH × Time	0.9184

References

1. Van Huis, A.; Van Itterbeeck, J.; Klunder, H.; Mertens, E.; Halloran, A.; Muir, G.; Vantomme, P. *Edible Insects: Future Prospects for Food and Feed Security*; FAO U.N.: Rome, Italy, 2013.

2. Akhtar, Y.; Isman, M. Insects as an Alternative Protein Source. In *Proteins in Food Processing*; Elsevier BV: Amsterdam, The Netherlands, 2018; pp. 263–288.

3. Hall, F.; Jones, O.G.; O'Haire, M.E.; Liceaga, A.M. Functional properties of tropical banded cricket (*Gryllodes sigillatus*) protein hydrolysates. *Food Chem.* **2017**, *224*, 414–422. [CrossRef]

4. Rumpold, B.A.; Schlüter, O.K. Nutritional composition and safety aspects of edible insects. *Mol. Nutr. Food Res.* **2013**, *57*, 802–823. [CrossRef]

5. Oonincx, D.G.A.B.; Van Itterbeeck, J.; Heetkamp, M.J.W.; Brand, H.V.D.; Van Loon, J.J.A.; Van Huis, A. An Exploration on Greenhouse Gas and Ammonia Production by Insect Species Suitable for Animal or Human Consumption. *PLoS ONE* **2010**, *5*, e14445. [CrossRef]

6. Oonincx, D.G.A.B.; De Boer, I.J.M. Environmental Impact of the Production of Mealworms as a Protein Source for Humans—A Life Cycle Assessment. *PLoS ONE* **2012**, *7*, e51145. [CrossRef]

7. Alexander, P.; Brown, C.; Arneth, A.; Dias, C.; Finnigan, J.; Moran, D.; Rounsevell, M.D. Could consumption of insects, cultured meat or imitation meat reduce global agricultural land use? *Glob. Food Secur.* **2017**, *15*, 22–32. [CrossRef]

8. Premalatha, M.; Abbasi, T.; Abbasi, T.; Abbasi, S. Energy-efficient food production to reduce global warming and ecodegradation: The use of edible insects. *Renew. Sustain. Energy Rev.* **2011**, *15*, 4357–4360. [CrossRef]

9. Ramos-Elorduy, J.; Moreno, J.M.P.; Prado, E.E.; Perez, M.A.; Otero, J.L.; De Guevara, O.L. Nutritional Value of Edible Insects from the State of Oaxaca, Mexico. *J. Food Compos. Anal.* **1997**, *10*, 142–157. [CrossRef]

10. Bukkens, S.G. The nutritional value of edible insects. *Ecol. Food Nutr.* **1997**, *36*, 287–319. [CrossRef]

11. Bußler, S.; Rumpold, B.A.; Jander, E.; Rawel, H.M.; Schlüter, O.K. Recovery and techno-functionality of flours and proteins from two edible insect species: Meal worm (*Tenebrio molitor*) and black soldier fly (*Hermetia illucens*) larvae. *Heliyon* **2016**, *2*, e00218. [CrossRef]

12. Gmuer, A.; Guth, J.N.; Hartmann, C.; Siegrist, M. Effects of the degree of processing of insect ingredients in snacks on expected emotional experiences and willingness to eat. *Food Qual. Prefer.* **2016**, *54*, 117–127. [CrossRef]

13. Hartmann, C.; Shi, J.; Giusto, A.; Siegrist, M. The psychology of eating insects: A cross-cultural comparison between Germany and China. *Food Qual. Prefer.* **2015**, *44*, 148–156. [CrossRef]

14. Hartmann, C.; Siegrist, M. Becoming an insectivore: Results of an experiment. *Food Qual. Prefer.* **2016**, *51*, 118–122. [CrossRef]

15. Schösler, H.; De Boer, J.; Boersema, J.J. Can we cut out the meat of the dish? Constructing consumer-oriented pathways towards meat substitution. *Appetite* **2012**, *58*, 39–47. [CrossRef] [PubMed]

16. Tan, H.S.G.; Fischer, A.R.; Tinchan, P.; Stieger, M.; Steenbekkers, L.; Van Trijp, H. Insects as food: Exploring cultural exposure and individual experience as determinants of acceptance. *Food Qual. Prefer.* **2015**, *42*, 78–89. [CrossRef]

17. Fombong, F.T.; Van Der Borght, M.; Broeck, J.V. Influence of Freeze-Drying and Oven-Drying Post Blanching on the Nutrient Composition of the Edible Insect Ruspolia differens. *Insects* **2017**, *8*, 102. [CrossRef]

18. Melgar-Lalanne, G.; Hernández-Álvarez, A.; Salinas-Castro, A. Edible Insects Processing: Traditional and Innovative Technologies. *Compr. Rev. Food Sci. Food Saf.* **2019**, *18*, 1166–1191. [CrossRef]

19. Kröncke, N.; Böschen, V.; Woyzichovski, J.; Demtröder, S.; Benning, R. Comparison of suitable drying processes for mealworms (*Tenebrio molitor*). *Innov. Food Sci. Emerg. Technol.* **2018**, *50*, 20–25. [CrossRef]

20. Purschke, B.; Meinlschmidt, P.; Horn, C.; Rieder, O.; Jäger, H. Improvement of techno-functional properties of edible insect protein from migratory locust by enzymatic hydrolysis. *Eur. Food Res. Technol.* **2017**, *244*, 999–1013. [CrossRef]

21. Azagoh, C.; Ducept, F.; Garcia, R.; Rakotozafy, L.; Cuvelier, M.-E.; Keller, S.; Lewandowski, R.; Samir, M. Extraction and physicochemical characterization of *Tenebrio molitor* proteins. *Food Res. Int.* **2016**, *88*, 24–31. [CrossRef]

22. Chalamaiah, M.; Rao, G.N.; Rao, D.; Jyothirmayi, T. Protein hydrolysates from meriga (*Cirrhinus mrigala*) egg and evaluation of their functional properties. *Food Chem.* **2010**, *120*, 652–657. [CrossRef]

23. Ghribi, A.M.; Gafsi, I.M.; Sila, A.; Blecker, C.; Danthine, S.; Attia, H.; Bougatef, A.; Besbes, S. Effects of enzymatic hydrolysis on conformational and functional properties of chickpea protein isolate. *Food Chem.* **2015**, *187*, 322–330. [CrossRef] [PubMed]

24. Wang, J.; Zhang, J.; Li, L.; Wang, Y.; Wang, Q.; Zhai, Y.; You, H.; Hu, D. Association of rs12255372 in theTCF7L2 gene with type 2 diabetes mellitus: A meta-analysis. *Braz. J. Med. Biol. Res.* **2013**, *46*, 382–393. [CrossRef] [PubMed]

25. Chen, Z.; Li, Y.; Lin, S.; Wei, M.; Du, F.; Ruan, G. Development of continuous microwave-assisted protein digestion with immobilized enzyme. *Biochem. Biophys. Res. Commun.* **2014**, *445*, 491–496. [CrossRef] [PubMed]

26. Uluko, H.; Zhang, S.; Liu, L.; Tsakama, M.; Lu, J.; Lv, J. Effects of thermal, microwave, and ultrasound pretreatments on antioxidative capacity of enzymatic milk protein concentrate hydrolysates. *J. Funct. Foods* **2015**, *18*, 1138–1146. [CrossRef]

27. Abadía-García, L.; Castaño-Tostado, E.; Ozimek, L.; Romero-Gómez, S.; Ozuna, C.; Amaya-Llano, S.L. Impact of ultrasound pretreatment on whey protein hydrolysis by vegetable proteases. *Innov. Food Sci. Emerg. Technol.* **2016**, *37*, 84–90. [CrossRef]

28. Kadam, S.U.; Tiwari, B.K.; Álvarez, C.; O'Donnell, C.P. Ultrasound applications for the extraction, identification and delivery of food proteins and bioactive peptides. *Trends Food Sci. Technol.* **2015**, *46*, 60–67. [CrossRef]

29. Lin, S.; Guo, Y.; Liu, J.; You, Q.; Yin, Y.; Cheng, S. Optimized enzymatic hydrolysis and pulsed electric field treatment for production of antioxidant peptides from egg white protein. *Afr. J. Biotech.* **2011**, *10*, 11648.

30. Lin, S.; Jin, Y.; Liu, M.; Yang, Y.; Zhang, M.; Guo, Y.; Jones, G.; Liu, J.; Yin, Y. Research on the preparation of antioxidant peptides derived from egg white with assisting of high-intensity pulsed electric field. *Food Chem.* **2013**, *139*, 300–306. [CrossRef]

31. Mikhaylin, S.; Boussetta, N.; Vorobiev, E.; Bazinet, L. High Voltage Electrical Treatments to Improve the Protein Susceptibility to Enzymatic Hydrolysis. *ACS Sustain. Chem. Eng.* **2017**, *5*, 11706–11714. [CrossRef]

32. Marciniak, A.; Suwal, S.; Naderi, N.; Pouliot, Y.; Doyen, A. Enhancing enzymatic hydrolysis of food proteins and production of bioactive peptides using high hydrostatic pressure technology. *Trends Food Sci. Technol.* **2018**, *80*, 187–198. [CrossRef]

33. Peyrano, F.; Speroni, F.; Avanza, M. Physicochemical and functional properties of cowpea protein isolates treated with temperature or high hydrostatic pressure. *Innov. Food Sci. Emerg. Technol.* **2016**, *33*, 38–46. [CrossRef]

34. Wang, X.-S.; Tang, C.-H.; Li, B.-S.; Yang, X.-Q.; Li, L.; Ma, C.-Y. Effects of high-pressure treatment on some physicochemical and functional properties of soy protein isolates. *Food Hydrocoll.* **2008**, *22*, 560–567. [CrossRef]

35. Rivalain, N.; Roquain, J.; Demazeau, G. Development of high hydrostatic pressure in biosciences: Pressure effect on biological structures and potential applications in Biotechnologies. *Biotechnol. Adv.* **2010**, *28*, 659–672. [CrossRef] [PubMed]

36. Knudsen, J.; Otte, J.; Olsen, K.; Skibsted, L.H. Effect of high hydrostatic pressure on the conformation of β-lactoglobulin A as assessed by proteolytic peptide profiling. *Int. Dairy J.* **2002**, *12*, 791–803. [CrossRef]

37. Hemker, A.K.; Nguyen, L.T.; Karwe, M.; Salvi, D. Effects of pressure-assisted enzymatic hydrolysis on functional and bioactive properties of tilapia (*Oreochromis niloticus*) by-product protein hydrolysates. *LWT* **2020**, *122*, 109003. [CrossRef]

38. Liceaga-Gesualdo, A.; Li-Chan, E. Functional Properties of Fish Protein Hydrolysate from Herring (*Clupea harengus*). *J. Food Sci.* **1999**, *64*, 1000–1004. [CrossRef]

39. De Holanda, H.D.; Netto, F.M. Recovery of Components from Shrimp (*Xiphopenaeus kroyeri*) Processing Waste by Enzymatic Hydrolysis. *J. Food Sci.* **2006**, *71*, C298–C303. [CrossRef]

40. Synowiecki, J.; Al-Khateeb, N.A. Production, Properties, and Some New Applications of Chitin and Its Derivatives. *Crit. Rev. Food Sci. Nutr.* **2003**, *43*, 145–171. [CrossRef]

41. Yi, L.; Lakemond, C.; Sagis, L.M.C.; Eisner-Schadler, V.; Van Huis, A.; Van Boekel, M. Extraction and characterisation of protein fractions from five insect species. *Food Chem.* **2013**, *141*, 3341–3348. [CrossRef]

42. Boukil, A.; Perreault, V.; Chamberland, J.; Samir, M.; Pouliot, Y.; Doyen, A. High Hydrostatic Pressure-Assisted Enzymatic Hydrolysis Affect Mealworm Allergenic Proteins. *Molecules* **2020**, *25*, 2685. [CrossRef]

43. Latimer, G.W.; AOAC International. *Official Methods of Analysis*; AOAC International: Gaithersburg, MD, USA, 2012.

44. Janssen, R.H.; Vincken, J.-P.; Broek, L.A.M.V.D.; Fogliano, V.; Lakemond, C.M.M. Nitrogen-to-Protein Conversion Factors for Three Edible Insects: *Tenebrio molitor*, *Alphitobius diaperinus*, and *Hermetia illucens*. *J. Agric. Food Chem.* **2017**, *65*, 2275–2278. [CrossRef] [PubMed]

45. Tzompa-Sosa, D.A.; Yi, L.; Van Valenberg, H.J.; Van Boekel, M.A.; Lakemond, C.M. Insect lipid profile: Aqueous versus organic solvent-based extraction methods. *Food Res. Int.* **2014**, *62*, 1087–1094. [CrossRef]

46. Spinelli, J.; Lehman, L.; Wieg, D. Composition, Processing, and Utilization of Red Crab (*Pleuroncodes planipes*) as an Aquacultural Feed Ingredient. *J. Fish. Res. Board Can.* **1974**, *31*, 1025–1029. [CrossRef]

47. Church, F.C.; Swaisgood, H.E.; Porter, D.H.; Catignani, G.L. Spectrophotometric Assay Using o-Phthaldialdehyde for Determination of Proteolysis in Milk and Isolated Milk Proteins. *J. Dairy Sci.* **1983**, *66*, 1219–1227. [CrossRef]

48. Turgeon, S.; Bard, C.; Gauthier, S. Comparaison de trois méthodes pour la mesure du degré d'hydrolyse de protéines laitières modifiées enzymatiquement. *Can. Inst. Food Sci. Technol. J.* **1991**, *24*, 14–18. [CrossRef]

49. Morr, C.; German, B.; Kinsella, J.; Regenstein, J.; Van Buren, J.; Kilara, A.; Lewis, B.; Mangino, M. A Collaborative Study to Develop a Standardized Food Protein Solubility Procedure. *J. Food Sci.* **1985**, *50*, 1715–1718. [CrossRef]

50. Mezger, T.G. *The Rheology Handbook: For Users of Rotational and Oscillatory Rheometers*; Vincentz Network GmbH & Co KG.: Hannover, Germany, 2006.

51. Guo, F.; Xiong, Y.L.; Qin, F.; Jian, H.; Huang, X.; Chen, J. Surface Properties of Heat-Induced Soluble Soy Protein Aggregates of Different Molecular Masses. *J. Food Sci.* **2015**, *80*, C279–C287. [CrossRef]

52. Quinn, J.R.; Paton, D. A pratical measurement of water hydratation capacity of protein materials. *Cereal Chem.* **1979**, *56*, 38–40.

53. Haque, Z.U.; Mozaffar, Z. Casein hydrolysate. II. Functional properties of peptides. *Food Hydrocoll.* **1992**, *5*, 559–571. [CrossRef]

54. Pearce, K.N.; Kinsella, J.E. Emulsifying properties of proteins: Evaluation of a turbidimetric technique. *J. Agric. Food Chem.* **1978**, *26*, 716–723. [CrossRef]

55. Zielińska, E.; Baraniak, B.; Karaś, M.; Rybczyńska, K.; Jakubczyk, A. Selected species of edible insects as a source of nutrient composition. *Food Res. Int.* **2015**, *77*, 460–466. [CrossRef]

56. Zielińska, E.; Karaś, M.; Jakubczyk, A. Antioxidant activity of predigested protein obtained from a range of farmed edible insects. *Int. J. Food Sci. Technol.* **2016**, *52*, 306–312. [CrossRef]

57. Gbogouri, G.; Linder, M.; Fanni, J.; Parmentier, M. Influence of Hydrolysis Degree on the Functional Properties of Salmon Byproducts Hydrolysates. *J. Food Sci.* **2004**, *69*, 615–622. [CrossRef]

58. Panyam, D.; Kilara, A. Enhancing the functionality of food proteins by enzymatic modification. *Trends Food Sci. Technol.* **1996**, *7*, 120–125. [CrossRef]

59. Guan, H.; Diao, X.; Jiang, F.; Han, J.; Kong, B. The enzymatic hydrolysis of soy protein isolate by Corolase PP under high hydrostatic pressure and its effect on bioactivity and characteristics of hydrolysates. *Food Chem.* **2018**, *245*, 89–96. [CrossRef]

60. Stone, A.K.; Tanaka, T.; Nickerson, M.T. Protein quality and physicochemical properties of commercial cricket and mealworm powders. *J. Food Sci. Technol.* **2019**, *56*, 3355–3363. [CrossRef]

61. Zhao, X.; Vázquez-Gutiérrez, J.L.; Johansson, D.P.; Landberg, R.; Langton, M. Yellow Mealworm Protein for Food Purposes-Extraction and Functional Properties. *PLoS ONE* **2016**, *11*, e147791. [CrossRef]

62. Wessels, M.; Azzollini, D.; Fogliano, V. Frozen storage of lesser mealworm larvae (*Alphitobius diaperinus*) changes chemical properties and functionalities of the derived ingredients. *Food Chem.* **2020**, *320*. [CrossRef]

63. Omotoso, O.T. Nutritional quality, functional properties and anti-nutrient compositions of the larva of *Cirina forda* (Westwood) (Lepidoptera: Saturniidae). *J. Zhejiang Univ. Sci. B* **2006**, *7*, 51–55. [CrossRef]

64. Womeni, H.M.; Tiencheu, B.; Linder, M.; Nabayo, C.; Martial, E.; Tenyang, N.; Tchouanguep Mbiapo, F.; Villeneuve, P.; Fanni, J.; Parmentier, M. Nutritional value and effect of cooking, drying and storage process on some functional properties of *Rhynchophorus phoenicis*. *J. Life Sci. Pharma Res.* **2012**, *2*, 203–217.

65. Adler-Nissen, J. *Enzymic Hydrolysis of Food Proteins*; Elsevier: New York, NY, USA, 1986.

66. Sathe, S.K.; Zaffran, V.D.; Gupta, S.; Li, T. Protein Solubilization. *J. Am. Oil Chem. Soc.* **2018**, *95*, 883–901. [CrossRef]

67. Bao, Z.-J.; Zhao, Y.; Wang, X.-Y.; Chi, Y. Effects of degree of hydrolysis (DH) on the functional properties of egg yolk hydrolysate with alcalase. *J. Food Sci. Technol.* **2017**, *54*, 669–678. [CrossRef] [PubMed]

68. Mora, P.G.; Peñas, E.; Frías, J.; Gomez, R.; Martinez-Villaluenga, C. High-pressure improves enzymatic proteolysis and the release of peptides with angiotensin I converting enzyme inhibitory and antioxidant activities from lentil proteins. *Food Chem.* **2015**, *171*, 224–232. [CrossRef] [PubMed]

69. Qin, Z.; Guo, X.; Lin, Y.; Chen, J.; Liao, X.; Hu, X.; Wu, J. Effects of high hydrostatic pressure on physicochemical and functional properties of walnut (*Juglans regia* L.) protein isolate. *J. Sci. Food Agric.* **2012**, *93*, 1105–1111. [CrossRef]

70. Bamdad, F.; Bark, S.; Kwon, C.H.; Suh, J.-W.; Sunwoo, H. Anti-Inflammatory and Antioxidant Properties of Peptides Released from β-Lactoglobulin by High Hydrostatic Pressure-Assisted Enzymatic Hydrolysis. *Molecules* **2017**, *22*, 949. [CrossRef]

71. Boukil, A.; Suwal, S.; Chamberland, J.; Pouliot, Y.; Doyen, A. Ultrafiltration performance and recovery of bioactive peptides after fractionation of tryptic hydrolysate generated from pressure-treated β-lactoglobulin. *J. Membr. Sci.* **2018**, *556*, 42–53. [CrossRef]

72. In, M.-J.; Kim, N.C.; Chae, H.J.; Oh, N.-S. Effects of degree of hydrolysis and pH on the solubility of heme-iron enriched peptide in hemoglobin hydrolysate. *Biosci. Biotechnol. Biochem.* **2003**, *67*, 365–367. [CrossRef]

73. Jung, S.; Murphy, P.A.; Johnson, L.A. Physicochemical and Functional Properties of Soy Protein Substrates Modified by Low Levels of Protease Hydrolysis. *J. Food Sci.* **2005**, *70*, 180–187. [CrossRef]

74. Lamsal, B.; Jung, S.; Johnson, L. Rheological properties of soy protein hydrolysates obtained from limited enzymatic hydrolysis. *LWT* **2007**, *40*, 1215–1223. [CrossRef]

75. Tsumura, K.; Saito, T.; Tsuge, K.; Ashida, H.; Kugimiya, W.; Inouye, K. Functional properties of soy protein hydrolysates obtained by selective proteolysis. *LWT* **2005**, *38*, 255–261. [CrossRef]

76. Hines, M.E.; Foegeding, E.A. Interactions of .alpha.-lactalbumin and bovine serum albumin with .beta.-lactoglobulin in thermally induced gelation. *J. Agric. Food Chem.* **1993**, *41*, 341–346. [CrossRef]

77. Le, X.T.; Rioux, L.-E.; Turgeon, S.L. Formation and functional properties of protein–polysaccharide electrostatic hydrogels in comparison to protein or polysaccharide hydrogels. *Adv. Colloid Interface Sci.* **2017**, *239*, 127–135. [CrossRef] [PubMed]

78. Turgeon, S.L.; Beaulieu, M. Improvement and modification of whey protein gel texture using polysaccharides. *Food Hydrocoll.* **2001**, *15*, 583–591. [CrossRef]

79. Pomeranz, Y. *Functional Properties of Food Components*; Elsevier BV: Amsterdam, The Netherlands, 1985.

80. Totosaus, A.; Montejano, J.G.; Salazar, J.A.; Guerrero, I. A review of physical and chemical protein-gel induction. *Int. J. Food Sci. Technol.* **2002**, *37*, 589–601. [CrossRef]

81. Zhang, T.; Jiang, B.; Wang, Z. Gelation properties of chickpea protein isolates. *Food Hydrocoll.* **2007**, *21*, 280–286. [CrossRef]

82. Akpossan, R.; Digbeu, Y.; Koffi, M.; Kouadio, J.; Dué, E.; Kouame, P. Protein Fractions and Functional Properties of Dried *Imbrasia oyemensis* Larvae Full-Fat and Defatted Flours. *Int. J. Biochem. Res. Rev.* **2015**, *5*, 116–126. [CrossRef]

83. Valle, F.R.; Mena, M.H.; Bourges, H. An investigation into insect protein. *J. Food Process. Preserv.* **1982**, *6*, 99–110. [CrossRef]

84. Lomakina, K.; Míková, K. A study of the factors affecting the foaming properties of egg white—A review. *Czech. J. Food Sci.* **2011**, *24*, 110–118. [CrossRef]

85. Zielińska, E.; Karaś, M.; Baraniak, B. Comparison of functional properties of edible insects and protein preparations thereof. *LWT* **2018**, *91*, 168–174. [CrossRef]

86. Damodaran, S.; Paraf, A. *Food Proteins and Their Applications*; Informa UK Limited: London, UK, 2017.

87. Deng, Q.; Wang, L.; Wei, F.; Xie, B.J.; Huang, F.-H.; Huang, W.; Shi, J.; Huang, Q.; Tian, B.; Xue, S. Functional properties of protein isolates, globulin and albumin extracted from *Ginkgo biloba* seeds. *Food Chem.* **2011**, *124*, 1458–1465. [CrossRef]

88. Pacheco-Aguilar, R.; Mazorra-Manzano, M.A.; Ramírez-Suárez, J.C. Functional properties of fish protein hydrolysates from Pacific whiting (*Merluccius productus*) muscle produced by a commercial protease. *Food Chem.* **2008**, *109*, 782–789. [CrossRef] [PubMed]

89. Wu, W.U.; Hettiarachchy, N.S.; Qi, M. Hydrophobicity, solubility, and emulsifying properties of soy protein peptides prepared by papain modification and ultrafiltration. *J. Am. Oil Chem. Soc.* **1998**, *75*, 845–850. [CrossRef]

90. Chatsuwan, N.; Nalinanon, S.; Puechkamut, Y.; Lamsal, B.P.; Pinsirodom, P. Characteristics, Functional Properties, and Antioxidant Activities of Water-Soluble Proteins Extracted from Grasshoppers, *Patanga succincta* and *Chondracris roseapbrunner*. *J. Chem.* **2018**, *2018*, 6528312. [CrossRef]

91. Linder, M.; Fanni, J.; Parmentier, M. Functional Properties of Veal Bone Hydrolysates. *J. Food Sci.* **1996**, *61*, 712–716. [CrossRef]

92. Meinlschmidt, P.; Schweiggert-Weisz, U.; Brode, V.; Eisner, P. Enzyme assisted degradation of potential soy protein allergens with special emphasis on the technofunctionality and the avoidance of a bitter taste formation. *LWT* **2016**, *68*, 707–716. [CrossRef]

93. Zhai, J.; Day, L.; Aguilar, M.-I.; Wooster, T.J. Protein folding at emulsion oil/water interfaces. *Curr. Opin. Colloid Interface Sci.* **2013**, *18*, 257–271. [CrossRef]

94. Chobert, J.M.; Bertrand-Harb, C.; Nicolas, M.G. Solubility and emulsifying properties of caseins and whey proteins modified enzymically by trypsin. *J. Agric. Food Chem.* **1988**, *36*, 883–892. [CrossRef]

95. Lam, R.S.; Nickerson, M.T. The effect of pH and temperature pre-treatments on the physicochemical and emulsifying properties of whey protein isolate. *LWT* **2015**, *60*, 427–434. [CrossRef]

96. Zhang, T.; Jiang, B.; Mu, W.; Wang, Z. Emulsifying properties of chickpea protein isolates: Influence of pH and NaCl. *Food Hydrocoll.* **2009**, *23*, 146–152. [CrossRef]

Sample Availability: Samples of the compounds are not available from the authors.

Publisher's Note: MDPI stays neutral with regard to jurisdictional claims in published maps and institutional affiliations.

Article

High Hydrostatic Pressure-Assisted Enzymatic Hydrolysis Affect Mealworm Allergenic Proteins

Abir Boukil [1], Véronique Perreault [1], Julien Chamberland [1], Samir Mezdour [2], Yves Pouliot [1] and Alain Doyen [1,*]

[1] Department of Food Sciences, Institute of Nutrition and Functional Foods (INAF), Université Laval, Quebec City, QC G1V 0A6, Canada; abir.boukil.1@ulaval.ca (A.B.); veronique.perreault.5@ulaval.ca (V.P.); julien.chamberland@fsaa.ulaval.ca (J.C.); Yves.Pouliot@fsaa.ulaval.ca (Y.P.)

[2] AgroParisTech, UMR782 Paris-Saclay Food and Bioproduct Engineering (SayFood and Bioproduct Engineering), 1, rue des Olympiades, 91077 Massy, France; samir.mezdour@agroparistech.fr

* Correspondence: alain.doyen@fsaa.ulaval.ca; Tel.: +1+418-656-2131 (ext. 4054540)

Academic Editors: Przemyslaw Lukasz Kowalczewski and Antonio-José Trujillo

Received: 12 May 2020; Accepted: 4 June 2020; Published: 9 June 2020

Abstract: Edible insects have garnered increased interest as alternative protein sources due to the world's growing population. However, the allergenicity of specific insect proteins is a major concern for both industry and consumers. This preliminary study investigated the capacity of high hydrostatic pressure (HHP) coupled to enzymatic hydrolysis by Alcalase® or pepsin in order to improve the in vitro digestion of mealworm proteins, specifically allergenic proteins. Pressurization was applied as pretreatment before in vitro digestion or, simultaneously, during hydrolysis. The degree of hydrolysis was compared between the different treatments and a mass spectrometry-based proteomic method was used to determine the efficiency of allergenic protein hydrolysis. Only the Alcalase® hydrolysis under pressure improved the degree of hydrolysis of mealworm proteins. Moreover, the in vitro digestion of the main allergenic proteins was increased by pressurization conditions that were specifically coupled to pepsin hydrolysis. Consequently, HHP-assisted enzymatic hydrolysis represents an alternative strategy to conventional hydrolysis for generating a large amount of peptide originating from allergenic mealworm proteins, and for lowering their immunoreactivity, for food, nutraceutical, and pharmaceutical applications.

Keywords: high hydrostatic pressure; mealworm proteins; enzymatic hydrolysis; allergenic proteins; proteomics analysis

1. Introduction

Edible insects have garnered increased interest for human consumption due to their high nutritional value and low environmental impact when compared to conventional livestock. Moreover, edible insects were targeted as a potential alternative protein resource to address the problem of a global food crisis. In Western countries, mealworm (*Tenebrio molitor*) is of particular interest, being evidenced in recent years by several publications related to protein quality and the improvement of techno-functional properties [1–3]. However, recent studies described an allergenic risk that was related to the consumption of edible insects due to potential cross-reactivity with other arthropods, especially crustaceans [4–6]. Tropomyosin, arginine-kinase, but also other insect proteins, such as larval cuticle protein, myosin light, and heavy chain, as well as troponin, are potentially involved in allergenic reactions in cross-reactivity with other allergenic species (shrimp, prawn, and crab) [5,7,8]. Consequently, particularly for consumers with a food allergy to crustaceans, it is necessary to consider the risk of an allergic reaction after consuming edible insects.

Different food processing methods, such as boiling [9,10], autoclaving [11], extrusion [12], microwave [13], pulsed-electric field [14], ultrasound [15], and HHP [16,17], may alter the intrinsic structure of food proteins and, consequently, decrease their allergenic properties. These processes may be coupled to enzymatic hydrolysis to further decrease the allergenicity of a wide range of conventional food proteins and generate protein hydrolysates that could potentially be integrated into specific formulations for food-allergic patients [18]. Similar strategies for reducing protein allergenicity were applied to edible insect products. Specifically, Alcalase® and pepsin were used to decrease the protein allergenicity of a wide range of food matrices, including edible insects. Indeed, Hall et al. [19] demonstrated that increasing the degree of hydrolysis (DH) of cricket proteins after proteolysis by Alcalase® positively impacted their bioactive potential while lowering the reactivity to tropomyosin. Van Broekhoven, Bastiaan-Net, de Jong, and Wichers [5] showed that mealworm protein allergenicity decreased significantly after heat processing and in vitro digestion by pepsin. Additionally, the application of microwave-assisted enzymatic hydrolysis by Alcalase® to cricket protein was an efficient method for generating hypoallergenic peptide fractions [20].

High hydrostatic pressure (HHP), a non-thermal technology that applies isostatic pressure from 100 to 1,000 MPa, induces the modification of secondary, tertiary, and quaternary structures of proteins, causing protein unfolding due to the rupture of noncovalent bonds (hydrogen, hydrophobic, and ionic bonds) [21]. The use of HHP combined with diverse enzymes to reduce protein allergenicity has been previously reviewed [18]. However, HHP-assisted enzymatic hydrolysis to modify allergenic protein digestion from edible insects has not yet been reported. Consequently, the aims of this preliminary work were 1) to apply HHP in combination with enzymatic hydrolysis by Alcalase® or pepsin to generate protein hydrolysates from mealworm meal and 2) to evaluate different pressurization strategies (applied as a pretreatment before enzymatic hydrolysis or simultaneously with enzymatic hydrolysis) to potentially improve the in vitro digestion of mealworm allergenic proteins.

2. Materials and Methods

2.1. Raw Material and Preparation of Mealworm Samples

Mealworm (*Tenebrio molitor*) meal powder was purchased from Entomo Farms (Norwood, ON, Canada). The meal was suspended in a Tris-HCl solution (pH 7) at a concentration of 5% w/V and stirred for 60 min at room temperature. Finally, the insect suspension was stored overnight at 4 °C prior to enzymatic hydrolysis.

2.2. Proximate Composition of Mealworm Meals

The crude fat content was obtained after hexane extraction based on a Soxhlet method (AOAC 960.39). The moisture and ash contents were determined by the Association of Official Analytical Chemists (AOAC) methods 925.09 and 923.03, respectively. The crude protein content was obtained by the Dumas method (Rapid Micro N Cube, Elementar, Langenselbold, Germany) while using a protein-to-nitrogen conversion factor of 4.76 [22]. The chitin content was determined according to the method described by Spinelli et al. [23]. Proximate composition of mealworm meal on dry basis was 45.8 ± 0.2 protein, 15.5 ± 0.9 lipid, 5.2 ± 0.0 ash, and 7.7 ± 1.0 chitin.

2.3. High Hydrostatic Pressure-Assisted Enzymatic Hydrolysis of Mealworm Proteins

Enzymatic hydrolysis of mealworm proteins was performed prior to (pretreated) or during (simultaneous) pressurization while using a laboratory scale HHP system (Mini FoodLab FPG5620, Stansted Fluid Power LTD, Essex, UK) with a rate of pressurization of 50 MPa/min. The pressure system has a maximum capacity of 900 MPa and a glycol/water solution (30:70) was the pressure transfer fluid. Alcalase® (Subtilisin from *Bacillus licheniformis*, Sigma Aldrich, St. Louis, MO, USA) and pepsin (BD Difco, Franklin Lakes, NJ, USA) were the enzymes used for hydrolysis. The pressure and duration parameters applied during enzymatic hydrolysis for both pretreated and simultaneous conditions were 380 MPa for 1 min. The specific value of 380 MPa was chosen because Zhang et al. [24] demonstrated a significant decrease of Alcalase® activity when the pressure reached 400 MPa, which indicated that this level of pressurization induced partial denaturation of the enzyme. Similarly, Curl and Jansen [25] observed that the activity of pepsin that was retained when diluted in a buffer solution at pH 1.9 was close to 100% at 400 MPa, but decreased drastically at higher pressures.

For pretreatment experiments, the mealworm meal suspension was first pressure-treated at 380 MPa for 1 min and then hydrolyzed at atmospheric pressure (0.1 MPa) while using Alcalase® or pepsin. The choice of these enzymes was based to their ability to hydrolyse the chitin-protein complex from arthropods and, consequently, to improve the recovery of separate protein and chitin separately, as described by Le Roux et al. [26] and De Holanda and Netto [27], respectively. Moreover, Alcalase® and pepsin were already used to decrease protein allergenicity in edible insect matrices [5,19]. The hydrolysis parameters for Alcalase® were an enzyme/substrate (E/S) ratio of 0.03% w/w at pH 8.5 and 60 °C for 120 min. The pepsin hydrolysis conditions were an E/S ratio of 0.25% w/w at pH 2.0 and 40 °C for 240 min. Hydrolysis durations of 120 and 240 min were determined as optimal for inducing the separation of the chitin-protein complexes in initial edible insect meals (results not shown) and consequently improved the recovery of protein/peptide fraction in the hydrolysates. For the simultaneous treatment condition, Alcalase® or pepsin enzyme, at the same E/S ratio, pH, and temperature as the pretreated condition, were added to the insect meal suspensions (5% w/V) before pressurization. The mixture of mealworm suspension and enzyme was also pressure-treated at 380 MPa for 1 min. An external thermoregulation system was used to maintain the enzyme's optimal temperature (60 and 40 °C for Alcalase® and pepsin, respectively) during pressurization. At the end of pressure treatment, decompression was instantaneous and mealworm samples were recovered. The remaining duration of hydrolysis (total hydrolysis of 120 and 240 min for Alcalase® and pepsin, respectively) was carried out at 0.1 MPa (atmospheric pressure) with constant pH and temperature control. The pH was controlled during control and pretreatment conditions, but not during simultaneous treatment (1 min), since it was not possible to open the reactor pressure vessel. However, it was controlled after pressure treatment and until the end of the hydrolysis. The control condition consisted of mealworm suspension (5% w/V) that was digested at atmospheric pressure (0.1 MPa) for 120 and 240 min with Alcalase® and pepsin, respectively. During the hydrolysis step, and for all conditions (control, pretreated, and simultaneous digestions), a sample of mealworm hydrolysate was collected every 2 min for the first 10 min of hydrolysis and then every 30 min until the end of digestion. Samples that were collected during and at the end of hydrolysis were immediately immersed in a 90 °C water bath for 5 min to inactivate the enzyme and then centrifuged at 9000× *g* for 10 min. Supernatants, corresponding to soluble peptide fractions and potential non hydrolyzed soluble proteins, were recovered and stored at 5 °C until analysis.

2.4. Analysis

2.4.1. Determination of the Degree of Hydrolysis

The degree of hydrolysis (DH), which is the proportion of peptide bonds released during in vitro protein digestion, was calculated according to the method that was described by Church et al. [28] for mealworm hydrolysates collected during and at the end of enzymatic hydrolysis, for control, pretreated, and simultaneous conditions. Briefly, 150 μL of mealworm hydrolysates (5% w/V) diluted by a factor of 60 was added to 3 mL of o-phthaldialdehyde (OPA) reagent. The mixture was incubated at room temperature for 2 min, transferred to polystyrene cuvettes and the absorbance at 340 nm measured in an Agilent 8453 UV-Visible spectrophotometer (Agilent Technologies, Santa Clara, CA, USA). DL-leucine was used as a standard with concentrations ranging from 0.75 mM to 3 mM. All of the samples were analyzed in triplicate. The DH was calculated using Equation (1):

$$\text{DH (\%)} = \left[\frac{h}{h_{tot}} \right] \times 100 \tag{1}$$

where h_{tot} of mealworm was determined from the amino acid composition of the protein, as the sum of mmols of the individual amino acids per g. The h_{tot} used in this study was 8.64 meq/g. The values of (h) were obtained by reference to a standard curve of absorbance at 340 nm versus mg/L amino nitrogen (using L-leucine) [29].

2.4.2. Digestion Profiles of Mealworm Proteins

The degradation profiles of the mealworm proteins after both pressurization conditions (pretreated and simultaneous) were obtained by sodium dodecyl sulfate-polyacrylamide gel electrophoresis (SDS-PAGE) under reducing conditions and compared to the control samples (enzymatic hydrolysis without pressurization treatment). Fifteen microliters of sample from initial and final hydrolysates (t = 0, 120 and 240 min) of both insect species was first diluted in 50 μL of deionized water. A volume of 25 μL of sample buffer (5% 2-mercaptoethanol, 95% Laemmli buffer) (Bio-Rad, Mississauga, ON, Canada) was added to 25 μL of each diluted sample. Subsequently, the solutions (hydrolysate and sample buffer) were immersed in a boiling water bath for 10 min and cooled on ice before injecting 10 μL per well. Electrophoresis was performed using 4–20% TGX Stain-Free polyacrylamide gel (Bio-Rad, Mississauga, ON, Canada) at 15 mA for 1 h at room temperature. The proteins were then stained with Coomassie blue (1 g/L of Coomassie Brilliant Blue R-250, 10% acetic acid, 40% ethanol, and 50% water) and destained with a solution of 10% V/V methanol and 10% V/V acetic acid. The MW of insect proteins and peptides were estimated using MW markers (Precision Plus Protein™ 161-0373 All Blue Prestained Protein Standards, Bio-Rad, Mississauga, ON, Canada). Images of the gels were captured using the ChemiDoc™ MP Imaging System (ChemiDoc MP, Bio-Rad, Mississauga, ON, Canada).

2.4.3. Protein Identification by Mass Spectrometry

Mass spectrometry (MS) analysis was performed by the Proteomics Platform of the Research Center of the *Centre Hospitalier Universitaire* (CHU) of Quebec City (QC, Canada). First, final hydrolysates (120 or 240 min of hydrolysis) generated by Alcalase® and pepsin in vitro digestion for each condition (control, pretreatment, and simultaneous conditions) were desalted on an Oasis HLB column (Waters, Mississauga, ON, Canada) and the peptides were quantified at 205 nm while using a nanodrop (ThermoFisher, Waltham, MA, USA). Peptide samples (1 μg) were analyzed by Liquid Chromatography-Mass Spectrometry (LC-MS/MS) on an Ekspert NanoLC425 (Eksigent, Redwood City, CA, USA) coupled to a 5600+ Triple TOF mass spectrometer (Sciex, Framingham, MA, USA) that was equipped with a nanoelectrospray ion source. The peptides were trapped on a pepmap 5 mm × 0.3 mm (ThermoFisher, Waltham, MA, USA) cartridge at 4 μL/min then separated on a self-packed picofrit column (New Objective, Woburn, MA, USA) with Reprosil 3 μL, (120A C18, 15 cm × 0.075 mm internal diameter), (Dr Maisch, Ammerbuch, Germany). The peptides were eluted with a linear gradient from 8–35% solvent B (acetonitrile, 0.1% V/V formic acid) in 30 min, at 300 ηL/min. Mass spectra were acquired using a data dependent acquisition mode and Analyst software version 1.7 (Sciex, Framingham, MA, USA). Each full scan mass spectrum (400 to 1,250 m/z) was followed by collision-induced dissociation of the twenty most intense ions. Dynamic exclusion was set for a period of 20 s and a tolerance of 100 ppm.

Mascot Generic Format (MGF) peak list files were created using Protein Pilot software (version 4.5, Sciex, Concord, ON, Canada). The MGF sample files were then analyzed using Mascot software (version 2.5.1, Matrix Science, London, UK). Uniprot databases were used to detect contaminants and align protein sequences while using the *Tenebrionidae* family (24,496 entries) database, assuming no enzyme. A fragment ion mass tolerance of 0.100 Da and a parent ion tolerance of 0.100 Da were used. The deamidation of asparagine and glutamine, and oxidation of methionine, were specified in Mascot as variable modifications.

Scaffold software (version 4.8.4, Proteome Software Inc., Portland, OR, USA) was used to validate MS/MS-based peptide and protein identifications. Peptide identifications were accepted if they could be established at a probability greater than 95.0% by the Scaffold Local False Discovery Rate (FDR) algorithm. Protein identifications were also accepted if they could be established at a probability that is greater than 95.0% and if they contained at least two identified peptides. Protein probabilities were assigned by the Protein Prophet algorithm [30]. Proteins that contained similar peptides that could not be differentiated based on MS/MS analysis alone were grouped to satisfy the principles of parsimony.

2.5. Statistical Analysis

The pressurization assays, enzymatic hydrolysis, and proximate composition were performed in triplicate. The DH results were expressed as mean ± standard deviation (SD). These data were subjected to a one-way analysis of variance (ANOVA) using Sigma Plot 14.0 (Systat Software Inc., Chicago, IL, USA). A 95% confidence level was used for all analyses.

3. Results

3.1. Mealworm Protein Degradation during Enzymatic Hydrolysis

Figure 1 presents the degradation of mealworm soluble proteins that were recovered at the beginning, during, and at the end of the enzymatic hydrolysis for control and pressurized conditions (pretreated and simultaneous). Regarding protein profiles during Alcalase® hydrolysis (Figure 1A), a band, whose intensity increased as a function of hydrolysis duration, was detected in wells for control, pretreated, and simultaneous conditions. Whatever the condition, three distinctive bands that corresponded to proteins with molecular weights (MW) close to 10, 20, and 50 kDa were observed. The protein degradation profiles for hydrolysates after 10 and 120 min of Alcalase® hydrolysis for control, pretreated, and simultaneous conditions were similar. More specifically, the same bands, as listed for control at t = 0 min (10, 20, and 50 kDa), were detected, but heavy bands close to 10 kDa were observed after 10 and 120 min of hydrolysis, resulting in mostly peptide fragments. A band with MW between 150 and 250 kDa observed after 10 min of hydrolysis, and whose intensity was increasing at 120 min of hydrolysis, was detected for control and pretreatment conditions but was absent at t = 0 min. When compared to Alcalase®, non-distinctive bands were observed for the pepsin control at t = 0 min except for the band at 20 kDa (Figure 1B). However, while peptides with molecular weights close to 10 kDa MW were detected for all conditions, the bands were more intense for simultaneous conditions after 10 and 240 min of hydrolysis as compared to the pretreatment condition. Under the control conditions, peptides with molecular weight close to 10 kDa MW were more concentrated after 10 min of hydrolysis as compared to 240 min.

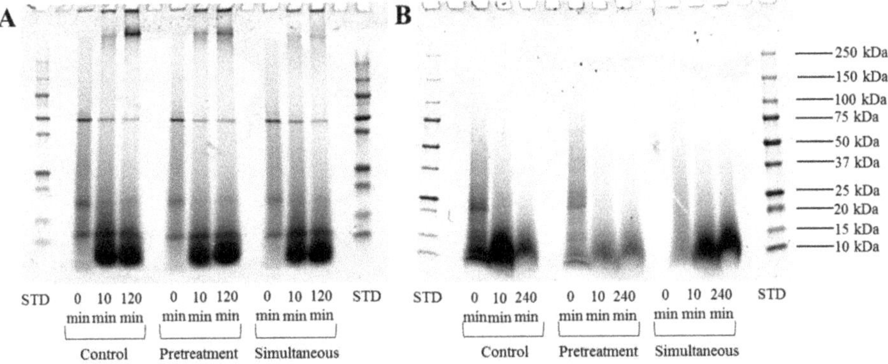

Figure 1. Protein degradation of mealworm proteins during enzymatic hydrolysis by Alcalase® (**A**) and pepsin (**B**) for control (0.1 MPa) and pressurization conditions (pretreated and simultaneous at 380 MPa for 1 min).

3.2. Effect of High Hydrostatic Pressure Coupled to Enzymatic Digestion on the Degree of Hydrolysis of Mealworm Proteins

Figure 2 shows the evolution of DH for mealworm proteins digested by Alcalase® and pepsin under three different conditions: without pressure (control), HHP applied as a pretreatment before hydrolysis, and HHP applied simultaneously during in vitro digestion. The degree of hydrolysis during in vitro digestion of mealworm protein by Alcalase® increased rapidly from 0 to 10 min and linearly until the end of hydrolysis (120 min), no matter the condition (control, pretreatment, and simultaneous conditions) (Figure 2A). The evolution of DH was similar for the control and pretreatment conditions with respective final values of 32.29 ± 1.90%, 37.82 ± 3.43%, and 29.12 ± 0.95% ($p > 0.05$). However, the degree of hydrolysis under HHP (simultaneous condition) was higher than control (21.75 ± 1.27%) and pretreated (22.20 ± 1.89%) conditions up to 60 min ($p < 0.05$) (27.15 ± 2.02%), but remained similar from 60 to 120 min ($p > 0.05$).

The evolution of the DH during mealworm protein hydrolysis by pepsin was similar to that of Alcalase®, with values of 35.20 ± 1.87%, 29.01 ± 3.55%, and 32.31 ± 1.09% for the control, pretreated, and simultaneous conditions, respectively ($p > 0.05$) (Figure 2B). When compared to hydrolysis by Alcalase®, the DH increased steadily over time and was lower at the end of hydrolysis (120 min for Alcalase® and 240 min for pepsin).

3.3. Determination of Mealworm Allergenic Protein Precursors from Hydrolysates

Table 1 presents the determination of mealworm allergenic protein precursors from peptides generated after 120 min of Alcalase® hydrolysis or 240 min of pepsin hydrolysis in vitro digestion. The total spectrum count (TSC) value, defined as the total number of spectra identified for a protein, is of particular interest, since this parameter is a semi-quantitative measure for a given protein abundance in proteomic studies [31]. Indeed, previous studies have demonstrated that the spectral counts of proteins correlate linearly with protein abundances in complex samples [32–34]. High throughput mass spectrometry identified a total of 110 proteins and close to 2800 spectrum counts. From these proteins, 19 allergenic proteins in mealworm meal were identified from the 161 unique peptides that were generated after Alcalase® or pepsin in vitro digestion of the mealworm proteins. The coverage and TSC ranged from 3 to 70% and 0 to 106, whatever the conditions (control, pretreatment, and simultaneous) and enzyme (Alcalase® and pepsin). The 19 allergenic proteins were identified according to the publication of Barre et al. in 2018 [35]. The majority of protein MWs ranged from 11 to 51 kDa (~74%), but five of the identified proteins ranged from 84 to 123 kDa. Most of the identified proteins had good peptide coverage, with ~31% of proteins having >10% of the sequence coverage and ~31% of proteins having >20% sequence coverage.

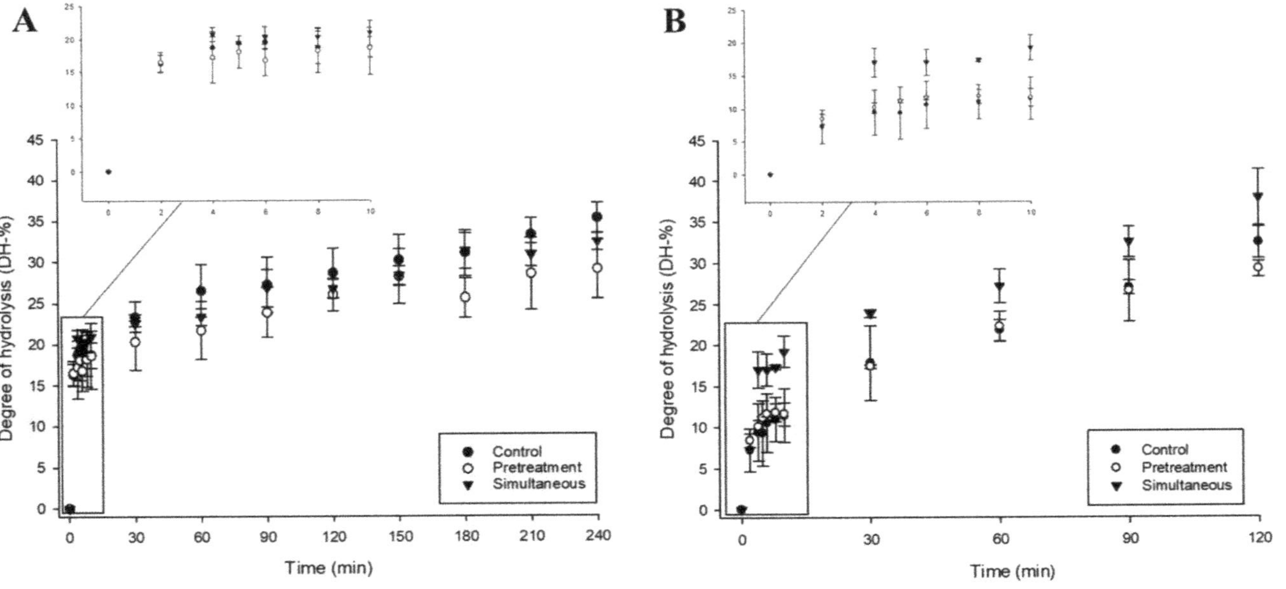

Figure 2. Degree of hydrolysis (%) of mealworm proteins digested by Alcalase® (**A**) and pepsin (**B**) for control (0.1 MPa), pretreated, and simultaneous conditions (380 MPa) (n = 3 ± SD).

Table 1. Mealworm allergenic protein precursors from peptides generated after Alcalase® and pepsin in vitro digestion of mealworm proteins.

Protein #	Identified Proteins *	MW (kDa)	UniProt ID	Number of Unique Peptides	Coverage (%)	Total Spectrum Count (TSC) **					
						Alcalase®			Pepsin		
						C [1]	P [2]	S [3]	C	P	S
1	Myosin heavy chain	123	A0A139WDZ4	15	7	2	4	4	21	31	19
2	Myosin heavy chain	141	A0A139WE70	8	6	2	4	6	9	22	10
3	Myosin heavy chain	103	A0A139WE10	2	9	0	0	0	20	0	21
4	Tropomyosin-1	40	A0A139WAN8	3	9	0	0	0	3	2	1
5	Actin-87E	42	D6WF19	28	39	24	31	51	42	106	71
6	Hexamerin 2	84	A0A288EPS5	7	9	4	6	7	11	28	22
7	Hexamerin 1	86	A0A288EIN5	7	11	0	0	8	7	13	11
8	Arginine kinase 1	40	A0A139WNX9	7	17	10	14	12	9	28	18
9	Troponin T	46	D6W953	3	3	0	1	0	3	6	7
10	Troponin C	17	D6WZP8	1	7	1	1	0	1	2	1
11	Tubulin beta chain	50	D6WSV2	10	16	1	1	1	10	15	17
12	Tubulin alpha chain	50	D6WBN7	5	13	1	0	1	5	10	10
13	Alpha-amylase	51	P56634	6	13	2	1	7	6	21	15
14	Larval cuticle protein A2B	12	P80682	21	70	13	10	16	23	58	52
15	Larval cuticle protein F1	15	Q9TXD9	13	48	13	9	27	18	42	26
16	Larval cuticle protein A1A	18	P80681	5	33	6	6	18	14	58	54
17	Larval cuticle protein A3A	14	P80683	8	58	16	9	21	17	55	59
18	Larval cuticle protein 8	11	D6WMB1	2	15	1	0	4	4	6	2
19	Larval / pupal cuticle protein H1C	21	P80686	10	33	0	0	2	14	29	19

* The probability of protein identification was over 95%. ** The peptide identification probability was ranging from 96 to 100%. [1] Control, [2] Pretreated, and [3] Simultaneous conditions.

Globally, the in vitro digestion of mealworm allergenic proteins by pepsin was more efficient than Alcalase®, since higher TSC values were obtained for all 19 allergenic proteins (Table 1). Regardless of the condition tested (control, pretreated, and simultaneous), allergenic proteins #5, 8, 14, 15, 16, and 17 were more susceptible to in vitro digestion by Alcalase®, as illustrated by high TSC values (TSC>10). Specifically, allergenic proteins #1, 2, and 6 were more resistant to hydrolysis by Alcalase® (TSC<10), whereas proteins #3, 4, 7, 9, 10, 11, 12, 18, and 19 were not digested by this enzyme (TSC values closed to 0). However, some differences were observed in TSC values as a function of the hydrolysis conditions. Indeed, while proteins #1, 2, 6, 8, 14, 18, and 19 were similarly digested by Alcalase®, as shown by the TSC detected for all conditions, the in vitro digestion of proteins #5, 13, 15, 16, and 17 was more efficient for the simultaneous treatment than for control and pretreated conditions. Only proteins #4, 9, 10 and 18 were resistant to pepsin hydrolysis, with TSC values < 10, regardless of the hydrolysis conditions (control and pressurization treatments). Protein #1 was not hydrolyzed by any treatment, since similar TSC values were obtained for all conditions. When compared to conventional hydrolysis at atmospheric pressure (0.1 MPa), the application of pressurization treatments (380 MPa for 1 min) before or simultaneous to hydrolysis improved the in vitro digestion of proteins #6, 11, 12, 13, 14, 15, 16, 17, and 19. The hydrolysis of proteins #2, 5, 8, 15, and 19 was enhanced by pretreatment, as compared to control and simultaneous conditions, but the hydrolysis of protein #3 was unaffected.

4. Discussion

The purpose of this study was to evaluate the effects of Alcalase® and pepsin hydrolysis under HHP processing conditions on the hydrolysis of allergenic proteins from mealworms. Our results provide the first evidence that HHP-assisted enzymatic hydrolysis by Alcalase® at 380 MPa for 1 min improved the degree of hydrolysis of mealworm proteins at an early stage of hydrolysis as well as in vitro digestion of allergenic proteins. Contrary to Alcalase®, HHP-assisted enzymatic hydrolysis by pepsin did not improve the degree of hydrolysis of mealworm proteins. However, the high pressure used as pretreatment condition enhanced the in vitro digestion of allergenic proteins by pepsin.

4.1. Effects of High Hydrostatic Pressure-Assisted Enzymatic Hydrolysis on Protein Degradation and Degree of Hydrolysis

Our results showed that Alcalase® hydrolysis of mealworm proteins under HHP (simultaneous condition-380 MPa for 1 min) improved the degree of hydrolysis by 24 and 22% after 60 min of hydrolysis when compared to the control and pretreated conditions, respectively (Figure 2), despite no major changes being observed in terms of protein degradation during hydrolysis (Figure 1). This non-correlation between SDS-PAGE profiles and degree of hydrolysis results could be explained by the generation of very low molecular weight peptides (largely < to 10 kDa), which could not be detected due to their migration outside the gel. Consequently, and specifically regarding the degree of hydrolysis, the result indicates that HHP treatment might have facilitated the mealworm protein conformational changes that are needed to increase the effectiveness of enzymatic digestion by providing Alcalase® access to the buried cleavage site at the very beginning of hydrolysis (until 10 min). Previous studies using various enzymes demonstrated that HHP-assisted enzymatic hydrolysis induced the exposure of new cleavage sites through protein unfolding, which enhanced enzyme activity, reduced hydrolysis time [36], and improved the degree of hydrolysis and concentration of peptides generated in protein hydrolysates [37–39]. In addition, the DH of mealworm proteins obtained was significantly increased (p < 0.05) during HHP-assisted enzymatic hydrolysis by Alcalase®. This increase was drastically improved as compared to the control (0.1 MPa) (Figure 2A), but was lower than similar studies [38,40,41], where DH increases ranged from 17% to 58% for pressure-treated protein hydrolysates as compared to the control (0.1 MPa). The effectiveness of HHP treatment on the enzymatic hydrolysis depends on specific parameters, such as substrate/enzyme ratio, pressure level, and treatment duration. Consequently, different hypotheses could explain the lack of influence of HHP on the in vitro digestion of mealworm protein. First, the duration of HHP treatment (1 min) was too short to induce optimal unfolding of mealworm proteins to improve Alcalase® access to buried sites. Second, irreversible protein aggregation occurring during the commercial-scale production of mealworm meals could decrease the efficiency of HHP and enzymatic hydrolysis. Indeed, the mealworm powders that were used for this study were roasted at approximately 107 °C and ground from fresh larvae [42]. When using the same mealworm powder from Entomo Farm, Stone et al. (2019) observed that the heating step of the insects during processing could result in protein denaturation, exposing hydrophobic groups and leading to protein aggregation [42]. Womeni et al. [43] demonstrated that roasting and grinding enhanced the aggregation of proteins, which makes the insect products unsuitable for food formulations due to their low solubility. A recent study published by Kröncke et al. [44] confirmed that oven drying of *T. molitor* larvae decreased the quality of proteins and reduced their solubility by 74% [44]. Consequently, the large amount of protein aggregates generated during production of mealworm meal rendered HHP ineffective for unfolding the proteins, leaving certain bonds in *T. molitor* proteins buried and inaccessible to Alcalase®, which decreased the proteolysis efficiency [45,46]. This second hypothesis is especially appealing given the large MW protein aggregates that were observed in the sodium dodecyl sulfate-polyacrylamide gel electrophoresis (SDS-PAGE) wells for control, pretreated, and simultaneous Alcalase® experimental conditions, and could explain why no significant DH was observed between the control and pretreatment conditions. Contrary to Alcalase® hydrolysis, coupling HHP and pepsin had no impact on the DH of mealworm proteins, despite the differences of protein degradation profiles observed in Figure 1. As mentioned for Alcalase®, it is difficult to correlate the results obtained in Figure 1; Figure 2, since very low molecular weight peptides could not be not detected in electrophoresis gels. The short duration of HHP treatment and the presence of protein aggregates in mealworm meal could explain the inability of HHP to improve protein digestion in vitro, as hypothesized for Alcalase®. Cleavage specificities are also important, since Alcalase® has broad specificity, hydrolyzing most peptide bonds. It preferentially hydrolyzes those containing aromatic amino acid residues whereas pepsin is more specific and cleaves peptide bonds following Phe or Tyr residues, as well as other hydrophobic amino acids. The recent study of Dai et al. [47] confirmed the importance of enzyme specificity during hydrolysis of *T. molitor* larva protein, since Alcalase® was the most efficient enzyme in terms of degree of hydrolysis compared to various other commercial enzymes (trypsin, Neutrase, papain, and pepsin) [47].

4.2. Digestion of Mealworm Allergenic Proteins by Pressurization Treatments

Recent publications demonstrated cross-reactivity between edible insects and other Arthropoda (crustaceans, mite), identifying different proteins that are involved in muscle contraction (actin, myosin, tropomyosin, troponin T and C, tubulin), in enzymatic pathways (arginine kinase 1, alpha-amylase) or part of the hemolymphatic system (hexamerin 1 and 2) as pan-allergens [48]. Globally, the results that are presented in Table 1 showed that pepsin was more efficient than Alcalase® for hydrolysis of allergenic mealworm proteins, despite the fact that, on the whole, the DH that was obtained with Alcalase® was higher (Figure 2). Our results also demonstrated that HHP, applied as a pretreatment before in vitro digestion or simultaneously with enzymatic hydrolysis, improved the in vitro digestion of specific mealworm allergenic proteins. These results are consistent with previous publications evaluating the effects of HHP coupled to enzymatic hydrolysis on the potential allergenicity of major protein allergens from different food matrices [49–52]. Indeed, after HHP-assisted enzymatic hydrolysis, these protein hydrolysates exhibited non-antigenic properties that were superior to those of proteins only treated with enzymatic hydrolysis. The production of hydrolysates with lower immunoreactivity from pressure-treated native protein is induced by increasing the protein susceptibility to enzymatic action by exposing new cleavage sites that allow for the proteases to reach otherwise buried hydrolysis sites.

However, our work also showed that HHP-assisted enzymatic hydrolysis had a limited effect on tropomyosin, myosin heavy chain, troponin, and tubulin proteolysis, as measured by TSC and enzyme efficiency when compared to the control condition. In the literature, the efficiency of Alcalase® and pepsin for the hydrolysis of these specific insect allergenic proteins has been scarcely reported.

Tropomyosin from shrimp, being classed as arthropods, was sensitive to HHP, since pressurization at 500 MPa for 10 min combined with thermal treatment at 55 °C decreased the protein allergenicity by 73.59% as compared to a boiling treatment [52]. However, tropomyosin is usually reported as heat stable and resistant to gastrointestinal digestion [53]. More specifically, it was demonstrated that pepsin could only slightly hydrolyze oyster tropomyosin, which demonstrated that tropomyosin has relatively good resistance to this enzyme. Nevertheless, Mejrhit et al. [54] found that shrimp tropomyosin IgE binding was decreased after heat and pepsin treatments. Hall, Johnson, and Liceaga [19] demonstrated that Alcalase® hydrolysis of cricket protein changed the binding characteristics of cricket tropomyosin to IgE, which indicated the susceptibility of tropomyosin to Alcalase® proteolysis. Similar results were obtained using house cricket *Acheta domesticus*, desert locust *Schistocerca gregaria* and yellow mealworm *T. molitor* [4]. To the best of our knowledge, there are no previous reports on the impact of Alcalase® or pepsin hydrolysis, coupled or not to HHP, on the vitro digestion of myosin heavy chain, troponin and tubulin from edible insects. However, Deng et al. [55] reported that pepsin was efficient for the proteolysis of myosin heavy chain and troponin from shrimp, while Alcalase® was often used for the hydrolysis of muscle proteins from meat and fish-based products [56]. Consequently, the low number of TSC for muscular mealworm proteins after enzymatic hydrolysis or the similar number of TSC obtained for control and pressurization conditions can be explained by irreversible protein aggregation induced by HHP, which drastically decreased the proteolytic activity of both Alcalase® and pepsin used at atmospheric pressure or in combination with HHP.

5. Conclusions

This preliminary study demonstrated that HHP-assisted enzymatic hydrolysis by Alcalase® improved the degree of hydrolysis of mealworm protein at the very beginning of digestion, despite no modification of protein degradation profiles. Moreover, pressurization treatments that were coupled to hydrolysis by Alcalase® and pepsin further increased the in vitro digestion of specific allergenic proteins. However, the presence of protein aggregates in the initial mealworm meal powder decreased the efficiency of enzymatic hydrolysis and pressurization treatment. Nevertheless, this work has produced encouraging results by combining HHP and enzymatic hydrolysis for the proteolysis of allergenic mealworm proteins. Besides, experiments are currently under way in order to understand the impact of HHP parameters on the denaturation and aggregation of edible insect proteins. Further research on native mealworm proteins extracted from fresh larvae and subject to minimal heat treatment is necessary to improve the combination of HHP and enzymatic hydrolysis.

Author Contributions: Conceptualization, A.B. and A.D.; Methodology, A.B., A.D.; Software, A.B.; Validation, A.B., S.M., Y.P., and A.D.; Formal Analysis, A.B.; Investigation, A.B. and A.D.; Resources, A.B., V.P., J.C., S.M., Y.P., and A.D.; Data Curation, A.B.; Writing—Original Draft Preparation, A.B. and A.D.; Writing—Review & Editing, A.B., V.P., J.C., S.M., Y.P. and A.D.; Visualization, A.B. and A.D.; Supervision, A.D.; Project Administration, A.D.; Funding Acquisition, A.D. All authors have read and agreed to the published version of the manuscript.

Funding: This research was funded by the Le Fonds de recherche du Québec—Nature et technologies (Grant # 2018-PR-208090 to Alain Doyen).

Acknowledgments: The authors thank Sylvie Bourassa and Victor Fourcassié for their assistance with proteomics analysis. The authors also thank Diane Gagnon and Dany Boisvert for technical support.

Conflicts of Interest: The authors declare no conflict of interest.

References

1. Yi, L.; Lakemond, C.; Sagis, L.M.; Eisner-Schadler, V.; Van Huis, A.; Van Boekel, M. Extraction and characterisation of protein fractions from five insect species. *Food Chem.* **2013**, *141*, 3341–3348. [CrossRef] [PubMed]

2. Purschke, B.; Meinlschmidt, P.; Horn, C.; Rieder, O.; Jäger, H. Improvement of techno-functional properties of edible insect protein from migratory locust by enzymatic hydrolysis. *Eur. Food Res. Technol.* **2017**, *244*, 999–1013. [CrossRef]

3. Mintah, B.K.; He, R.; Dabbour, M.; Xiang, J.; Agyekum, A.A.; Ma, H. Techno-functional attribute and antioxidative capacity of edible insect protein preparations and hydrolysates thereof: Effect of multiple mode sonochemical action. *Ultrason. Sonochem.* **2019**, *58*, 104676. [CrossRef] [PubMed]

4. Pali-Schöll, I.; Meinlschmidt, P.; Larenas-Linnemann, D.; Purschke, B.; Hofstetter, G.; Rodríguez-Monroy, F.A.; Einhorn, L.; Mothes-Luksch, N.; Jensen-Jarolim, E.; Jäger, H. Edible insects: Cross-recognition of IgE from crustacean- and house dust mite allergic patients, and reduction of allergenicity by food processing. *World Allergy Organ. J.* **2019**, *12*, 100006. [CrossRef]

5. Van Broekhoven, S.; Bastiaan-Net, S.; De Jong, N.W.; Wichers, H.J. Influence of processing and in vitro digestion on the allergic cross-reactivity of three mealworm species. *Food Chem.* **2016**, *196*, 1075–1083. [CrossRef]

6. Barre, A.; Caze-Subra, S.; Gironde, C.; Bienvenu, F.; Bienvenu, J.; Rougé, P. Entomophagie et risque allergique. *Rev. Française d'Allergol.* **2014**, *54*, 315–321. [CrossRef]

7. Ozawa, H.; Umezawa, K.; Takano, M.; Ishizaki, S.; Watabe, S.; Ochiai, Y. Structural and dynamical characteristics of tropomyosin epitopes as the major allergens in shrimp. *Biochem. Biophys. Res. Commun.* **2018**, *498*, 119–124. [CrossRef]

8. Broekman, H.C.; Knulst, A.; Jager, S.D.H.; Monteleone, F.; Gaspari, M.; De Jong, G.; Houben, G.F.; Verhoeckx, K. Effect of thermal processing on mealworm allergenicity. *Mol. Nutr. Food Res.* **2015**, *59*, 1855–1864. [CrossRef]

9. Liu, G.-M.; Cheng, H.; Nesbit, J.B.; Su, W.-J.; Cao, M.-J.; Maleki, S. Effects of boiling on the IgE-Binding properties of tropomyosin of shrimp (*Litopenaeus vannamei*). *J. Food Sci.* **2010**, *75*, T1–T5. [CrossRef]

10. De Gier, S.; Verhoeckx, K. Insect (food) allergy and allergens. *Mol. Immunol.* **2018**, *100*, 82–106. [CrossRef]

11. Villa, C.; Moura, M.B.M.V.; Costa, J.; Mafra, I. Immunoreactivity of lupine and soybean allergens in foods as affected by thermal processing. *Foods* **2020**, *9*, 254. [CrossRef] [PubMed]

12. Chen, L.; Phillips, R.D. Effects of twin-screw extrusion of peanut flour on in vitro digestion of potentially allergenic peanut proteins. *J. Food Prot.* **2005**, *68*, 1712–1719. [CrossRef] [PubMed]

13. Ketnawa, S.; Liceaga, A.M. Effect of microwave treatments on antioxidant activity and antigenicity of fish frame protein hydrolysates. *Food Bioprocess Technol.* **2016**, *10*, 582–591. [CrossRef]

14. Yang, W.; Tu, Z.; Wang, H.; Zhang, L.; Gao, Y.; Li, X.; Tian, M. Immunogenic and structural properties of ovalbumin treated by pulsed electric fields. *Int. J. Food Prop.* **2017**, *20*, S3164–S3176. [CrossRef]

15. Wang, C.; Xie, Q.; Wang, Y.; Fu, L. Effect of ultrasound treatment on allergenicity reduction of milk casein via colloid formation. *J. Agric. Food Chem.* **2020**, *68*, 4678–4686. [CrossRef]

16. Kato, T.; Katayama, E.; Matsubara, S.; Omi, Y.; Matsuda, T. Release of allergenic proteins from rice grains induced by high hydrostatic pressure. *J. Agric. Food Chem.* **2000**, *48*, 3124–3129. [CrossRef]

17. Li, H.; Zhu, K.; Zhou, H.; Peng, W. Effects of high hydrostatic pressure treatment on allergenicity and structural properties of soybean protein isolate for infant formula. *Food Chem.* **2012**, *132*, 808–814. [CrossRef]

18. Dong, X.; Wang, J.; Raghavan, V. Critical reviews and recent advances of novel non-thermal processing techniques on the modification of food allergens. *Crit. Rev. Food Sci. Nutr.* **2020**, 1–15. [CrossRef]

19. Hall, F.; Johnson, P.; Liceaga, A.M. Effect of enzymatic hydrolysis on bioactive properties and allergenicity of cricket (*Gryllodes sigillatus*) protein. *Food Chem.* **2018**, *262*, 39–47. [CrossRef]

20. Hall, F.; Liceaga, A.M. Effect of microwave-assisted enzymatic hydrolysis of cricket (*Gryllodes sigillatus*) protein on ACE and DPP-IV inhibition and tropomyosin-IgG binding. *J. Funct. Foods* **2020**, *64*, 103634. [CrossRef]

21. Butz, P.; Tauscher, B. Emerging technologies: Chemical aspects. *Food Res. Int.* **2002**, *35*, 279–284. [CrossRef]

22. Janssen, R.H.; Vincken, J.-P.; Broek, L.A.V.D.; Fogliano, V.; Lakemond, C. Nitrogen-to-protein conversion factors for three edible insects: *Tenebrio molitor*, *Alphitobius diaperinus*, and *Hermetia illucens*. *J. Agric. Food Chem.* **2017**, *65*, 2275–2278. [CrossRef] [PubMed]

23. Spinelli, J.; Lehman, L.; Wieg, D. Composition, Processing, and utilization of red crab (*Pleuroncodes planipes*) as an aquacultural feed ingredient. *J. Fish. Res. Board Can.* **1974**, *31*, 1025–1029. [CrossRef]

24. Zhang, T.; Jiang, B.; Miao, M.; Mu, W.; Li, Y. Combined effects of high-pressure and enzymatic treatments on the hydrolysis of chickpea protein isolates and antioxidant activity of the hydrolysates. *Food Chem.* **2012**, *135*, 904–912. [CrossRef]

25. Curl, A.L.; Jansen, E.F. Effect of high pressures on trypsin and chymotrypsin. *J. Biol. Chem.* **1950**, *184*, 45–54.

26. Le Roux, K.; Berge, J.P.; Baron, R.; Leroy, E.; Arhaliass, A. Extraction of Chitins in a Single Step by Enzymatic Hydrolysis in an Acid Medium. U.S. Patent Application No. 14/122,427, 10 April 2014.

27. De Holanda, H.D.; Netto, F.M. Recovery of components from shrimp (*Xiphopenaeus kroyeri*) processing waste by enzymatic hydrolysis. *J. Food Sci.* **2006**, *71*, C298–C303. [CrossRef]

28. Church, F.C.; Swaisgood, H.E.; Porter, D.H.; Catignani, G.L. Spectrophotometric assay using o-phthaldialdehyde for determination of proteolysis in milk and isolated milk proteins. *J. Dairy Sci.* **1983**, *66*, 1219–1227. [CrossRef]

29. Hall, F.; Jones, O.G.; O'Haire, M.E.; Liceaga, A.M. Functional properties of tropical banded cricket (*Gryllodes sigillatus*) protein hydrolysates. *Food Chem.* **2017**, *224*, 414–422. [CrossRef]

30. Nesvizhskii, A.I.; Keller, A.; Kolker, E.; Aebersold, R. A statistical model for identifying proteins by tandem mass spectrometry. *Anal. Chem.* **2003**, *75*, 4646–4658. [CrossRef]

31. Lundgren, D.H.; Hwang, S.; Wu, L.; Han, D.K. Role of spectral counting in quantitative proteomics. *Expert Rev. Proteom.* **2010**, *7*, 39–53. [CrossRef]

32. Liu, H.; Sadygov, R.G.; Yates, J.R. A model for random sampling and estimation of relative protein abundance in shotgun proteomics. *Anal. Chem.* **2004**, *76*, 4193–4201. [CrossRef] [PubMed]

33. Qian, W.-J.; Jacobs, J.M.; Camp, D.G.; Monroe, M.E.; Moore, R.J.; Gritsenko, M.A.; Calvano, S.E.; Lowry, S.F.; Xiao, W.; Moldawer, L.L.; et al. Comparative proteome analyses of human plasma followingin vivo lipopolysaccharide administration using multidimensional separations coupled with tandem mass spectrometry. *Proteomics* **2005**, *5*, 572–584. [CrossRef]

34. Old, W.M.; Meyer-Arendt, K.; Aveline-Wolf, L.; Pierce, K.G.; Mendoza, A.; Sevinsky, J.R.; Resing, K.A.; Ahn, N.G. Comparison of label-free methods for quantifying human proteins by shotgun proteomics. *Mol. Cell. Proteom.* **2005**, *4*, 1487–1502. [CrossRef] [PubMed]

35. Barre, A.; Simplicien, M.; Cassan, G.; Benoist, H.; Rougé, P. Food allergen families common to different arthropods (mites, insects, crustaceans), mollusks and nematods: Cross-reactivity and potential cross-allergenicity. *Rev. Française d'Allergol.* **2018**, *58*, 581–593. [CrossRef]

36. Garcia-Mora, P.; Peñas, E.; Frias, J.; Gomez, R.; Martinez-Villaluenga, C. High-pressure improves enzymatic proteolysis and the release of peptides with angiotensin I converting enzyme inhibitory and antioxidant activities from lentil proteins. *Food Chem.* **2015**, *171*, 224–232. [CrossRef]

37. Franck, M.; Perreault, V.; Suwal, S.; Marciniak, A.; Bazinet, L.; Doyen, A. High hydrostatic pressure-assisted enzymatic hydrolysis improved protein digestion of flaxseed protein isolate and generation of peptides with antioxidant activity. *Food Res. Int.* **2019**, *115*, 467–473. [CrossRef]

38. Nazir, M.A.; Mu, T.; Zhang, M. Preparation and identification of angiotensin I-converting enzyme inhibitory peptides from sweet potato protein by enzymatic hydrolysis under high hydrostatic pressure. *Int. J. Food Sci. Technol.* **2019**, *55*, 482–489. [CrossRef]

39. Boukil, A.; Suwal, S.; Chamberland, J.; Pouliot, Y.; Doyen, A. Ultrafiltration performance and recovery of bioactive peptides after fractionation of tryptic hydrolysate generated from pressure-treated β-lactoglobulin. *J. Membr. Sci.* **2018**, *556*, 42–53. [CrossRef]

40. Ahmed, J.; Mulla, M.; Al-Ruwaih, N.; Arfat, Y.A. Effect of high-pressure treatment prior to enzymatic hydrolysis on rheological, thermal, and antioxidant properties of lentil protein isolate. *Legum. Sci.* **2019**, *1*, 10. [CrossRef]

41. Al-Ruwaih, N.; Ahmed, J.; Mulla, M.F.; Arfat, Y.A. High-pressure assisted enzymatic proteolysis of kidney beans protein isolates and characterization of hydrolysates by functional, structural, rheological and antioxidant properties. *LWT* **2019**, *100*, 231–236. [CrossRef]

42. Stone, A.K.; Tanaka, T.; Nickerson, M. Protein quality and physicochemical properties of commercial cricket and mealworm powders. *J. Food Sci. Technol.* **2019**, *56*, 3355–3363. [CrossRef] [PubMed]

43. Womeni, H.M.; Tiencheu, B.; Linder, M.; Nabayo, C.; Martial, E.; Tenyang, N.; Tchouanguep Mbiapo, F.; Villeneuve, P.; Fanni, J.; Parmentier, M. Nutritional value and effect of cooking, drying and storage process on some functional properties of Rhynchophorus phoenicis. *Int. J. Life Pharma Res.* **2012**, *2*, 203–219.

44. Kröncke, N.; Böschen, V.; Woyzichovski, J.; Demtröder, S.; Benning, R. Comparison of suitable drying processes for mealworms (*Tenebrio molitor*). *Innov. Food Sci. Emerg. Technol.* **2018**, *50*, 20–25. [CrossRef]

45. Mulvihill, D.; Fox, P. Proteolysis of α s1-casein by chymosin: Influence of pH and urea. *J. Dairy Res.* **1977**, *44*, 533–540. [CrossRef]

46. Agboola, S.O.; Dalgleish, D.G. Enzymatic hydrolysis of milk proteins used for emulsion formation. 1. Kinetics of protein breakdown and storage stability of the emulsions. *J. Agric. Food Chem.* **1996**, *44*, 3631–3636. [CrossRef]

47. Dai, C.; Ma, H.; Luo, L.; Yin, X. Angiotensin I-converting enzyme (ACE) inhibitory peptide derived from *Tenebrio molitor* (L.) larva protein hydrolysate. *Eur. Food Res. Technol.* **2013**, *236*, 681–689. [CrossRef]

48. Leni, G.; Tedeschi, T.; Faccini, A.; Pratesi, F.; Folli, C.; Puxeddu, I.; Migliorini, P.; Gianotten, N.; Jacobs, J.; Depraetere, S.; et al. Shotgun proteomics, in-silico evaluation and immunoblotting assays for allergenicity assessment of lesser mealworm, black soldier fly and their protein hydrolysates. *Sci. Rep.* **2020**, *10*. [CrossRef]

49. Chicón, R.; Belloque, J.; Alonso, E.; Martín-Álvarez, P.J.; López-Fandiño, R. Hydrolysis under high hydrostatic pressure as a means to reduce the binding of β-lactoglobulin to immunoglobulin E from human sera. *J. Food Prot.* **2008**, *71*, 1453–1459. [CrossRef]

50. Peñas, E.; Préstamo, G.; Polo, F.; Gomez, R. Enzymatic proteolysis, under high pressure of soybean whey: Analysis of peptides and the allergen Gly m 1 in the hydrolysates. *Food Chem.* **2006**, *99*, 569–573. [CrossRef]

51. Zhou, H.; Wang, C.; Ye, J.; Tao, R.; Chen, H.; Cao, F. Effects of enzymatic hydrolysis assisted by high hydrostatic pressure processing on the hydrolysis and allergenicity of proteins from ginkgo seeds. *Food Bioprocess Technol.* **2016**, *9*, 839–848. [CrossRef]

52. Long, F.; Yang, X.; Wang, R.; Hu, X.; Chen, F. Effects of combined high pressure and thermal treatments on the allergenic potential of shrimp (*Litopenaeus vannamei*) tropomyosin in a mouse model of allergy. *Innov. Food Sci. Emerg. Technol.* **2015**, *29*, 119–124. [CrossRef]

53. Palmer, L. Edible Insects as a Source of Food Allergens. Master's Thesis, University of Nebraska-Lincoln, Lincoln, NE, USA, 2016.

54. Mejrhit, N.; Azdad, O.; Chda, A.; El Kabbaoui, M.; Bousfiha, A.; Bencheikh, R.; Tazi, A.; Aarab, L. Evaluation of the sensitivity of Moroccans to shrimp tropomyosin and effect of heating and enzymatic treatments. *Food Agric. Immunol.* **2017**, *28*, 969–980. [CrossRef]

55. Deng, S.-G.; Lutema, P.C.; Gwekwe, B.; Li, Y.; Akida, J.S.; Pang, Z.; Huang, Y.; Dang, Y.; Wang, S.; Chen, M.; et al. Bitter peptides increase engulf of phagocytes in vitro and inhibit oxidation of myofibrillar protein in peeled shrimp (*Litopenaeus vannamei*) during chilled storage. *Aquac. Rep.* **2019**, *15*, 100234. [CrossRef]

56. Muguruma, M.; Ahhmed, A.M.; Katayama, K.; Kawahara, S.; Maruyama, M.; Nakamura, T. Identification of pro-drug type ACE inhibitory peptide sourced from porcine myosin B: Evaluation of its antihypertensive effects in vivo. *Food Chem.* **2009**, *114*, 516–522. [CrossRef]

Article

Effects of Hexane on Protein Profile, Solubility and Foaming Properties of Defatted Proteins Extracted from *Tenebrio molitor* Larvae

Alexia Gravel [1], Alice Marciniak [1], Manon Couture [2] and Alain Doyen [1,*]

[1] Department of Food Sciences, Institute of Nutrition and Functional Foods (INAF), Université Laval, Quebec City, QC G1V 0A6, Canada; alexia.gravel.1@ulaval.ca (A.G.); alice.marciniak.1@ulaval.ca (A.M.)

[2] Department of Biochemistry, Microbiology and Bio-Informatics, Institute of Integrative Biology and Systems (IBIS), Université Laval, Quebec City, QC G1V 0A6, Canada; manon.couture@bcm.ulaval.ca

* Correspondence: alain.doyen@fsaa.ulaval.ca; Tel.: +1-418-656-2131 (ext. 405454)

Abstract: Inclusion of edible insects in human diets is increasingly promoted as a sustainable source of proteins with high nutritional value. While consumer acceptability remains the main challenge to their integration into Western food culture, the use of edible insects as meal and protein concentrate could decrease neophobia. The defatting of edible insects, mostly done with hexane, is the first step in producing protein ingredients. However, its impact on protein profiles and techno-functionality is still unclear. Consequently, this study compares the protein profiles of hexane-defatted and non-hexane-defatted yellow mealworm (*Tenebrio molitor*) meals and protein extracts, and evaluates the impact of hexane on protein solubility and foaming properties. Results showed that profiles for major proteins were similar between hexane-defatted and non-defatted samples, however some specific content differences (e.g., hexamerin 2) were observed and characterized using proteomic tools. Protein solubility was markedly lower for *T. molitor* meals compared to protein extracts. A large increase in the foaming capacity was observed for defatted fractions, whereas foam stability decreased similarly in all fractions. Consequently, although the hexane-defatting step was largely studied to produce edible insect protein ingredients, it is necessary to precisely understand its impact on their techno-functional properties for the development of food formulations.

Keywords: *Tenebrio molitor*; hexane defatting step; protein concentrates; protein profile and identification; protein solubility; foaming capacity

Citation: Gravel, A.; Marciniak, A.; Couture, M.; Doyen, A. Effects of Hexane on Protein Profile, Solubility and Foaming Properties of Defatted Proteins Extracted from *Tenebrio molitor* Larvae. *Molecules* **2021**, *26*, 351. https://doi.org/10.3390/molecules26020351

Received: 15 December 2020
Accepted: 9 January 2021
Published: 12 January 2021

Publisher's Note: MDPI stays neutral with regard to jurisdictional claims in published maps and institutional affiliations.

1. Introduction

As food security challenges related to worldwide population growth are anticipated in the coming years, an interest in insects as an emerging protein source for human consumption is being promoted. More specifically, Western countries have shown great interest in yellow mealworm (*Tenebrio molitor*) due to its high protein content and composition of essential lipids, as well as the low environmental impact of farming insects [1]. Nevertheless, food neophobia is the major challenge, which negatively affects the social acceptability of this alternative food resource [2,3]. It was suggested that the integration of edible insects as meal and protein concentrates or isolates in different food formulations could enhance consumer acceptability [2–5]. Consequently, numerous studies have investigated the nutritional and functional properties (e.g., solubility, foaming, gelling and emulsions) [6–9] of insect protein powders as a function of processing methods, mainly blanching and drying, in order to optimize the ingredient quality [10,11].

Defatting of raw food materials and by-products using organic solvents is a frequent method for producing protein-enriched ingredients [12–14]. Hexane is the most frequently used solvent for the production of defatted insect meal and protein extracts [15–17] despite its environmental, economic and safety disadvantages [18]. These drawbacks have motivated the study of different extraction solvents (ethanol, methanol, etc.) [19,20], as well as more sustainable options (supercritical CO_2 lipid extraction) [19,21]. Nevertheless, hexane remains the most popular method to remove lipid from the solid insect matrix for its high efficiency and availability [18]. Until recently, very few studies were available describing the effects of solvent and defatting parameters on the composition of protein extracts and their techno-functional properties. Kim et al. (2020) evaluated the impact of different organic solvents on the functional properties of defatted proteins extracted from *Protaetia brevitarsis* larvae. These authors determined that hexane-defatted protein fractions had better amino acid composition, protein solubility and functional properties compared to the same fractions defatted with methanol and ethanol [20]. Borremans et al. (2020) compared the functional properties of full-fat and hexane-defatted *T. molitor* meals and demonstrated that the defatted meal's foaming capacity was equivalent to that of egg albumen, indicating its potential as alternative foaming agents in food formulation [22]. However, the detailed effects of hexane-defatting on the protein profile of *T. molitor* meals and protein extracts, their solubility and foaming properties were never evaluated. Consequently, the aims of this study were to compare and determine the impact of hexane-defatting on the proximate composition of *T. molitor* meal and protein extract fractions. Specific emphasis was given to the modification of protein profiles by proteomic tools. Protein solubility and foaming properties of defatted and non-defatted meals and protein extracts were determined.

2. Materials and Methods

2.1. Raw Materials

Living mealworm larvae (*T. molitor*) at age of 73 days were kindly provided by Neoxis (Saint-Flavien, Québec, Canada). First, mealworm larvae were separated from feed substrate and frass residues by passing through a coarse filter. Next, larvae were killed by freezing at $-30\ °C$ without starvation. Frozen mealworm larvae were freeze-dried, ground with a household compact blender (Magic Bullet, Los Angeles, CA, USA) and passed through an 800 μm mesh sieve to produce *T. molitor* meal (TMI). This meal was stored at $-20\ °C$ before further analysis and processing. All experiments were performed in triplicate using three separate insect batches.

2.2. Methods

2.2.1. Lipid Extraction

Half of the TMI meal from each insect batch was stored at $-20\ °C$ while the other half was defatted using the Soxhlet method as described by Tzompa-Sosa et al. (2014) [23]. Briefly, 10 g of TMI were weighed into a cellulose cartridge (Fisher Scientific catalog number 12-101-100) and the lipids were extracted with hexane for 6 h. The defatted *T. molitor* meal (TMD) was recovered, vacuum-dried in an oven at 50 °C for 5 h and stored at $-20\ °C$ before further analysis and processing.

2.2.2. Protein Extraction

Soluble proteins from the TMI and TMD fractions were recovered as described by Yi et al. (2013), with modifications [24]. First, TMI and TMD fractions were solubilized in a 9.5 mM ascorbic acid solution in a 1:8 (w/v) ratio and magnetically stirred overnight at 10 °C. The suspensions were then centrifuged at $15,000 \times g$ for 30 min at 4 °C and the supernatant containing the soluble proteins was filtered three times through a Whatman® Grade 4 filter paper in a Büchner funnel. Both filtrates were freeze-dried, producing two different fractions: a hexane-defatted soluble protein (HDSP) and a non-hexane-defatted soluble protein (NHSP) extract from *T. molitor*.

The detailed methodology used to generate the different fractions is illustrated in Figure 1.

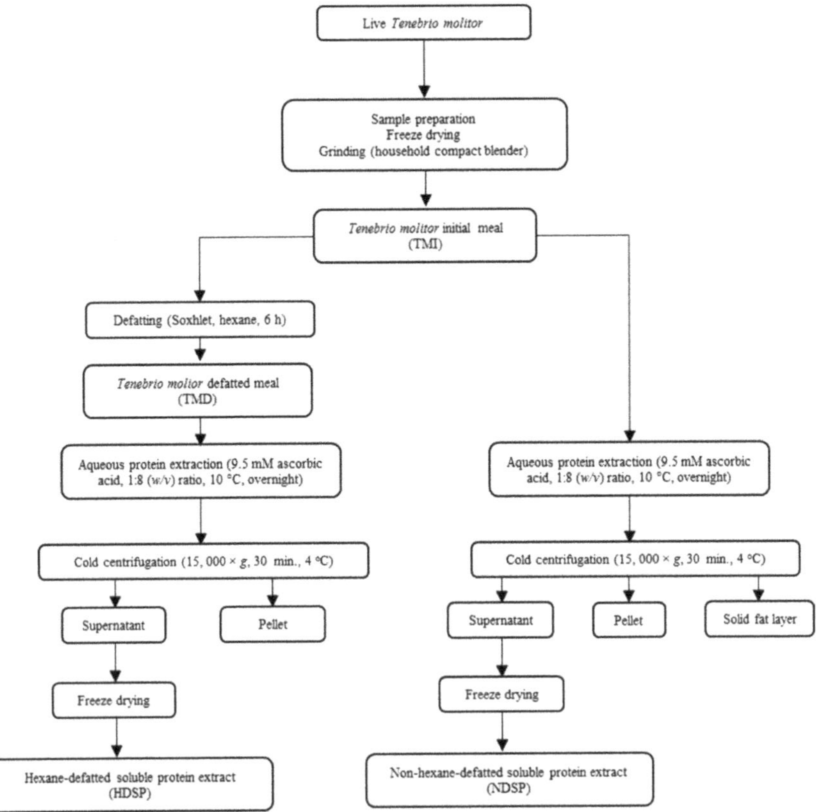

Figure 1. Experimental design for the production of defatted and non-defatted mealworm meals and protein extracts.

2.3. Analysis

2.3.1. Proximate Composition

The proximate composition was determined for the four *T. molitor* fractions (TMI, TMD, HDSP, and NDSP). The protein content was measured using the Dumas Method [19,25] (Elementar rapid Micro N cube, Langenselbold, Germany), with a nitrogen-to-protein conversion factor of 4.76 for TMI and TMD, and 5.60 for HDSP and NDSP, accounting for the high chitin content of TMI and TMD fractions and therefore allowing for a more accurate protein content determination, as suggested by Janssen et al. (2017) [26]. The lipid content was determined using the Mojonnier method (AOAC 925.32). Moisture and ash content were determined by official methods AOAC 950.46 (A) and AOAC 920.153, respectively [27]. The method described by Spinelli et al. (1974) was used to determine the chitin content of the four fractions [28].

2.3.2. Protein Profile

The two-dimensional (2D) gel analysis was carried out according to Kumar et al. (2017), with modifications [29]. Briefly, 2 mg of proteins from the four *T. molitor* fractions were mixed with the sample buffer containing 7 M urea, 2 M thiourea, 4% CHAPS (w/v), 1% Triton-X-100 (v/v), 20 mM Tris, 1% DTT (w/v) and 0.5% pH 3-10 IPG buffer (v/v). A volume of 250 μL of each sample solution (TMI, TMD, HDSP and NHSP) was loaded on a 13 cm GE Healthcare Immobiline® DryStrip (pH 3–10) and first-dimension isoelectric focusing (IEF) was performed using the Ettan IPGphor 3 IEF system (GE Healthcare, Piscataway, NJ, USA) at 20 °C, 50 µA/strip, 30 V for 12 h (rehydration), 100 V for 1 h, 500 V for 1 h, 1000 V for 1 h, 5000 V for 1 h and then 8000 V until 16,000 Vh was reached. The first-dimension strips were kept at −20 °C until used for the second dimension. For the second dimension, each strip was soaked in 20 mL of equilibration buffer containing 6 M urea, 2 M Milli-Q H_2O, 50 mM Tris-HCl pH 8.8, 30% glycerol (*v/v*), 2% SDS (*v/v*), 2% DTT (*w/v*) and traces of bromophenol blue for 15 min at 20 °C. SDS-PAGE was carried out at 10 °C using 17.5 cm × 16.8 cm × 1 mm 15% polyacrylamide gels topped with 5% polyacrylamide stacking gels. The migration was performed at 25 µA for each gel. The running buffer consisted of 1X tris-glycine SDS solution (Bio-Rad, Hercules, CA, USA). The molecular weights (MW) of mealworm proteins were estimated by using a MW marker (Prestained Protein Standards Precision Plus Protein™ All Blue, Bio-Rad, cat. #1610373, Hercules, CA, USA). Proteins were fixed by soaking the gels in a 50% MeOH (*v/v*) and 10% acetic acid (*v/v*) solution for 12 h. Gels were then stained by soaking in GelCode™ Blue Stain Reagent (Thermo Fisher Scientific, Whaltham, MA, United-States) for 24 h, rinsed in Milli-Q water and scanned on a ChemiDoc™ MP Imaging System (Bio-Rad, Hercules, CA, USA). The 2-D gel image analysis was performed using ImageJ software according to the procedure of Natale et al. (2011), with modifications [30]. Briefly, gel images were aligned using the same image as reference with the bUnwarpJ plugin and repetitions for TMI, TMD, HDSP, and NDSP were stacked and summed to give a single gel image representative of all 3 repetitions.

2.3.3. Protein Identification by Mass Spectrometry

Mass Spectrometry

Mass spectrometry experiments on selected 2D gel spots were performed by the Proteomics platform of the CHU de Quebec Research Center, Quebec, Canada. Samples were analyzed by nanoLC/MSMS using a Dionex UltiMate 3000 nanoRSLC chromatography system (Thermo Fisher Scientific, San Jose, CA, USA) connected to an Orbitrap Fusion mass spectrometer (Thermo Fisher Scientific, San Jose, CA, USA). Peptides were trapped at 20 µL/min in loading solvent (2% acetonitrile, 0.05% TFA) on a 5 mm × 300 µm C18 pepmap cartridge pre-column (Thermo Fisher Scientific/Dionex Softron GmbH, Germering, Germany) over 5 min. The pre-column was then switched online with a Pepmap Acclaim column (Thermo Fisher Scientific, San Jose, CA, USA)—a 50 cm × 75 µm internal diameter separation column—and the peptides were eluted with a linear gradient from 5–40% solvent B (A: 0.1% formic acid, B: 80% acetonitrile, 0.1% formic acid) over 35 min, at 300 nL/min. Mass spectra were acquired using a data dependent acquisition mode with Thermo XCalibur software version 4.1.50. Full scan mass spectra (350 to 1800 m/z) were acquired in the orbitrap using an AGC target of 4e5, a maximum injection time of 50 ms and a resolution of 120,000. Internal calibration was done using lock mass on the m/z 445.12003 siloxane ion. Each MS scan was followed by MSMS fragmentation of the most intense ions for a total cycle time of 3 s (top speed mode). The selected ions were isolated using the quadrupole analyzer in a window of 1.6 m/z and fragmented by Higher energy Collision-induced Dissociation (HCD) with 35% of collision energy. The resulting fragments were detected by the linear ion trap at rapid scan rate with an AGC target of 1e4 and a maximum injection time of 50 ms. Dynamic exclusion of previously fragmented peptides was set for a period of 20 s and a tolerance of 10 ppm.

Database Searching

Mascot Generic Format (MGF) peak list files were created using Proteome Discoverer 2.3 software (Thermo Fisher Scientific, San Jose, CA, USA). The MGF sample files were then analyzed using Mascot (Matrix Science, London, UK; version 2.5.1). Mascot was set up to search a contaminant database and Uniprot *Tenebrionidae* database (40,907 entries), assuming that the digestion enzyme was trypsin. Mascot was searched with a fragment ion mass tolerance of 0.60 Da and a parent ion tolerance of 10.0 ppm. Carbamidomethyl of cysteine was specified in Mascot as a fixed modification. Deamidation of asparagine and glutamine and oxidation of methionine were specified in Mascot as variable modifications. Two missed cleavages were allowed.

Criteria for Protein Identification

Scaffold (version Scaffold_4.8.4, Proteome Software Inc., Portland, OR, USA) was used to validate MS/MS-based peptide and protein identifications. A false discovery rate of 1% was used for peptide and protein. Proteins that contained similar peptides and could not be differentiated based on MS/MS analysis alone were grouped to satisfy the principles of parsimony. The total spectrum count value, which corresponds to the total number of spectra identified for a protein, was used to identify selected protein spots from the 2D gels, as this parameter was reported to be a semi-quantitative measure for a given protein abundance in proteomic studies [31,32].

2.4. Protein Solubility

Suspensions of TMI, TMD, HDSP, and NDSP fractions (1% *w/v*) were prepared in Milli-Q water. The pH of suspensions was adjusted to 5, 7 or 9 with 1 M NaOH or 1 M HCl. The suspensions were centrifuged at $20,000 \times g$ for 30 min at 20 °C and filtered through a Whatman® Grade 1 filter paper. The filtrates were freeze-dried and the recovered fractions were analyzed for their protein content by the Dumas method with a 4.76 and 5.60 nitrogen-to-protein conversion factor, respectively for the meal larvae (TMI and TMD) and for the protein extracts (HDSP and NDSP) as described previously [26]. The protein solubility was calculated using Equation (1) [33]:

$$Solubility(\%) = \frac{protein\ content\ in\ supernatant}{total\ protein\ content\ in\ sample} \times 100 \tag{1}$$

2.5. Foaming Properties

Foaming capacity (*FC*) and foam stability (*FS*) were determined as described by Zielińska et al. (2018) with some modifications [7]. A volume of 30 mL of a 3% (*w/v*) TMI, TMD, HDSP, or NDSP sample was whipped using a hand mixer (KitchenAid, KHM512IB, Benton Charter Township, MI, USA) at a speed of approximately 800 rpm for 10 min and immediately transferred into a cylinder. The total volumes of foam were read at time 0 (V_0), and then every 15 min until 90 min following (V_t). Foaming capacity was calculated using Equation (2) and *FS* were calculated using Equation (3) [34]:

$$FC(\%) = \frac{V_0 - 30}{30} \times 100 \tag{2}$$

$$FS(\%) = \frac{V_t - 30}{V_0 - 30} \times 100 \tag{3}$$

2.6. Statistical Analysis

The proximate composition, functional properties experiments, and 2D gels were all performed in triplicate. Proximate composition data were subjected to a one-way analysis of variance (ANOVA) and functional properties experiments data were subjected to a two-way ANOVA using the Statistical Analysis System (SAS) University Edition, SAS® Studio 3.5 software. The differences of the means between the samples were determined using the Tukey test. *p*-values < 0.05 were considered statistically significant.

3. Results

3.1. Proximate Composition of Meals and Protein Extracts

Table 1 presents the proximate compositions of the four *T. molitor* fractions. The total solids contents were quite similar for all fractions (meals and protein extracts) with values ranging from 90.3 (TMI) to 95.2% (TMD). The defatting step applied to TMI meal was efficient, since very low lipid concentrations (0.3–0.4%) were obtained in TMD and HDSP. Of all meals and protein extracts, TMI had the lowest protein content with a value of 38.6%. The defatting step largely improved protein recovery in defatted fractions with respective values of 59.1% and 62.7% for TMD and HDSP samples. The defatting step also helped to improve protein content in the extracts since the protein recovery in HDSP (62.7%) was increased by 9% compared to NDSP (54.0%) (Table 1). In addition, chitin content was higher in TMD (7.8%) compared to TMI (5.0%), and after aqueous solubilization, no chitin was detected in defatted (HDSP) and non-hexane-defatted (NDSP) soluble protein extracts.

Table 1. Proximate composition of *T. molitor* fractions at different processing stages expressed on a dry basis.

Processing Stage	Protein	Lipid	Chitin	Ash	Moisture	Total Solids
	% (*w/w*)					
TMI	38.6 ± 1.4 [a]	28.5 ± 1.0 [a]	5.0 ± 0.6 [a]	3.4 ± 0.1 [a]	9.7 ± 3.2 [a]	90.3 ± 3.2 [a]
TMD	59.1 ± 0.9 [b,c]	0.4 ± 0.2 [b]	7.8 ± 0.6 [b]	4.8 ± 0.1 [a]	4.8 ± 0.1 [b]	95.2 ± 0.1 [b]
HDSP	62.7 ± 3.0 [c]	0.3 ± 0.1 [b]	0.0 ± 0.0 [c]	8.4 ± 1.1 [b]	7.3 ± 0.8 [a,b]	92.7 ± 0.8 [a,b]
NDSP	54.0 ± 5.0 [b]	0.5 ± 0.1 [b]	0.0 ± 0.0 [c]	10.3 ± 1.3 [b]	9.3 ± 1.2 [a,b]	90.7 ± 1.2 [a,b]

Different letters in the same column indicate significant differences ($p < 0.05$).

3.2. Profiles and Characterization of Proteins in Meals and Protein Extracts

Figure 2 presents the protein profiles of TMI, TMD, HDSP, and NDSP fractions obtained after 2D SDS-PAGE analysis. The 11 main spots (#1 to #11) visualized in 2D gels were excised for protein characterization by proteomic tools (Table 2). As observed for 2D gels (Figure 2), proteomic results confirmed that *T. molior* fractions (TMI, TMD, HDSP, and NDSP) have very similar protein profiles consisting of muscular (e.g., actin-87E-like protein) and hemolymph proteins (e.g., hexamerin 2, 12 kDa hemolymph protein b) as well as proteins associated with various metabolic activities (e.g., α-amylase, chitinase, arginine kinase). However, some differences in spot intensities were observed between gels. While the cockroach allergen-like protein (spot #2) and the 28 kDa desiccation stress protein (spot #9) were present in all fractions with similar relative intensities, spot #3, identified as a cockroach allergen-like protein, was present at high intensity for each *T. molitor* fraction except for TMI. Similarly, the actin-87E-like protein (spot #1) was present in all fractions, except for NDSP where the intensity of the protein spot was low. Furthermore, 12 kDa hemolymph protein b (spot #4 and #5), α-amylase (spot #8), the melanin-inhibiting protein (spot #10), as well as chitinase (spot #11) were predominantly abundant in soluble protein extracts NDSP and HDSP. Conversely, an arginine kinase fragment (spot #7) could only be found in whole meals TMI and TMD. Finally, the 86 kDa early-staged encapsulation inducing protein and hexamerin 2 (spot #6) were only abundant in HDSP. It is worth noting that some identified proteins were associated only with other insects from the *Tenebrionidae* family, such as *Tribolium castaneum*, since they were absent from the Uniprot database for *T. molitor* (Table 2).

3.3. Protein Solubility

Figure 3 shows the protein solubility of TMI, TMD, NDSP, and HDSP fractions at pH 5, 7, and 9. Both the type of fraction and the pH had a significant impact on protein solubility ($p < 0.0001$). Similar solubilities were obtained for HDSP and NDSP at pH 7 and 9, whereas the solubility of HDSP was calculated to be lower than NDSP at pH 5. As expected, and whatever the pH value, protein fractions recovered from HDSP and NDSP (soluble fraction) were more soluble than TMI and TMD. A similar increase in solubility was observed for TMI and TMD when the pH increased, except at pH 9 where the solubility of TMI was higher than TMD. Globally, results showed that fractions had minimum and maximum solubilities at pH 5 and 9, respectively.

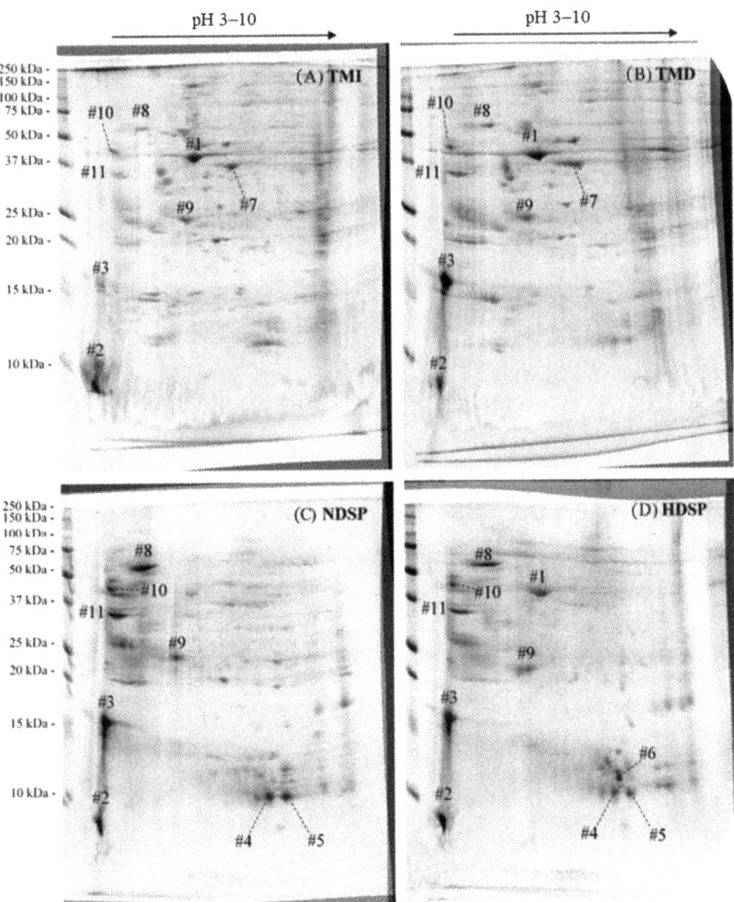

Figure 2. Two-dimensional electrophoresis of *T. molitor* proteins for (**A**) TMI, (**B**) TMD, (**C**) NDSP, and (**D**) HDSP. Proteins composing each numbered spot were characterized by proteomics (Table 2).

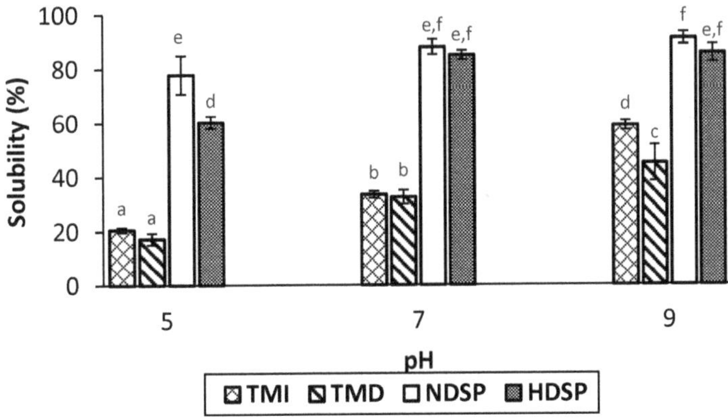

Figure 3. Protein solubility at pH 5, 7 and 9 of TMI, TMD, NDSP, and HDSP fractions. Different letters indicate significant differences ($p < 0.05$).

Table 2. Proteins identified by proteomics in *T. molitor* 2D gels.

#	Identified Proteins	Molecular Weight MW (kDa)	UniProt ID	Number of Unique Peptides	Coverage (%)	Total Spectrum Count	Most Abundant in Fractions
1	Actin-87E-like protein (*Tribolium castaneum*)	42	D6WF19	39	77	736	TMI, TMD, and HDSP
2	Cockroach allergen-like protein (*Tenebrio molitor*)	65	Q7YZB8	5	11	53	TMI, TMD, NDSP, and HDSP
3				9	15	158	TMD, NDSP, and HDSP
4	12 kDa hemolymph protein b (*Tenebrio molitor*)	14	Q7YWD7	2	81	297	NDSP and HDSP
5				2	81	271	
6	86 kDa early-staged encapsulation inducing protein (*Tenebrio molitor*)	91	Q9Y1W5	21	27	93	HDSP
	Hexamerin 2 (*Tenebrio molitor*)	85	Q95PI7	12	20	81	
7	Arginine kinase (Fragment) (*Tenebrio molitor*)	27	A0A0U4ARJ8	8	85	414	TMI and TMD
8	α-amylase (*Tenebrio molitor*)	51	P56634	30	80	731	NDSP and HDSP
9	28 kDa desiccation stress protein (*Tenebrio molitor*)	25	Q27013	21	49	202	TMI, TMD, NDSP, and HDSP
10	Melanin-inhibiting protein (*Tenebrio molitor*)	40	Q4LE89	25	64	225	NDSP and HDSP
11	Chitinase (*Tenebrio molitor*)	40	Q7YZB9	11	38	156	NDSP and HDSP

3.4. Foaming Properties

Foaming capacity and foam stability of TMI, TMD, NDSP, and HDSP are presented in Figures 4 and 5. Low foaming capacities of 13% and 41% were measured for TMI and NDSP, respectively. However, the foaming capacities for TMD and HDSP were drastically improved (546% and 629%, respectively), even if HDSP exhibited the highest foaming capacity. Foam stability values were similar for all insect fractions ($p > 0.05$) and decreased similarly as a function of time for both defatted and non-defatted fractions. More specifically, 90 min after whipping, a decrease to about 61% of initial foam volume was obtained.

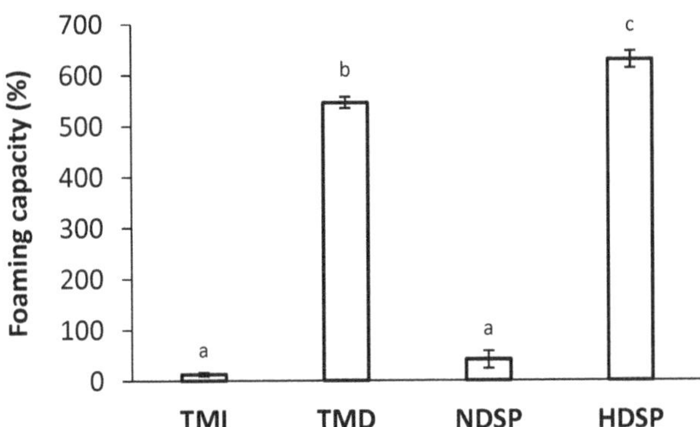

Figure 4. Foaming capacity of TMI, TMD, NDSP, and HDSP. Different letters indicate a significant difference ($p < 0.05$).

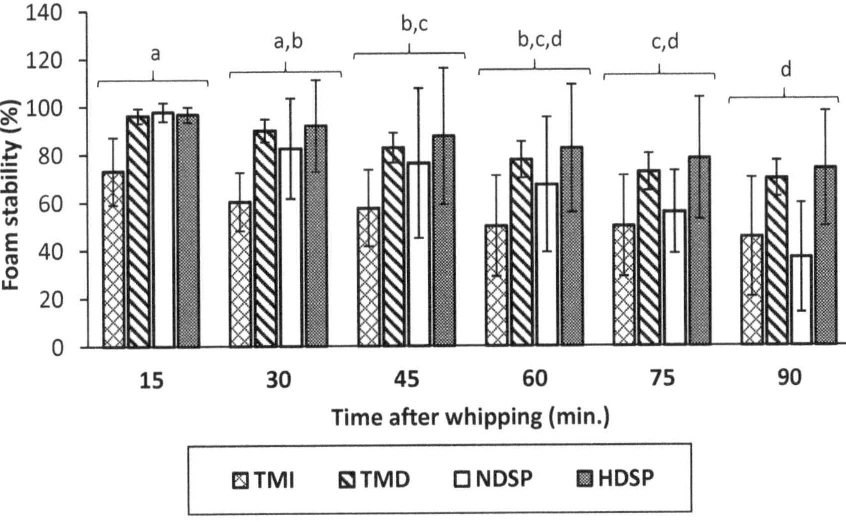

Figure 5. Foam stability values as a function of time after whipping of TMI, TMD, NDSP, and HDSP. Different letters indicate significant differences ($p < 0.05$).

4. Discussion

A growing number of studies are becoming available regarding the impact of processing methods on protein techno-functionalities [15]. However, no information is specifically available concerning the impact of the defatting step on mealworm protein profiles and functionalities. Consequently, the purpose of this study was to compare the protein profiles of hexane-defatted and non-hexane-defatted mealworm meals and protein extracts and to evaluate the impact of hexane on protein solubility and foaming capacity. Our results showed that defatting with hexane had little impact on protein profile. However, the 86 kDa early-staged encapsulation inducing protein and hexamerin 2 were more abundant in hexane-defatted protein extracts than in non-hexane-defatted protein extracts. Moreover, compared to non-defatted meals and protein extracts, solubility was reduced and foaming properties were largely improved for defatted fractions.

The initial fat content of 28.5% was within the normal range of similar studies [16,19,35] and residual lipids in hexane-defatted fractions did not exceed 0.4%, which is comparable to the results obtained by Choi et al. (2017) for defatted *T. molitor* meals [16]. The low-fat content of NDSP (0.5%) indicates that the cold centrifugation step following aqueous solubilization of proteins allows lipid to be extracted as efficiently as the hexane-defatting step by the formation of an easily removable solid fat layer. However, unlike the hexane-defatting step, cold centrifugation can only be performed after aqueous protein extraction, which resulted in a lower protein content in the soluble NDSP fraction compared to HDSP, since some lipid binding proteins might have been removed along the lipid fraction during the cold centrifugation step in the process to obtain the NDSP protein extract (Table 1).

Our results showed that the initial *T. molitor* (TMI) protein content of 38.6% was close to that of Purschke et al. (2019), who obtained 39% for dried *T. molitor* larvae [36]. As previously reported, hexane-defatting increased the protein content of TMI compared to TMD by removing lipids [19,37]. Our results also confirmed an increase in the protein content of soluble protein extracts of HDSP compared to NDSP (62.7% and 54.0%, respectively) after subsequent aqueous protein extraction. This result could be explained by the fact that protein-lipid interactions limit protein solubility during protein extraction [14,38].

Two-dimensional gel electrophoresis coupled to proteomic identification of the resulting spots was performed with all four fractions to understand the effects of the processing methods, especially the defatting step, on protein abundance and profile. Both proteomic and 2D electrophoresis results were consistent with previously published work on the characterization of *T. molitor* proteins. These studies identified many pan-allergens, such as muscle proteins (actin, myosin, tropomyosin) [15,32,39], hemolymph proteins (hexamerin 1 and 2) [15,32,39], and proteins associated with various metabolic activities (arginine kinase, α-amylase) [39–41], to be the most abundant in *T. molitor*. Overall, the protein profiles were quite similar for the four fractions although specific differences in the abundance of major proteins were observed. Interestingly, the biggest differences were observed between whole protein meals (TMI and TMD) and soluble protein extracts (NDSP and HDSP), rather than between hexane-defatted and non-hexane-defatted fractions. Indeed, five protein spots corresponding to the 12 kDa hemolymph protein (spot #4 and #5), α-amylase (spot #8), melanin-inhibiting protein (spot #10), and chitinase (spot #11) were predominantly abundant in extracted proteins fractions. These proteins, which are involved in different metabolic processes, are highly soluble in hemolymph and other aqueous solutions [42], which could explain the difference observed between the meal and extracted proteins fractions. Some differences in the protein profile were also detected between hexane-defatted and non-hexane-defatted soluble extracts. For example, actin-87E-like protein of *Tribolium castaneum* (spot #1), 86 kDa early-staged encapsulation inducing protein and Hexamerin 2 (spot #6) were present only in HDSP and not in NDSP. While the lack of knowledge about the structure and function of insect proteins make it difficult to formulate a hypothesis or draw any conclusion about the roles of these proteins [15], they are undoubtedly soluble. Furthermore, insect hexamerins, which have been studied in more depth, can provide some insight into the mechanisms that are active during the production of *T. molitor*. Indeed, hexamerins are synthesized and stored in the fat body of the insect in its larval form [43]. The fat body is a heterogenous organ consisting of cells containing multiple lipid droplets [44]. Consequently, their high abundance in HDSP fractions can be explained by the fact that those proteins are imprisoned in adipocyte cells and are only released during hexane-defatting, thus making them more abundant in HDSP. As many proteins are synthesized in the fat body, this hypothesis could also explain the abundance of 86 kDa early-staged encapsulation inducing protein in HDSP compared to NDSP [44]. Finally, it should be noted that the cockroach allergen-like protein was identified as the major constituent of both protein spots #2 and #3, whereas protein spot #3 in TMI had very low intensity. Indeed, some TMI proteins did not migrate on the gel during the second dimension, as can be observed by the presence of an intense band at 250 kDa (Figure 2A).

Proteins were probably unable to migrate in TMI fractions, since they were aggregated with other components of the cuticular matrix, such as chitin and lipids [45,46]. This phenomenon of protein aggregates was also observed by Boukil et al. (2020) for non-processed *T. molitor* meal [32].

The solubility of *T. molitor* proteins was minimal at pH 5 and increased at pH values of 7 and 9 (Figure 3). Our results were consistent with the previous work of Bußler et al. (2016) and Azagoh et al. (2016) who observed that the lowest protein solubility was obtained between pH 3 and 5, corresponding to the average pI. Indeed, at this pH value, the protein-water interactions are at their minimum, thereby inducing their insolubility. Interestingly, the same authors reported maximal edible insect protein solubility at alkaline pH values ranging from 7 to 12, depending on the fraction [37,47]. Borremans et al. (2020) observed that the solubility of hexane-defatted *T. molitor* meals (i.e., TMD) decreased compared to full-fat meals (i.e., TMI) at pH values above 8 [9,22], as obtained in our study (Figure 3). However, the same tendency was not repeatable for soluble protein extracts, since the pH increase did not affect the protein solubility of NDSP compared to HDSP. Interestingly, this phenomenon was observed at pH 5 since HDSP solubility was lower than that of NDSP. This could be related to a change in the average pI value due to the variation in protein molecular structure and protein content caused by the defatting process [48], evidenced by the profile differences observed in Figure 2. Overall, these results suggest that little protein denaturation occurred during the defatting process, as very few differences were observed, especially at neutral pH, between whole protein meals (TMI and TMD) and soluble protein extracts (NDSP and HDSP), respectively [49].

Our results also showed the highest foam capacity for hexane-defatted protein fractions TMD and HDSP. Yi et al. (2013), Zielińska et al. (2018), and Stone et al. (2019) reported that non-defatted *T. molitor* had poor foaming properties, which was confirmed by our results for TMI and NDSP (Figure 4) [7,8,24]. Borremans et al. (2020) evaluated the foaming properties of hexane-defatted *T. molitor* meal and showed increased foam capacity from 94% in the control sample to 540% for the defatted mealworms, which is similar to what is observed for TMD [22]. Lastly, foaming capacity of HDSP (629%) happened to be higher than the one of egg albumen (575%) which is commonly used as a foaming agent in food formulation. Akpossan et al. (2015) and Kim et al. (2020) also determined that the foaming capacity of defatted *Imbrasia oyemensis* meals and *Protaetia brevitarsis* protein extracts were superior to that of full-fat meals and non-defatted protein extracts, respectively [20,50]. Kim et al. (2020) partly attributed these results to the high-fat content of the non-defatted insect protein extracts, but that is not the case in this study, according to the proximate composition of NDSP, which still showed poor foaming capacity [20]. On the other hand, as experimented and discussed by Mishyna et al. (2019) for *Schistocerca gregaria* and *Apis mellifera*, higher foaming capacity could be attributed to the higher protein content of defatted fractions as well as the alteration of protein intrinsic molecular properties (e.g., partial protein unfolding) during the transformation process [9].

5. Conclusions

This study demonstrated that mealworm larvae processing methods, including defatting, modified the meal and protein extract's protein profile and altered solubility and foaming capacity. Specifically, hexane-defatting of *T. molitor* meal before aqueous extraction of proteins increased the total protein content and abundance of the 86 kDa early-staged encapsulation inducing protein and hexamerin 2 of hexane-defatted protein extracts compared to the non-hexane-defatted extracts. Hexane-defatted *T. molitor* meals and protein extracts also had highly increased foaming capacity. Furthermore, *T. molitor* protein fractions exhibited the highest solubility at pH values above 7, which could facilitate their incorporation into various food systems. While further research on the effect of different processing steps (e.g., grinding, drying, protein extraction etc.) on insect protein quality and techno-functional properties is required to improve their use as a protein source in food formulations, this study provides insights on how hexane-defatted *T. molitor* protein ingredients could be incorporated into different foods, especially as foams, meringues, cakes, and mousses.

Author Contributions: Conceptualization, A.G., M.C., and A.D.; Methodology, A.G., M.C., and A.D.; Validation, A.G., A.M., M.C., and A.D.; Formal Analysis, A.G.; Investigation, A.G. and A.D.; Writing—Original Draft Preparation, A.G. and A.D.; Writing—Review & Editing, A.G., A.M., M.C., and A.D.; Visualization, A.G. and A.D.; Supervision, A.D.; Project Administration, A.D.; Funding Acquisition, A.D. All authors have read and agreed to the published version of the manuscript.

Funding: This research was funded by the Fonds de recherche du Québec—Nature et technologies (Grant # 2018-PR-208090 to Alain Doyen).

Data Availability Statement: Data contained within the article are available from the authors.

Acknowledgments: The authors thank Sylvie Bourassa and Victor Fourcassié for their assistance with proteomics analysis. The authors also thank Diane Gagnon, Véronique Perreault, Pascal Lavoie and Dany Boisvert for technical support.

Conflicts of Interest: The authors declare no conflict of interest.

Sample Availability: Samples of the compounds are not available from the authors.

References

1. Oonincx, D.G.; van Itterbeeck, J.; Heetkamp, M.J.; van den Brand, H.; van Loon, J.J.; van Huis, A. An exploration on greenhouse gas and ammonia production by insect species suitable for animal or human consumption. *PLoS ONE* **2010**, *5*, e14445. [CrossRef] [PubMed]

2. La Barbera, F.; Verneau, F.; Amato, M.; Grunert, K. Understanding Westerners' disgust for the eating of insects: The role of food neophobia and implicit associations. *Food Qual. Prefer.* **2018**, *64*, 120–125. [CrossRef]

3. Gere, A.; Székely, G.; Kovács, S.; Kókai, Z.; Sipos, L. Readiness to adopt insects in Hungary: A case study. *Food Qual. Prefer.* **2017**, *59*, 81–86. [CrossRef]

4. Verbeke, W. Profiling consumers who are ready to adopt insects as a meat substitute in a Western society. *Food Qual. Prefer.* **2015**, *39*, 147–155. [CrossRef]

5. Schosler, H.; de Boer, J.; Boersema, J.J. Can we cut out the meat of the dish? Constructing consumer-oriented pathways towards meat substitution. *Appetite* **2012**, *58*, 39–47. [CrossRef]

6. Williams, J.P.; Williams, J.R.; Kirabo, A.; Chester, D.; Peterson, M. Chapter 3—Nutrient Content and Health Benefits of Insects. In *Insects as Sustainable Food Ingredients*; Dossey, A.T., Morales-Ramos, J.A., Rojas, M.G., Eds.; Academic Press: Cambridge, MA, USA, 2016; pp. 61–84.

7. Zielińska, E.; Karaś, M.; Baraniak, B. Comparison of functional properties of edible insects and protein preparations thereof. *LWT* **2018**, *91*, 168–174. [CrossRef]

8. Stone, A.K.; Tanaka, T.; Nickerson, M.T. Protein quality and physicochemical properties of commercial cricket and mealworm powders. *J. Food Sci. Technol.* **2019**, *56*, 3355–3363. [CrossRef]

9. Mishyna, M.; Martinez, J.I.; Chen, J.; Benjamin, O. Extraction, characterization and functional properties of soluble proteins from edible grasshopper (*Schistocerca gregaria*) and honey bee (*Apis mellifera*). *Food Res. Int.* **2019**, *116*, 697–706. [CrossRef]

10. Gravel, A.; Doyen, A. The use of edible insect proteins in food: Challenges and issues related to their functional properties. *Innov. Food Sci. Emerg. Technol.* **2020**, *59*. [CrossRef]

11. Cappelli, A.; Cini, E.; Lorini, C.; Oliva, N.; Bonaccorsi, G. Insects as food: A review on risks assessments of Tenebrionidae and Gryllidae in relation to a first machines and plants development. *Food Control.* **2020**, *108*, 106877. [CrossRef]

12. L'hocine, L.; Boye, J.I.; Arcand, Y. Composition and Functional Properties of Soy Protein Isolates Prepared Using Alternative Defatting and Extraction Procedures. *J. Food Sci.* **2006**, *71*, C137–C145. [CrossRef]

13. Capellini, M.C.; Novais, J.S.; Monteiro, R.F.; Veiga, B.Q.; Osiro, D.; Rodrigues, C.E.C. Thermal, structural and functional properties of rice bran defatted with alcoholic solvents. *J. Cereal Sci.* **2020**, *95*, 103067. [CrossRef]

14. Teh, S.-S.; Bekhit, A.E.-D.; Carne, A.; Birch, J. Effect of the defatting process, acid and alkali extraction on the physicochemical and functional properties of hemp, flax and canola seed cake protein isolates. *J. Food Meas. Charact.* **2013**, *8*, 92–104. [CrossRef]

15. Yi, L.; Van Boekel, M.A.J.S.; Boeren, S.; Lakemond, C.M.M. Protein identification and in vitro digestion of fractions from *Tenebrio molitor*. *Eur. Food Res. Technol.* **2016**, *242*, 1285–1297. [CrossRef]

16. Choi, B.D.; Wong, N.A.K.; Auh, J.H. Defatting and Sonication Enhances Protein Extraction from Edible Insects. *Korean J. Food Sci. Anim. Resour.* **2017**, *37*, 955–961. [PubMed]

17. Son, Y.J.; Choi, S.Y.; Hwang, I.K.; Nho, C.W.; Kim, S.H. Could Defatted Mealworm (*Tenebrio molitor*) and Mealworm Oil Be Used as Food Ingredients? *Foods* **2020**, *9*, 40. [CrossRef] [PubMed]

18. Russin, T.A.; Boye, J.I.; Arcand, Y.; Rajamohamed, S.H. Alternative Techniques for Defatting Soy: A Practical Review. *Food Bioprocess. Technol.* **2010**, *4*, 200–223. [CrossRef]

19. Laroche, M.; Perreault, V.; Marciniak, A.; Gravel, A.; Chamberland, J.; Doyen, A. Comparison of Conventional and Sustainable Lipid Extraction Methods for the Production of Oil and Protein Isolate from Edible Insect Meal. *Foods* **2019**, *8*, 572. [CrossRef]

20. Kim, T.K.; Yong, H.I.; Kim, Y.B.; Jung, S.; Kim, H.W.; Choi, Y.S. Effects of organic solvent on functional properties of defatted proteins extracted from *Protaetia brevitarsis* larvae. *Food Chem* **2020**, *336*, 127679. [CrossRef]

21. Purschke, B.; Stegmann, T.; Schreiner, M.; Jäger, H. Pilot-scale supercritical CO_2 extraction of edible insect oil from *Tenebrio molitor* L. larvae-Influence of extraction conditions on kinetics, defatting performance and compositional properties. *Eur. J. Lipid Sci. Technol.* **2017**, *119*. [CrossRef]

22. Borremans, A.; Bussler, S.; Sagu, S.T.; Rawel, H.; Schluter, O.K.; Leen, V.C. Effect of Blanching Plus Fermentation on Selected Functional Properties of Mealworm (*Tenebrio molitor*) Powders. *Foods* **2020**, *9*, 917. [CrossRef]

23. Tzompa-Sosa, D.A.; Yi, L.; van Valenberg, H.J.F.; van Boekel, M.A.J.S.; Lakemond, C.M.M. Insect lipid profile: Aqueous versus organic solvent-based extraction methods. *Food Res. Int.* **2014**, *62*, 1087–1094. [CrossRef]

24. Yi, L.; Lakemond, C.M.; Sagis, L.M.; Eisner-Schadler, V.; van Huis, A.; van Boekel, M.A. Extraction and characterisation of protein fractions from five insect species. *Food Chem* **2013**, *141*, 3341–3348. [CrossRef]

25. Simonne, A.H.; Simonne, E.H.; Eitenmiller, R.R.; Mills, H.A.; Cresman, C. Could the Dumas method replace the Kjeldahl digestion for nitrogen and crude protein determinations in foods? *J. Sci. Food Agric.* **1997**, *73*, 39–45. [CrossRef]

26. Janssen, R.H.; Vincken, J.P.; van den Broek, L.A.; Fogliano, V.; Lakemond, C.M. Nitrogen-to-Protein Conversion Factors for Three Edible Insects: *Tenebrio molitor*, Alphitobius diaperinus, and Hermetia illucens. *J. Agric. Food Chem.* **2017**, *65*, 2275–2278. [CrossRef]

27. AOAC International. *Official Methods of Analysis*, 17th ed.; AOAC International: Gaithersburg, MD, USA, 2017.

28. Spinelli, J.; Lehman, L.; Wieg, D. Composition, Processing, and Utilization of Red Crab (Pleuroncodes planipes) as an Aquacultural Feed Ingredient. *J. l'Office Rech. Pêcheries Can.* **1974**, *31*, 1025–1029. [CrossRef]

29. Kumar, M.; Singh, R.; Meena, A.; Patidar, B.S.; Prasad, R.; Chhabra, S.K.; Bansal, S.K. An Improved 2-Dimensional Gel Electrophoresis Method for Resolving Human Erythrocyte Membrane Proteins. *Proteom. Insights* **2017**, *8*, 1178641817700880. [CrossRef]

30. Natale, M.; Maresca, B.; Abrescia, P.; Bucci, E.M. Image Analysis Workflow for 2-D Electrophoresis Gels Based on Image. *J. Proteom. Insights* **2011**, *4*. [CrossRef]

31. Lundgren, D.H.; Hwang, S.-I.; Wu, L.; Han, D.K. Role of spectral counting in quantitative proteomics. *Expert Rev. Proteom.* **2010**, *7*, 39–53. [CrossRef]

32. Boukil, A.; Perreault, V.; Chamberland, J.; Mezdour, S.; Pouliot, Y.; Doyen, A. High Hydrostatic Pressure-Assisted Enzymatic Hydrolysis Affect Mealworm Allergenic Proteins. *Molecules* **2020**, *25*, 2685. [CrossRef]

33. Hall, F.G.; Jones, O.G.; O'Haire, M.E.; Liceaga, A.M. Functional properties of tropical banded cricket (Gryllodes sigillatus) protein hydrolysates. *Food Chem* **2017**, *224*, 414–422. [CrossRef] [PubMed]

34. Guo, F.; Xiong, Y.L.; Qin, F.; Jian, H.; Huang, X.; Chen, J. Surface properties of heat-induced soluble soy protein aggregates of different molecular masses. *J. Food Sci.* **2015**, *80*, C279–C287. [CrossRef] [PubMed]

35. Zielińska, E.; Baraniak, B.; Karaś, M.; Rybczyńska, K.; Jakubczyk, A. Selected species of edible insects as a source of nutrient composition. *Food Res. Int.* **2015**, *77*, 460–466. [CrossRef]

36. Purschke, B.; Mendez Sanchez, Y.D.; Jäger, H. Centrifugal fractionation of mealworm larvae (*Tenebrio molitor* L.) for protein recovery and concentration. *LWT* **2018**, *89*, 224–228. [CrossRef]

37. Bußler, S.; Rumpold, B.A.; Jander, E.; Rawel, H.M.; Schluter, O.K. Recovery and techno-functionality of flours and proteins from two edible insect species: Meal worm (*Tenebrio molitor*) and black soldier fly (*Hermetia illucens*) larvae. *Heliyon* **2016**, *2*, e00218.

38. Lam, A.C.Y.; Can Karaca, A.; Tyler, R.T.; Nickerson, M.T. Pea protein isolates: Structure, extraction, and functionality. *Food Rev. Int.* **2018**, *34*, 126–147. [CrossRef]

39. Barre, A.; Pichereaux, C.; Velazquez, E.; Maudouit, A.; Simplicien, M.; Garnier, L.; Bienvenu, F.; Bienvenu, J.; Burlet-Schiltz, O.; Auriol, C.; et al. Insights into the Allergenic Potential of the Edible Yellow Mealworm (*Tenebrio molitor*). *Foods* **2019**, *8*, 515. [CrossRef]

40. Verhoeckx, K.C.; van Broekhoven, S.; den Hartog-Jager, C.F.; Gaspari, M.; de Jong, G.A.; Wichers, H.J.; van Hoffen, E.; Houben, G.F.; Knulst, A.C. YiHouse dust mite (Der p 10) and crustacean allergic patients may react to food containing Yellow mealworm proteins. *Food Chem. Toxicol.* **2014**, *65*, 364–373. [CrossRef]

41. Francis, F.; Doyen, V.; Debaugnies, F.; Mazzucchelli, G.; Caparros, R.; Alabi, T.; Blecker, C.; Haubruge, E.; Corazza, F. Limited cross reactivity among arginine kinase allergens from mealworm and cricket edible insects. *Food Chem.* **2019**, *276*, 714–718. [CrossRef]

42. Salgado, V.L.; Sparks, T.C. 6.5—The Spinosyns: Chemistry, Biochemistry, Mode of Action, and Resistance. In *Comprehensive Molecular Insect Science*; Gilbert, L.I., Ed.; Elsevier: Amsterdam, The Netherlands, 2005; pp. 137–173.

43. Kanost, M.R. Chapter 117—Hemolymph. In *Encyclopedia of Insects*, 2nd ed.; Resh, V.H., Cardé, R.T., Eds.; Academic Press: Cambridge, MA, USA, 2009; pp. 446–449.

44. Arrese, E.L.; Soulages, J.L. Insect fat body: Energy, metabolism, and regulation. *Annu. Rev. Entomol.* **2010**, *55*, 207–225. [CrossRef]

45. Andersen, S.O. Characteristic properties of proteins from pre-ecdysial cuticle of larvae and pupae of the mealworm *Tenebrio molitor*. *Insect Biochem. Mol. Biol.* **2002**, *32*, 1077–1087. [CrossRef]

46. Andersen, S.O.; Rafn, K.; Roepstorff, P. Sequence studies of proteins from larval and pupal cuticle of the yellow meal worm, *Tenebrio molitor*. *Insect Biochem. Mol. Biol.* **1997**, *27*, 121–131. [CrossRef]

47. Azagoh, C.; Ducept, F.; Garcia, R.; Rakotozafy, L.; Cuvelier, M.E.; Keller, S.; Lewandowski, R.; Mezdour, S. Extraction and physicochemical characterization of *Tenebrio molitor* proteins. *Food Res. Int.* **2016**, *88*, 24–31. [CrossRef] [PubMed]
48. Galves, C.; Stone, A.K.; Szarko, J.; Liu, S.; Shafer, K.; Hargreaves, J.; Siarkowski, M.; Nickerson, M.T. Effect of pH and defatting on the functional attributes of safflower, sunflower, canola, and hemp protein concentrates. *Cereal Chem.* **2019**, *96*, 1036–1047. [CrossRef]
49. Zayas, J.F. Solubility of Proteins. In *Functionality of Proteins in Food*; Zayas, J.F., Ed.; Springer: Berlin/Heidelberg, Germany, 1997; pp. 6–75.
50. Akpossan, R.; Digbeu, Y.; Koffi, M.; Kouadio, J.; Dué, E.; Kouamé, P. Protein Fractions and Functional Properties of Dried Imbrasia oyemensis Larvae Full-Fat and Defatted Flours. *Int. J. Biochem. Res. Rev.* **2015**, *5*, 116–126. [CrossRef]

Article

Nutritional, Physiochemical, and Antioxidative Characteristics of Shortcake Biscuits Enriched with *Tenebrio molitor* Flour

Ewelina Zielińska and Urszula Pankiewicz *

Department of Analysis and Evaluation of Food Quality, University of Life Sciences in Lublin, Skromna Str. 8, 20-704 Lublin, Poland; ewelina.zielinska@up.lublin.pl
* Correspondence: urszula.pankiewicz@up.lublin.pl

Academic Editors: Przemyslaw Lukasz Kowalczewski, Anubhav Pratap Singh and David Kitts
Received: 9 November 2020; Accepted: 27 November 2020; Published: 30 November 2020

Abstract: Edible insects, due to their high nutritional value, are a good choice for traditional food supplementation. The effects of partial replacement of wheat flour and butter with mealworm flour (*Tenebrio molitor*) on the quality attributes of shortcake biscuits were studied. The approximate composition was analyzed, along with the physical properties and color. Moreover, the antioxidant properties, starch digestibility, and glycemic index were determined in vitro. The protein and ash contents in biscuits supplemented with mealworm flour increased, while the carbohydrates content decreased. The increasing insect flour substitution decreased the lightness (L*) and yellowness (b*) but increased the redness (a*), total color difference (ΔE), and browning index (BI). The spread factor for the sample with the highest proportion of mealworm flour was significantly higher than the other biscuits. Furthermore, higher additions of mealworm flour increased the antioxidant activity of the biscuits and contributed to an increase in the content of slowly digested starch, with a decrease in the content of rapidly digested starch. Therefore, the results of the research are promising and indicate the possibility of using edible insects to enrich food by increasing the nutritional and health-promoting values.

Keywords: edible insects; entomophagy; shortcake biscuits with insects; mealworm; novel protein

1. Introduction

Entomophagy is commonly practiced by 2 billion people mainly in Asia, Africa, and South America [1]. In recent years, insect consumption has also been promoted in areas where it is not a tradition, such as Europe and North America, because of the high nutritional value and low environmental impact associated with eating insects [2]. Edible insects are known to be good sources of proteins, lipids, certain vitamins, and minerals, such as calcium, iron, or zinc. Importantly, insects are good sources of essential amino acids and polyunsaturated fatty acids [3].

Despite the advantages of using insects in human nutrition, the barrier to wider consumption is negative consumer perception. Insects in Western countries have several bad associations, namely with dirtiness, poverty, and diseases [4]. Visual impressions are very important because they are the consumer's first chance to form an opinion of the product; therefore, promoting the consumption of whole insects is not popular in Western societies. Sensory aspects of food are very important for consumer acceptance, so it is a good idea to design insect-based foods with minimal negative sensory characteristics [5]. Moreover, it has been confirmed that the addition of insects to food products in an invisible way allows for the acceptance of these products [6,7]. Insect-based foods started with the addition of insects into familiar food products such as bread [8,9], frankfurters [10], pastes [11,12], bakery products [13,14], and snacks bars [15] or cereal-based snacks [16]. In particular, cereal-based foods such as bread, biscuits, bakery products, and pasta are very popular and highly accepted worldwide, and thus research on how to enrich them with insect flours would be a good starting point. Additionally, insect flours could be sources of protein in gluten-free products when substitutes for gluten protein are sought [17]. Based on previous research, we can generally conclude that the addition of insects to enriched food products increased the contents of protein and ash and decreased the carbohydrate content [18]. However, the addition of insects can also enrich products with valuable fatty acids. Edible insects are certainly rich in unsaturated fatty acids, with levels comparable to those of poultry and fish, however containing more polyunsaturated fatty acids (PUFAs). One of the insects that is abundant in unsaturated fatty acids is the mealworm (*Tenebrio molitor*). The unsaturated fatty acid content in *T. molitor* amounts to 74.64% [3,19]. There is a growing number of papers discussing the use of insect flour, many of which focus on mealworms. For example, supplementation of maize tortillas with mealworm powder contributed to an increase in the contents of protein and essential amino acids [20]. In turn, the presence of mealworm flour in wheat bread increased the dough's stability and tenacity, as well as the crumb density [9]. This type of supplementation also affected the nutritional value of the bread—it increased the protein, fat, and ash contents [14]. For cookies supplemented with mealworm flour, the moisture, carbohydrate, protein, fat, ash, and mineral contents were higher compared with control cookies [21]. Other changes in products supplemented with mealworm flour include visual aspects. In one study, the lightness and yellowness of the muffins decreased with an increase in the mealworm powder concentration, however the redness increased [22]. An attempt was made to replace pork meat with mealworms in Frankfurters. The sausages formulated with a combination of 40% pork meat and 10% mealworm were similar in terms of cooking loss, emulsion stability, protein solubility, and overall acceptability to regular control frankfurters, maintaining the overall quality [10].

Furthermore, edible insects, in addition to their high nutritional value, have nutraceutical properties, such as strong antioxidant properties [23]. Our previous studies showed that after the in vitro digestion and absorption process, edible insects, including *T. molitor*, still have high antiradical activity [24–26]. Therefore, their addition to food products can increase the nutritional and pro-health value. Moreover, among the edible insects that are popular in Europe, *T. molitor* is characterized by its reasonably high fat content, with an appropriate fatty acid profile [3]. Based on the composition of mealworm meal, we can conclude that it should be used in the supplementation of traditional foods rich in flour and fat. An example is a shortcake, which is the basis of many food products (not only confectionery products, but also baking tarts and salty pastries). As such, in our research, we decided to replace some of the flour and butter in shortcake biscuits with ground mealworms. We examine the extent to which replacing some of the flour and butter with insect flour could influence the nutritional, physiochemical, and antioxidative properties of food products.

2. Results and Discussion

2.1. Physical Properties

The physical properties of the biscuits were determined by weight, diameter, thickness, spread factor, and apparent density. The physical properties of the biscuits are shown in Table 1. The biscuit weights ranged from 10.88 g to 13.41 g, with sample modification 1 (M1) presenting the lowest weight value (10.88 g) as compared to the control (C) (12.37 g). This decrease may be associated with the weaker water-binding capacity of mealworm flour than wheat flour, which may be responsible for moisture loss during the baking process, resulting in the lower weight of the biscuit. Moreover, in modification 1, as a result of the proportion of ingredients applied to the dough, the smallest amount of water was introduced in this batch. The smallest amounts of butter and flour were also used, with these ingredients being replaced by mealworm flour, which contains less moisture. The diameters of the biscuits were not significantly different ($p < 0.05$) and varied from 4.94 mm to 4.98 mm, whereas for the thickness, only sample M1 was significantly lower than the others (6.89 mm) ($p < 0.05$), which is correlated with the low weight of this sample. Therefore, the spread factor for this sample was significantly higher than the other biscuits (7.18) ($p < 0.05$). Undoubtedly, this was a result of the largest amount of mealworm flour being added to this batch. The spread ratio is a measure of cookie quality. A higher spread ratio is desirable for better cookies, which indicates higher product performance [27]. On this basis, it can be concluded that a higher addition of insect flour may increase the product's yield. Similar results were noted for cookies prepared from amaranth flour—the obtained spread factor was 7.95 [28]. The spread ratios of the other samples did not differ significantly from each other ($p < 0.05$). For the studied biscuits, we also calculated the apparent density, which was significantly higher for biscuits with higher amounts of mealworm flour (M1 and modification 2 (M2)) than the others. Although there is not much research on insect supplementation of bakery products, cookies are a very popular matrix for assessing supplementation with other functional additives. Certain functional additives, such as the insect flour used in this study, cause similar changes in the physical properties of bakery products. For example, cookies enriched with dephytinized oat bran [29] and novel cookies with a sclerotium of edible mushroom [30] had higher spread ratios.

Table 1. Physical properties of the biscuits.

Samples	Weight (g)	Diameter (cm)	Thickness (mm)	Spread Ratio	Apparent Density (g/cm^3)
C	12.37 ± 0.92 [b]	4.98 ± 0.06 [a]	8.04 ± 0.5 [a]	6.19 ± 0.3 [b]	0.197 ± 0.01 [b]
M1	10.88 ± 0.82 [c]	4.94 ± 0.05 [a]	6.89 ± 0.47 [b]	7.18 ± 0.5 [a]	0.204 ± 0.01 [a]
M2	13.41 ± 0.94 [a]	4.97 ± 0.07 [a]	8.33 ± 0.32 [a]	5.96 ± 0.2 [b]	0.207 ± 0.01 [a]
M3	12.39 ± 0.9 [b]	4.96 ± 0.05 [a]	8.11 ± 0.31 [a]	6.12 ± 0.2 [b]	0.196 ± 0.02 [b]

C—control sample; M1—modification 1; M2—modification 2; M3—modification 3. Mean values with different letters in each column are significantly ($p < 0.05$) different.

2.2. Nutrient Composition

The approximate compositions of the studied biscuits and mealworm flour are reported in Table 2. Mealworms are a good source of compounds of high nutritional value, such as proteins, lipids, and particularly polyunsaturated fatty acids and minerals, such as iron, magnesium, and zinc [3]. Edible insects are known as high-protein food products, and the protein content of the mealworms was determined to be 54.6% of the dry weight (d.w.). Furthermore, insects are rich in minerals—the ash content was determined to be 3.89% d.w. The substitution of wheat flour with mealworm flour changed the nutritional value of the products. The protein content of the biscuits with the addition of mealworm flour ranged from 10.82% to 13.52% of dry weight. This effect was expected because mealworm flour is the richest in protein among all ingredients in the recipe, so the increase in its content causes a proportional increase in protein content in the end product. In turn, the fat content was at a similar level in all biscuits ($p < 0.05$). In this case, the fat content that was removed from the dough in the form of butter was almost completely replaced in the form of fat from mealworms, and the estimated fat contents introduced into the dough from the ingredients in all modification batches were similar. Biscuits with the highest amounts of mealworm flour (M1 and M2) were found to be the richest in ash, with levels equivalent to the microelement contents (0.63 and 0.7, respectively). Considering possible ash content deviations, sample M3 was characterized by having the lowest value due to the having the lowest amount of added mealworm flour. Generally, there was a progressive increase in the protein and ash contents, with a decrease in the carbohydrates content as the concentration of mealworm flour increased. In turn, the moisture content decreased as the mealworm flour concentration increased in the biscuits. The water contained in the dough was derived only from water contained in the ingredients used for its production—no additional water was used. This proportional change in the moisture content was, therefore, caused by the replacement of the flour and butter with mealworm flour, which has a lower moisture content than these ingredients, meaning the water content of the dough was also reduced. The nutritional compositions of the products are important, however the moisture content is a significant quality factor affecting preservation, packaging, and transport convenience [31]. Although the proportions of protein and carbohydrates had changed, no effects on the energy values of the tested biscuits were found ($p < 0.05$). This is because protein and carbohydrates have the same conversion factor values as energy values [32]. A difference in energy value could only be caused by a significant difference in fat content.

Table 2. Nutritional values of the biscuits.

Sample	Protein (% d.w.)	Fat (% d.w.)	Ash (% d.w.)	Carbohydrates (% d.w.)	Moisture (%)	Energy Value (kcal/100g d.w.)	Energy Value (kJ/100g d.w.)
C	9.09 ± 0.46 [c]	27.03 ± 1.48 [a]	0.28 ± 0.04 [b]	63.6 ± 1.63 [a]	6.4 ± 0.23 [a]	534 ± 4.3 [a]	2236 ± 12.8 [a]
M1	13.52 ± 0.6 [a]	27.17 ± 0.39 [a]	0.63 ± 0.06 [a]	58.69 ± 1.85 [b]	4.33 ± 0.09 [c]	533 ± 4.5 [a]	2233 ± 13.3 [a]
M2	11.97 ± 0.5 [b]	26.97 ± 1.69 [a]	0.7 ± 0.1 [a]	60.36 ± 1.49 [ab]	5.3 ± 0.16 [b]	532 ± 4.9 [a]	2227 ± 12 [a]
M3	10.82 ± 0.5 [b]	28.47 ± 0.36 [a]	0.44 ± 0.05 [b]	60.27 ± 1.72 [ab]	5.97 ± 0.2 [a]	541 ± 3.9 [a]	2262 ± 12.7 [a]

C—control sample; M1—modification 1; M2—modification 2; M3—modification 3, Values followed by a different superscript in a column differ significantly ($p < 0.05$).

2.3. Color Measurements

The surface colors of the shortcake biscuits in terms of the L*, a*, b*, ΔE, and browning index (BI) values can be found in Table 3. The appearances of the biscuits are shown in Figure 1. Increasing the amount of insect flour decreased the lightness (L*) and yellowness (b*) while increasing the redness (a*), total color difference (ΔE), and browning index. Such results were expected, because mealworm flour is darker than the wheat flour used in these biscuits, hence supplementation with mealworm flour will give products a darker color, because usually the color of a baked product is directly dependent on the colors of the raw materials used. On the other hand, the protein content in mealworm flour was higher than in wheat flour, resulting in a higher degree of Maillard reaction with increased surface redness [33]. Similar results for color darkening were observed by researchers for muffins enriched with mealworm [22] and cricket powder [18], cookies enriched with mealworm powder [21], bread supplemented with insect flour [8,14], and pasta enriched with cricket powder [17]. The highest color difference in the control samples was seen in sample M1, which contained the greatest amount of mealworm flour. As Pauter et al. [18] suggested, consumers tend to see darker bakery products as healthier and containing more fiber or whole grains. Therefore, this color change may increase consumer interest in this type of biscuit.

Table 3. Color determinants of biscuits.

Sample	L*	a*	b*	ΔE	BI
C	34.44 ± 2.97 [a]	5.77 ± 1.63 [b]	8.61 ± 0.94 [a]	-	11.80
M1	20.96 ± 0.96 [d]	8.64 ± 1.31 [a]	2.14 ± 0.64 [c]	15.23	26.67
M2	26.45 ± 0.56 [c]	6.57 ± 1.01 [ab]	5.62 ± 0.81 [b]	8.56	17.18
M3	29.46 ± 0.97 [b]	7.12 ± 0.8 [ab]	7.46 ± 0.3 [ab]	5.28	16.75

C—control sample; M1—modification 1; M2—modification 2; M3—modification 3. Values followed by a different superscript in a column differ significantly ($p < 0.05$). * mean in terms of lightness (L*) and color (a*—redness; b*—yellowness).

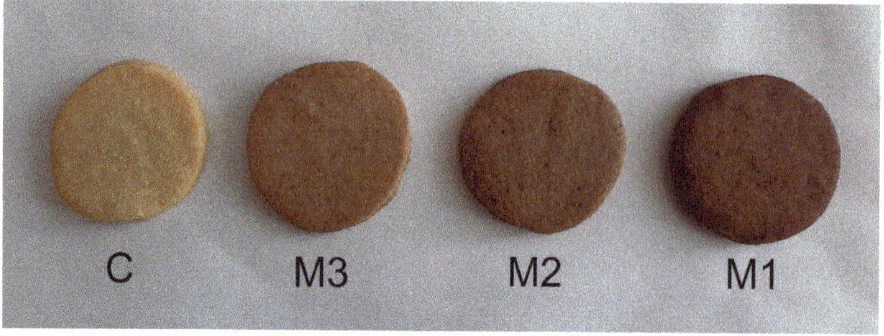

Figure 1. Shortcake biscuits prepared with different amounts of mealworm flour and wheat flour. C—control sample; M1—modification 1; M2—modification 2; M3—modification 3.

2.4. Antioxidant Properties

Measuring the antioxidant capacity of food products is of increasing interest because it can provide a wide variety of information on factors such as the oxidation resistance, the quantitative contribution of antioxidants, or the antioxidant effects that can occur in the body at the time of consumption. The antioxidant activity of the sample extracts was evaluated by assessing the ability of the extracts to inhibit 2,2-diphenyl-1-picrylhydrazyl (DPPH$^\bullet$) and 2,2′-azino-bis(3-ethylbenzothiazoline-6-sulfonic acid (ABTS$^{\bullet+}$). The antioxidant properties of the mealworm flour and the biscuits prepared from it are shown in Figure 2. The highest antiradical activity values against both ABTS$^{\bullet+}$ and DPPH$^\bullet$ were observed for the mealworm flour (0.67 and 2.70 mM TE, respectively). Navarro del Hierro et al. [34] studied the DPPH$^\bullet$ scavenging activity of mealworm extracts and also confirmed their strong antioxidant properties. As expected, the partial substitution of wheat flour with mealworm flour significantly ($p < 0.05$) increased the free radical scavenging capacity, as reflected by DPPH$^\bullet$ and ABTS$^{\bullet+}$ scavenging activity values. Therefore, the antioxidant activity levels of the biscuits increases as the concentration of mealworm flour in the recipe increased. Higher free radical scavenging activity was observed for DPPH$^\bullet$ than ABTS$^{\bullet+}$. Food fortification has gained increased interest among consumers. Many products are enriched to increase their antioxidant potential. For example, cookies produced with 5% to 20% camu camu (*Myrciaria dubia*) coproduct powder as a replacement for wheat flour had higher antioxidant potential than control cookies [35]. In turn, cookies supplemented with flaxseed in amounts of 5% and 30% exhibited DPPH radical scavenging activity values of 7.93% and 12.25% as compared to 5.5% for the control [31]. Food enrichment with unconventional protein sources is becoming increasingly popular. For example, the incorporation of microalgae in amounts ranging from 2% to 6% into cookies led to a significant increase in the antioxidant capacity [36]. As unconventional sources of protein, insects may, therefore, have good potential to enrich food and improve its nutraceutical value. The antioxidant activity of bread enriched with cricket flour as measured by the DPPH and ABTS scavenging activity showed was significantly higher than standard dough [37]. Many bioactive compounds have been identified in insects (e.g., chitins, polyphenols, antioxidant enzymes, peptides, proteins, etc.) [23,24,26,38,39]. As high-protein products, insects are, therefore, potential sources of bioactive proteins and peptides. Numerous amino acid sequences from insects have been identified, which have been associated with in vitro bioactive properties [23,26,40–42]. Proteins and peptides are also involved in the antioxidant properties of insects [25,26,43], and depending on the species of edible insects these proteins might change the DPPH$^\bullet$ and ABTS$^{\bullet+}$ radical scavenging activity. Such changes might be depend on the molecular weight of protein or peptide in question, as well as the amino acid composition [44]. Moreover, based on the high activity of DPPH$^\bullet$ in terms of radical scavenging, the obtained results suggest that mealworm proteins contain amino acids or peptides that act as electron donors and can react with free radicals to transform them into more stable compounds.

It has been confirmed that during in vitro digestion of the mealworm, peptides with strong antioxidant properties are released. Their activity has been confirmed based on the activity of the synthesized peptide sequences. The identified peptides also showed higher activity against DPPH$^\bullet$ than ABTS$^{\bullet+}$. The following peptides were identified in the hydrolysates of *T. molitor*: NYVADGLG from raw insect cuticle protein, AAAPVAVAK from boiled insect cuticle protein, YDDGSYKPH from baked insect ADFb protein, and AGDDAPR from isolated protein [26]. Moreover, the results of the presented study show that the heat treatment used for the insects increased the antiradical activity of the peptide fractions, with baking yielding particularly good results [25].

2.5. Rapidly and Slowly Digested Starch Contents and In Vitro Glycemic Index (GI) Values

Rapidly digested starch (RDS) is defined as the part of the starch that is digested within 20 min of food intake and which causes a rapid increase in blood glucose levels. Slowly digested starch (SDS) is the part of the starch that is completely digested in the small intestine, but at a slower rate than RDS, i.e., within 20–120 min of ingestion [45].

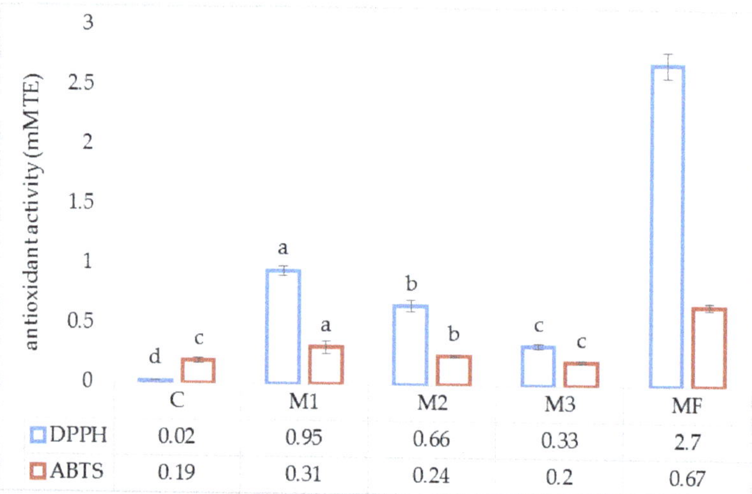

	C	M1	M2	M3	MF
DPPH	0.02	0.95	0.66	0.33	2.7
ABTS	0.19	0.31	0.24	0.2	0.67

Figure 2. Antioxidant properties of biscuits. C—control sample; M1—modification 1; M2—modification 2; M3—modification 3; MF—mealworm flour. Values followed by a different superscript in a column differ significantly ($p < 0.05$).

Based on the obtained results, it can be concluded that the addition of flour from mealworm affected the changes in rapidly and slowly digested starch contents (Table 4). All samples were found to have more rapidly than slowly digested starch, but the control and modification 3 (M3) samples contained three times more RDS than SDS, whereas the M1 sample contained only about 1.8 times more RDS than SDS. Generally, samples M1 and M2, which contained the highest amounts of mealworm flour, had significantly lower contents of RDS and higher contents of SDS than control, which is desirable for consumers. The dietary benefits attributed to SDS are associated with a slower postprandial rise in blood glucose and glycemia maintenance for longer periods compared to RDS, which results in a rapid rise and then a rapid fall in blood glucose, often to below the initial value due to the increased insulin levels. The prolonged absorption of glucose after consumption of products rich in SDS inhibits the release of free fatty acids from adipose tissue. Reducing their inflow to the liver promotes the faster removal of glucose from the cardiovascular system and consequently leads to a decrease in the serum concentration. These features of SDS give products rich in this component low glycemic index values [46].

Table 4. Rapidly and slowly digested starch contents and in vitro glycemic index values for the biscuits.

Sample	RDS (mg Glucose/g Sample)	SDS (mg Glucose/g Sample)	In Vitro GI
C	223.28 ± 9.85 [a]	73.55 ± 2.69 [b]	39.85 ± 0.12 [a]
M1	181.66 ± 9.61 [b]	102.95 ± 6.12 [a]	39.83 ± 0.1 [a]
M2	191.8 ± 5.58 [b]	91.01 ± 6.75 [a]	39.84 ± 0.08 [a]
M3	218.52 ± 1.85 [a]	72.10 ± 4.89 [b]	39.85 ± 0.15 [a]

C—control sample; M1—modification 1; M2—modification 2; M3—modification 3; RDS—rapidly digested starch; SDS—slowly digested starch. Values followed by a different superscript in a column differ significantly ($p < 0.05$).

The effect of the mealworm flour addition on starch digestibility in biscuits has been documented. The digestibility of starch present in food, especially when subjected to different types of heat treatment, is influenced by the other ingredients present in the food. The other ingredients can include proteins; interactions between proteins and starch are crucial [47]. The addition of edible insects as high-protein foods to bakery products can, therefore, lead increase interactions and also increase the content of slowly digested starch.

An in vitro starch hydrolysis method was used in this study to estimate the metabolic glycemic response to food products. The in vitro glycemic index values of the biscuits are shown in Table 4. The addition of mealworm flour to the biscuits did not affect the in vitro glycemic index values. So far, no effects of supplementation with edible insects on starch composition, digestibility, and in vitro glycemic index values have been shown in the literature. Further studies on these issues are, therefore, needed to increase the knowledge of food supplementation with edible insects.

3. Materials and Methods

3.1. Chemicals, Reagents, and Biscuit Ingredients

Trolox (6-hydroxy-2,5,7,8-tetramethylchroman-2-carboxylic acid), ABTS (2,2′-azino-bis(3-ethylbenzothiazoline-6-sulfonic acid), DPPH (2,2-diphenyl-1-picrylhydrazyl), dinitrosalicylic acid, pepsin from porcine gastric mucosa (cat no. 10080, 250 U/Mg), pancreatin from porcine pancreas (cat no. P1750), amyloglucosidase (cat no. 10115, 70 U/Mg), and invertase (cat no. I4504, 300 U/Mg) were purchased from Sigma-Aldrich Company, Ltd. (St Louis, MO, USA). The D-glucose assay kit (GOPOD format) was obtained from Megazyme International (Wicklow, Ireland). All other chemicals used were of analytical grade.

Wheat flour (type 500), sugar, eggs, and butter (80% fat, including 55% saturated fatty acids) were purchased from a local market. The *Tenebrio molitor* mealworms (Linnaeus, Coleoptera: Tenebrionidae) (larvae) were obtained from a local breeder (Lublin, Poland). The insects were fasted for two days to empty their digestive tract, then they were frozen for 24 h at −18 °C and lyophilized. Afterward, the insects were ground in a laboratory grinder (IKA A11 basic) to obtain flour. In order to obtain a uniform particle size, the flour was passed through a 20 mesh sieve, then the flour was analyzed (protein: 54.6 ± 2.2% d.w.; fat: 29.47 ± 1.95% d.w.; ash: 3.89 ± 0.8% d.w.; carbohydrates: 12.04 ± 0.96% d.w.; moisture: 4.72 ± 0.12%; energetic value: 532 ± 4.2 kcal/100g d.w. or 2223 ± 13.5 kJ/100g d.w). All analyses were performed within two weeks of the preparation of the flour, during which time it was stored in an airtight container at −18 °C. All chemicals and reagents used were of analytical grade.

3.2. Shortcake Biscuit Preparation

The biscuits were prepared as three variants with different amounts of added insect flour, which partially replaced the wheat flour and butter. Control biscuits were prepared without the addition of insects. The shortcake recipe contained commonly used ingredients, including wheat flour (300 g), butter (150 g), eggs (60 g), and sugar (70 g) [48]. The cold ingredients were then kneaded manually into dough, cooled, then sheeted manually on a dough sheeter to a uniform thickness of 0.5 cm, cut into round shapes measuring 5 cm in diameter, and baked at 180 °C for 25 min in a preheated oven. The biscuits were placed randomly to minimize the impact of their placement on the tray on their characteristics during baking. The biscuits were then allowed to cool at room temperature and subjected to physical and color measurements. Next, the biscuits were frozen at −18 °C so that additional analyses could be performed within two weeks. The quantities of ingredients for each biscuit variant are presented in Table 5.

Table 5. Recipe for shortcake biscuits.

Ingredients	C	M1	M2	M3
Wheat flour (g)	300	270	280	285
Butter (g)	150	135	140	142.5
Sugar (g)	70	70	70	70
Eggs (pcs)	1	1	1	1
Mealworm flour (g)	-	30	20	15

C—control sample; M1—modification method 1; M2—modification method 2; M3—modification method 3.

3.3. Physical Properties

Fifteen biscuits were randomly selected from each batch and were analyzed in terms of weight, thickness, and diameter. The spread ratio was calculated by finding the ratio of a biscuit's diameter versus its thickness [30]. The apparent density was calculated by evaluating the weight-to-volume ratio [49].

3.4. Nutrient Composition

The insect flour and biscuit samples were assessed for their moisture, ash, fat, and protein contents ($N \times 6.25$) by employing standard methods of analysis [50]. The carbohydrate content was determined by the following formula: 100 − (weight in grams (protein + fat + ash) in 100 g of the dry weight of edible insects). The conversion method was used for the determination of the nutritional value [33].

3.5. Color Measurements

Fifteen biscuits were randomly selected from each batch and color measurements were performed in triplicate. The color of the biscuits was measured using an EnviSense colorimeter NH310 (EnviSense, Lublin, Poland). Color differences were recorded in the CIE L*a*b* scale in terms of lightness (L*) and color (a*—redness; b*—yellowness). Additionally, the total color difference (ΔE) was calculated using the following formula:

$$\Delta E = \sqrt{\Delta L^{*2} + \Delta a^{*2} + \Delta b^{*2}} \tag{1}$$

where ΔL^*, Δa^*, and Δb^* are differences in the L*, a*, and b* values, respectively, between the reference sample and the test sample.

The browning index (BI, Equations (2) and (3)) was calculated using the measured L*, a*, and b* values as follows [51]:

$$BI = \frac{100(x - 0.31)}{0.17} \tag{2}$$

where x:

$$x = \frac{(a^* + 1.75L^*)}{(5.645L^* + a^* - 0.012b^*)} \tag{3}$$

3.6. Antioxidant Properties

3.6.1. Extraction of Bioactive Compounds

The biscuits samples (1 g) were ground in a laboratory grinder and shaken with 10 mL of 4:1 ethanol/water (*v/v*) for 120 min in a laboratory shaker. Next, the samples were centrifuged at 3000 g for 10 min. The supernatant was stored at −18 °C until the next day for further analysis.

3.6.2. DPPH Radical Scavenging Activity

The DPPH assay was estimated according to Brand-Williams, Cuvelier, and Berset [52], with slight modification. A 0.1 mL volume of the sample was mixed with 0.9 mL of a 6 µM solution of DPPH• in 75% methanol. The absorbance was measured after 3 min of reaction at 515 nM. Here, 75% methanol was used as a blank. The scavenging effect was calculated using the formula:

$$\text{Scavenging activity (\%)} = [1 - (A\,\text{sample}/A\,\text{control})] \times 100 \tag{4}$$

where A sample is the absorbance of the mixture of sample and DPPH•; A control is the absorbance of the control (DPPH• solution).

The results were expressed as Trolox equivalent antioxidant activity (TEAC) values (mM Trolox).

3.6.3. ABTS Radical Scavenging Activity

The ABTS assay was determined according to Re et al. [53] with slight modifications regarding the quantities of the antioxidant solution. The radical solution was prepared with ABTS and potassium persulfate, diluted in water to a final concentration of 2.45 mM, and left in the dark for 16 h to allow for radical development. The solution was diluted to reach the absorbance measures around 0.7 at 734 nM. Then, 2.95 mL of the ABTS$^{\bullet+}$ solution was mixed with 0.05 mL of each sample. The absorbance was measured after 3 min of the reaction at 734 nM. Deionized water was used as a blank. The scavenging effect was calculated according to the equation:

$$\text{Scavenging activity } (\%) = [1 - (A \text{ sample}/A \text{ control})] \times 100 \tag{5}$$

where A sample is the absorbance of the mixture of sample and ABTS$^{\bullet+}$; A control is the absorbance of the control (ABTS$^{\bullet+}$ solution).

The results were expressed as Trolox equivalent antioxidant activity (TEAC) values (mM Trolox).

3.7. In Vitro Digestion of Biscuits

The in vitro digestibility of the starch was determined according to the method of Monro et al. [45] with slight modifications. Briefly, 1 g of the sample was combined with 30 mL of water and 0.8mL of 1M HCl to reach a pH of 2.5, then 1mL of 10% pepsin solution in 0.05M HCl was added. The reaction was carried out for 30 min at 37 °C with stirring (130 rpm) to complete gastric digestion. Next, 2mL of 1M NaHCO$_3$ and 5mL of 0.1 M phosphate buffer (pH 6) were added, followed by 4.6 mg amyloglucosidase and 5mL of 2.5% pancreatin in 0.1 M phosphate buffer (pH 6) to start the small intestinal phase. The tubes were filled with distilled water to a volume of 55 mL. Digesta aliquots measuring 1.0 mL were removed after the gastric phase and at 20, 30, 60, 90, 120, and 180 min from the start of amylolysis, then were each subsequently added to 4mL absolute ethanol in a tube and mixed.

3.8. Rapidly and Slowly Digested Starch Contents

The rapidly digested starch (RDS) and slowly digested starch (SDS) levels were determined by measuring reducing sugars released during in vitro digestion of biscuits according to the procedure described by Soong, Tan, Leong, and Henry [54]. Briefly, sugars released during in vitro digestion were evaluated as monosaccharides using a dinitrosalicylic acid (DNS) colorimetric method with slight modifications. To complete the depolymerization into monosaccharides, 50 µL of the supernatant was mixed with 0.25 mL of acetate buffer containing 0.4% invertase and 1% amyloglucosidase. Incubation was carried out at room temperature for 30 min. Next, 0.75 mL of DNS mixture containing 0.5 mg/mL glucose, 4 M NaOH, and DNS reagent at a 1:1:5 ratio was added and heated for 15 min at 95–100 °C, cooled, and then diluted with 4 mL distilled water. The absorbance of the samples and standards was read at 530 nM against the blank.

The amount of sugars released was calculated in mg of glucose/g of each sample using the absorbance values. The rapidly digested starch (RDS) was evaluated as the amount of reducing sugars determined in the sample aliquot after 20 min from the start of pancreatic digestion, whereas slowly digested starch (SDS) was calculated as the difference between the amount of reducing sugars measured at 120 min and RDS.

3.9. In Vitro Glycemic Index (GI)

The in vitro GI of the biscuits was determined by evaluation of the in vitro starch digestibility according to the method described by Reis and Abu-Ghannam [55] with slight modifications, using the digestion procedure described in Section 3.7. Aliquots of 1 mL of the hydrolyzed solution were taken at 10, 20, 30, 60, 90, 120, and 180 min intervals, respectively. They were mixed with 4 mL of absolute ethanol to deactivate the enzymes. The glucose content of the hydrolysates was determined using the GOPOD method. The values were expressed as mg glucose/g sample. Glucose content was plotted as a function of time and the areas under the hydrolysis curves (AUC) were calculated. The hydrolysis index (HI) for each sample was calculated as the ratio between the AUC of the sample and the AUC for the reference food (white bread). The HI values were normalized for the total carbohydrate available in each sample and reference and expressed as percentages. The GI was predicted according to the equation described by Goñi, Garcia-Alonso, and Saura-Calixto [56]:

$$GI\ (\%) = 39.71 + 0.549 \times HI \tag{6}$$

3.10. Statistical Analysis

All assays were performed in triplicate. All data are presented as means plus standard deviation. Statistical analyses were carried out using Statistica (version 13.0, StatSoft, Krakow, Poland). Tukey's test was used to compare the groups. The differences between the mean values were found to be statistically significant at a *p* values of less than 0.05.

4. Conclusions

This study was undertaken to identify the potential for edible insects to increase the nutritional and antioxidant properties of shortcake biscuits. The results revealed that the substitution of wheat flour with mealworm flour changed the nutritional value of the products. There was a progressive increase in the protein and ash contents of biscuits as the concentration of mealworm flour increased. In turn, the fat content and energy value did not differ significantly.

Increased levels of mealworm flour in the biscuits led to a darker appearance. Additionally, the biscuit with the greatest amount of mealworm flour was characterized by the highest spread factor. Regarding the antioxidant activity, mealworms were found to have high antioxidant potential, as evident from the higher free radical scavenging activity of biscuits enriched in mealworm flour in comparison to control. Moreover, it is very beneficial that the addition of mealworm flour to biscuits caused an increase in slowly digested starch, with a decrease in rapidly digested starch.

The results showed that edible insects might be a good nutritive and bioactive additive to traditionally eaten food, which was confirmed based on fortification of the shortcake dough. This type of dough can be used as an intermediate product to make various types of food, including not only confectionery products, but also savory products. A further step would be to investigate the effects of the addition of edible insects to other food matrices.

Author Contributions: Conceptualization, E.Z. and U.P.; methodology, E.Z. and U.P.; software, E.Z.; validation, E.Z.; formal analysis, E.Z.; investigation, E.Z.; resources, E.Z.; data curation, E.Z.; writing—original draft preparation, E.Z.; writing—review and editing, E.Z. and U.P.; supervision, U.P.; project administration, E.Z.; funding acquisition, E.Z. All authors have read and agreed to the published version of the manuscript.

Funding: This research was funded by the Ministry of Science and Higher Education in Poland, grant number VKA/MN-3/TŻ/20.

Conflicts of Interest: The authors declare no conflict of interest. The sponsors had no role in the design, execution, interpretation, or writing of the study.

References

1. Food and Argriculture Organization of the United Nations. *Edible Insects. Future Prospects for Food and Feed Security*; Food and Argriculture Organization of the United Nations Edible Insects: Rome, Italy, 2013; Volume 171, ISBN 9789251075951.
2. Smetana, S.; Palanisamy, M.; Mathys, A.; Heinz, V. Sustainability of insect use for feed and food: Life Cycle Assessment perspective. *J. Clean. Prod.* **2016**, *137*, 741–751. [CrossRef]
3. Zielińska, E.; Baraniak, B.; Karaś, M.; Rybczyńska, K.; Jakubczyk, A. Selected species of edible insects as a source of nutrient composition. *Food Res. Int.* **2015**, *77*, 460–466. [CrossRef]
4. Looy, H.; Dunkel, F.V.; Wood, J.R. How then shall we eat? Insect-eating attitudes and sustainable foodways. *Agric. Human Values* **2014**, *31*, 131–141. [CrossRef]
5. Mishyna, M.; Chen, J.; Benjamin, O. Sensory attributes of edible insects and insect-based foods – Future outlooks for enhancing consumer appeal. *Trends Food Sci. Technol.* **2020**, *95*, 141–148. [CrossRef]
6. Hartmann, C.; Siegrist, M. Becoming an insectivore: Results of an experiment. *Food Qual. Prefer.* **2016**, *51*, 118–122. [CrossRef]

7. Tan, H.S.G.; van den Berg, E.; Stieger, M. The influence of product preparation, familiarity and individual traits on the consumer acceptance of insects as food. *Food Qual. Prefer.* **2016**, *52*, 222–231. [CrossRef]

8. Haber, M.; Mishyna, M.; Martinez, J.J.I.; Benjamin, O. The influence of grasshopper (Schistocerca gregaria) powder enrichment on bread nutritional and sensorial properties. *LWT* **2019**, *115*, 108395. [CrossRef]

9. Cappelli, A.; Oliva, N.; Bonaccorsi, G.; Lorini, C.; Cini, E. Assessment of the rheological properties and bread characteristics obtained by innovative protein sources (Cicer arietinum, Acheta domesticus, Tenebrio molitor): Novel food or potential improvers for wheat flour? *LWT* **2020**, *118*. [CrossRef]

10. Choi, Y.S.; Kim, T.K.; Choi, H.D.; Park, J.D.; Sung, J.M.; Jeon, K.H.; Paik, H.D.; Kim, Y.B. Optimization of replacing pork meat with yellow worm (Tenebrio molitor L.) for frankfurters. *Korean J. Food Sci. Anim. Resour.* **2017**, *37*, 617–625. [CrossRef]

11. De Smet, J.; Lenaerts, S.; Borremans, A.; Scholliers, J.; Van Der Borght, M.; Van Campenhout, L. Stability assessment and laboratory scale fermentation of pastes produced on a pilot scale from mealworms (Tenebrio molitor). *LWT* **2019**, *102*, 113–121. [CrossRef]

12. Smarzyński, K.; Sarbak, P.; Musiał, S.; Jezowski, P.; Piatek, M.; Kowalczewski, P.T. Nutritional analysis and evaluation of the consumer acceptance of pork pâté enriched with cricket powder-preliminary study. *Open Agric.* **2019**, *4*, 159–163. [CrossRef]

13. Delicato, C.; Schouteten, J.J.; Dewettinck, K.; Gellynck, X.; Tzompa-Sosa, D.A. Consumers' perception of bakery products with insect fat as partial butter replacement. *Food Qual. Prefer.* **2020**, *79*, 103755. [CrossRef]

14. González, C.M.; Garzón, R.; Rosell, C.M. Insects as ingredients for bakery goods. A comparison study of H. illucens, A. domestica and T. molitor flours. *Innov. Food Sci. Emerg. Technol.* **2019**, *51*, 205–210. [CrossRef]

15. Adámek, M.; Adámková, A.; Mlček, J.; Borkovcová, M.; Bednářová, M. Acceptability and sensory evaluation of energy bars and protein bars enriched with edible insect. *Potravin. Slovak J. Food Sci.* **2018**, *12*, 431–437. [CrossRef]

16. Severini, C.; Azzollini, D.; Albenzio, M.; Derossi, A. On printability, quality and nutritional properties of 3D printed cereal based snacks enriched with edible insects. *Food Res. Int.* **2018**, *106*, 666–676. [CrossRef] [PubMed]

17. Duda, A.; Adamczak, J.; Chelminska, P.; Juszkiewicz, J.; Kowalczewski, P. Quality and nutritional/textural properties of durum wheat pasta enriched with cricket powder. *Foods* **2019**, *8*, 46. [CrossRef] [PubMed]

18. Pauter, P.; Różańska, M.; Wiza, P.; Dworczak, S.; Grobelna, N.; Sarbak, P.; Kowalczewski, P. Effects of the replacement of wheat flour with cricket powder on the characteristics of muffins. *Acta Sci. Pol. Technol. Aliment.* **2018**, *17*, 227–233. [CrossRef]

19. Tzompa-Sosa, D.A.; Yi, L.; van Valenberg, H.J.F.; van Boekel, M.A.J.S.; Lakemond, C.M.M. Insect lipid profile: Aqueous versus organic solvent-based extraction methods. *Food Res. Int.* **2014**, *62*, 1087–1094. [CrossRef]

20. Aguilar-Miranda, E.D.; Lo´pez, M.G.; Lo´pez, L.; Escamilla-Santana, C.; Barba De, A.P.; Rosa, L.A. Characteristics of Maize Flour Tortilla Supplemented with Ground Tenebrio molitor Larvae. *J. Agric. Food Chem.* **2002**. [CrossRef]

21. Min, K.-T.; Kang, M.-S.; Kim, M.-J.; Lee, S.-H.; Han, J.-S.; Kim, A.-J. Manufacture and Quality Evaluation of Cookies prepared with Mealworm (Tenebrio molitor) Powder. *Korean J. Food Nutr.* **2016**, *29*. [CrossRef]

22. Hwang, S.-Y.; Choi, S.-K. Quality Characteristics of Muffins Containing Mealworm(Tenebrio molitor). *Korean J. Culin. Res.* **2015**, *21*, 104–115.

23. Nongonierma, A.B.; FitzGerald, R.J. Unlocking the biological potential of proteins from edible insects through enzymatic hydrolysis: A review. *Innov. Food Sci. Emerg. Technol.* **2017**, *43*, 239–252. [CrossRef]

24. Zielińska, E.; Karaś, M.; Jakubczyk, A. Antioxidant activity of predigested protein obtained from a range of farmed edible insects. *Int. J. Food Sci. Technol.* **2017**, *52*, 306–312. [CrossRef]

25. Zielińska, E.; Baraniak, B.; Karaś, M. Antioxidant and anti-inflammatory activities of hydrolysates and peptide fractions obtained by enzymatic hydrolysis of selected heat-treated edible insects. *Nutrients* **2017**, *9*, 970. [CrossRef] [PubMed]

26. Zielińska, E.; Baraniak, B.; Karaś, M. Identification of antioxidant and anti-inflammatory peptides obtained by simulated gastrointestinal digestion of three edible insects species (Gryllodes sigillatus, Tenebrio molitor, Schistocerca gragaria). *Int. J. Food Sci. Technol.* **2018**, *53*, 2542–2551. [CrossRef]

27. Mudgil, D.; Barak, S.; Khatkar, B.S. Cookie texture, spread ratio and sensory acceptability of cookies as a function of soluble dietary fiber, baking time and different water levels. *LWT-Food Sci. Technol.* **2017**, *80*, 537–542. [CrossRef]

28. Chauhan, A.; Saxena, D.C.; Singh, S. Total dietary fibre and antioxidant activity of gluten free cookies made from raw and germinated amaranth (Amaranthus spp.) flour. *LWT-Food Sci. Technol.* **2015**, *63*, 939–945. [CrossRef]

29. Baumgartner, B.; Özkaya, B.; Saka, I.; Özkaya, H. Functional and physical properties of cookies enriched with dephytinized oat bran. *J. Cereal Sci.* **2018**, *80*, 24–30. [CrossRef]

30. Kolawole, F.L.; Akinwande, B.A.; Ade-Omowaye, B.I.O. Physicochemical properties of novel cookies produced from orange-fleshed sweet potato cookies enriched with sclerotium of edible mushroom (Pleurotus tuberregium). *J. Saudi Soc. Agric. Sci.* **2020**, *19*, 174–178. [CrossRef]

31. Kaur, M.; Singh, V.; Kaur, R. Effect of partial replacement of wheat flour with varying levels of flaxseed flour on physicochemical, antioxidant and sensory characteristics of cookies. *Bioact. Carbohydr. Diet. Fibre* **2017**, *9*, 14–20. [CrossRef]

32. European Parliment REGULATION (EU) No 1169/2011 OF THE EUROPEAN PARLIAMENT AND OF THE COUNCIL of 25 October 2011 on the provision of food information to consumers. *Off. J. Eur. Union* **2011**, 18–63.

33. Sozer, N.; Cicerelli, L.; Heiniö, R.L.; Poutanen, K. Effect of wheat bran addition on invitro starch digestibility, physico-mechanical and sensory properties of biscuits. *J. Cereal Sci.* **2014**, *60*, 105–113. [CrossRef]

34. Navarro del Hierro, J.; Gutiérrez-Docio, A.; Otero, P.; Reglero, G.; Martin, D. Characterization, antioxidant activity, and inhibitory effect on pancreatic lipase of extracts from the edible insects Acheta domesticus and Tenebrio molitor. *Food Chem.* **2020**, *309*, 125742. [CrossRef] [PubMed]

35. das Chagas, E.G.L.; Vanin, F.M.; dos Santos Garcia, V.A.; Yoshida, C.M.P.; de Carvalho, R.A. Enrichment of antioxidants compounds in cookies produced with camu-camu (Myrciaria dubia) coproducts powders. *LWT* **2020**, 110472. [CrossRef]

36. Batista, A.P.; Niccolai, A.; Fradinho, P.; Fragoso, S.; Bursic, I.; Rodolfi, L.; Biondi, N.; Tredici, M.R.; Sousa, I.; Raymundo, A. Microalgae biomass as an alternative ingredient in cookies: Sensory, physical and chemical properties, antioxidant activity and in vitro digestibility. *Algal Res.* **2017**, *26*, 161–171. [CrossRef]

37. Nissen, L.; Samaei, S.P.; Babini, E.; Gianotti, A. Gluten free sourdough bread enriched with cricket flour for protein fortification: Antioxidant improvement and Volatilome characterization. *Food Chem.* **2020**, *333*, 127410. [CrossRef]

38. Mlcek, J.; Borkovcova, M.; Bednarova, M. Biologically active substances of edible insects and their use in agriculture, veterinary and human medicine—A review. *J. Cent. Eur. Agric.* **2014**, *15*, 225–237. [CrossRef]

39. Zielińska, E.; Karaś, M.; Jakubczyk, A.; Zieliński, D.; Baraniak, B. Edible Insects as Source of Proteins. In *Bioactive Molecules in Food*; Springer: Cham, Switzerland, 2019; pp. 389–441. ISBN 9783319545288.

40. Vercruysse, L.; Smagghe, G.; Herregods, G.; Van Camp, J. ACE inhibitory activity in enzymatic hydrolysates of insect protein. *J. Agric. Food Chem.* **2005**, *53*, 5207–5211. [CrossRef]

41. Cito, A.; Botta, M.; Francardi, V.; Dreassi, E. Insects as source of angiotensin converting enzyme inhibitory peptides. *J. Insects as Food Feed* **2017**, *3*, 231–240. [CrossRef]

42. Tao, M.; Wang, C.; Liao, D.; Liu, H.; Zhao, Z.; Zhao, Z. Purification, modification and inhibition mechanism of angiotensin I-converting enzyme inhibitory peptide from silkworm pupa (Bombyx mori) protein hydrolysate. *Process Biochem.* **2017**, *54*, 172–179. [CrossRef]

43. Zielińska, E.; Karaś, M.; Baraniak, B.; Jakubczyk, A. Evaluation of ACE, α-glucosidase, and lipase inhibitory activities of peptides obtained by in vitro digestion of selected species of edible insects. *Eur. Food Res. Technol.* **2020**, 1–9. [CrossRef]

44. Chatsuwan, N.; Nalinanon, S.; Puechkamut, Y.; Lamsal, B.P.; Pinsirodom, P. Characteristics, Functional Properties, and Antioxidant Activities of Water-Soluble Proteins Extracted from Grasshoppers, Patanga succincta and Chondracris roseapbrunner. *J. Chem.* **2018**, *2018*. [CrossRef]

45. Monro, J.A.; Wallace, A.; Mishra, S.; Eady, S.; Willis, J.A.; Scott, R.S.; Hedderley, D. Relative glycaemic impact of customarily consumed portions of eighty-three foods measured by digesting in vitro and adjusting for food mass and apparent glucose disposal. *Br. J. Nutr.* **2010**, *104*, 407–417. [CrossRef]

46. Jenkins, D.J.; Kendall, C.W.; Augustin, L.S.; Franceschi, S.; Hamidi, M.; Marchie, A.; Axelsen, M. Glycemic index: Overview of implications in health and disease. *Am. J. Clin. Nutr.* **2002**, *76*, 266S–273S. [CrossRef] [PubMed]

47. López-Barón, N.; Gu, Y.; Vasanthan, T.; Hoover, R. Plant proteins mitigate in vitro wheat starch digestibility. *Food Hydrocoll.* **2017**, *69*, 19–27. [CrossRef]

48. Kaźmierczak, M. *Technologie Produkcji Cukierniczej. Wyroby Cukiernicze*; WSiP: Warsaw, Poland, 2019; ISBN 9788302147197.

49. Marzec, A.; Kowalska, H.; Gasowski, W. Właściwości mechaniczne ciastek biszkoptowych o zróżnicowanej porowatości. *Acta Agrophys.* **2010**, *16*, 359–368.

50. AOAC International. *Official Methods of Analysis of AOAC International*; AOAC International: Gaithersburg, MD, USA, 2010; pp. 2–4.

51. Sarıçoban, C.; Yılmaz, M. Modelling the Effects of Processing Factors on the Changes in Colour Parameters of Cooked Meatballs Using Response Surface Methodology. *World Appl. Sci. J.* **2010**, *9*, 14–22.

52. Brand-Williams, W.; Cuvelier, M.E.; Berset, C. Use of a free radical method to evaluate antioxidant activity. *LWT-Food Sci. Technol.* **1995**, *28*, 25–30. [CrossRef]

53. Re, R.; Pellegrini, N.; Proteggente, A.; Pannala, A.; Yang, M.; Rice-Evans, C. Antioxidant activity applying an improved ABTS radical cation decolorization assay. *Free Radic. Biol. Med.* **1999**, *26*, 1231–1237. [CrossRef]

54. Soong, Y.Y.; Tan, S.P.; Leong, L.P.; Henry, J.K. Total antioxidant capacity and starch digestibility of muffins baked with rice, wheat, oat, corn and barley flour. *Food Chem.* **2014**, *164*, 462–469. [CrossRef]

55. Reis, S.F.; Abu-Ghannam, N. Antioxidant capacity, arabinoxylans content and invitro glycaemic index of cereal-based snacks incorporated with brewer's spent grain. *LWT-Food Sci. Technol.* **2014**, *55*, 269–277. [CrossRef]

56. Goñi, I.; Garcia-Alonso, A.; Saura-Calixto, F. A starch hydrolysis procedure to estimate glycemic index. *Nutr. Res.* **1997**, *17*, 427–437. [CrossRef]

Sample Availability: Samples of the insects hydrolysates and extracts are available from the authors.

Publisher's Note: MDPI stays neutral with regard to jurisdictional claims in published maps and institutional affiliations.

Article

Low-Field NMR Study of Shortcake Biscuits with Cricket Powder, and Their Nutritional and Physical Characteristics

Krzysztof Smarzyński [1], Paulina Sarbak [1], Przemysław Łukasz Kowalczewski [2], Maria Barbara Różańska [2], Iga Rybicka [3], Katarzyna Polanowska [2], Monika Fedko [4], Dominik Kmiecik [2], Łukasz Masewicz [5], Marcin Nowicki [6], Jacek Lewandowicz [7], Paweł Jeżowski [8], Miroslava Kačániová [9,10], Mariusz Ślachciński [8], Tomasz Piechota [11] and Hanna Maria Baranowska [5,*]

1. Students' Scientific Club of Food Technologists, Poznań University of Life Sciences, 31 Wojska Polskiego St., 60-624 Poznań, Poland; krzysztof.smarzynski@gmail.com (K.S.); paulina.sarbak@onet.pl (P.S.)
2. Department of Food Technology of Plant Origin, Poznań University of Life Sciences, 31 Wojska Polskiego St., 60-624 Poznań, Poland; przemyslaw.kowalczewski@up.poznan.pl (P.Ł.K.); maria.rozanska@up.poznan.pl (M.B.R.); katarzyna.polanowska@up.poznan.pl (K.P.); dominik.kmiecik@up.poznan.pl (D.K.)
3. Department of Technology and Instrumental Analysis, Poznań University of Economics and Business, Al. Niepodległości 10, 61-875 Poznań, Poland; iga.rybicka@ue.poznan.pl
4. Department of Gastronomy Science and Functional Food, Poznań University of Life Sciences, 31 Wojska Polskiego St., 60-634 Poznań, Poland; monika.fedko@up.poznan.pl
5. Department of Physics and Biophysics, Poznań University of Life Sciences, 38/42 Wojska Polskiego St., 60-637 Poznań, Poland; lukasz.masewicz@up.poznan.pl
6. Institute of Agriculture, University of Tennessee, 370 Plant Biotechnology Building, 2505 EJ Chapman Drive, Knoxville, TN 37996-4560, USA; mnowicki@utk.edu
7. Department of Production Management and Logistics, Poznan University of Technology, 2 Jacka Rychlewskiego St., 60-965 Poznań, Poland; jacek.lewandowicz@put.poznan.pl
8. Institute of Chemistry and Technical Electrochemistry, Poznan University of Technology, Berdychowo 4, 60-965 Poznań, Poland; pawel.jezowski@put.poznan.pl (P.J.); mariusz.slachcinski@put.poznan.pl (M.Ś.)
9. Department of Fruit Sciences, Viticulture and Enology, Faculty of Horticulture and Landscape Engineering, Slovak University of Agriculture in Nitra, Tr. A. Hlinku 2, 949 76 Nitra, Slovakia; miroslava.kacaniova@gmail.com
10. Department of Bioenergy and Food Technology, Institute of Food Technology and Nutrition, University of Rzeszow, Cwiklinskiej 1, 35-601 Rzeszow, Poland
11. Department of Agronomy, Poznań University of Life Sciences, 11 Dojazd St., 60-631 Poznań, Poland; tomasz.piechota@up.poznan.pl
* Correspondence: hanna.baranowska@up.poznan.pl

Citation: Smarzyński, K.; Sarbak, P.; Kowalczewski, P.Ł.; Różańska, M.B.; Rybicka, I.; Polanowska, K.; Fedko, M.; Kmiecik, D.; Masewicz, Ł.; Nowicki, M.; et al. Low-Field NMR Study of Shortcake Biscuits with Cricket Powder, and Their Nutritional and Physical Characteristics. *Molecules* **2021**, *26*, 5417. https://doi.org/10.3390/molecules26175417

Academic Editor: Davide Bertelli

Received: 28 July 2021
Accepted: 1 September 2021
Published: 6 September 2021

Publisher's Note: MDPI stays neutral with regard to jurisdictional claims in published maps and institutional affiliations.

Abstract: The growing human population renders challenges for the future supply of food products with high nutritional value. Here, we enhanced the functional and nutritional value of biscuits, a popular sweet snack, by replacing the wheat flour with 2%, 6%, or 10% (w/w) cricket powder. Consumer acceptance ratings for reference and 2% augmented cookies were comparable, whereas the higher levels of enhancement received inferior consumer scores. This relatively small change in biscuit recipe provided significant and nutritionally desirable enhancements in the biscuits, observed in a series of analyses. An increase in the protein content was observed, including essential amino acids, as well as minerals and fat. This conversion also affected the physical properties of the biscuits, including hardness, and water molecular dynamics measured by ^{1}H NMR. Cricket powder-augmented biscuits join the line of enhanced, functionally superior food products. This and similar food augmentation provide a viable scenario to meet the human food demands in the future.

Keywords: *Acheta domesticus*; amino acids composition; cookies with insects; edible insects; fatty acids; nutritional value; minerals; ^{1}H NMR; water dynamics

1. Introduction

The growing awareness of consumers regarding proper nutrition makes them look for food that is right for them. A balanced diet ensures that the demand for macronutrients is met in the right proportions. It is important that the consumed food meets the energy and physiological needs in the right portions, but at the same time ensuring a sufficient amount of micronutrients necessary for the proper functioning of our body [1–4]. Numerous recent studies have focused on the enrichment of food products in various bioactive compounds [5]. Improving the nutritional value was also attempted by decreasing the fat and sugar content or increasing the protein content [6–8]. The changing demands of consumers looking for a "healthier" snack led to attempts to improve its nutritional value and functional properties. Even though biscuits are not regarded as a healthy choice, they are eagerly consumed around the world. The market of biscuits is constantly growing, which was particularly observed in last months of COVID-19, e.g., in the United States [9,10]. Additionally, in the United Kingdom, the average consumption of confectionery products remains at the high level of 123–137 g per person per week in 2008–2019 [11]. Therefore, biscuits, being one of the world's most popular staple sweets, are considered a convenient food matrix for modification of their recipe by incorporation of various ingredients. Improvements in their nutritional value is achieved by adding whole grains or raw materials rich in dietary fiber, as well as by increasing the content of protein or minerals [12–14].

The FAO-estimated population growth to 9 billion in 2050 poses new challenges for food producers [15]. One of the main challenges will be to provide not only the right amount of food, but also an adequate supply of protein. The application of an unconventional source of protein—cricket powder (CP)—seems a promising approach to food for fortification with protein, vitamins, minerals such as Ca, Mg, K, Fe, Cu, Mn, and Zn, and dietary fiber [16–18]. The replacement of wheat flour with CP affects changes in the quality and digestibility of the product's protein as well as the desirable essential amino acids profile [19]. Our previous investigations were conducted on enhancing the nutritional value of various food products: muffins, gluten-free bread, pasta, and pork pâtés by their supplementation with CP [20–23]. Interestingly, texture analysis showed that in the case of gluten-free bread, the replacement of starch by CP in the amount of up to 6% resulted in a reduction in firmness, likely due to the emulsifying properties of cricket proteins [22]. A similar observation was also found for muffins [20]. Moreover, the addition of CP reduced cooking losses and caused a significant increase in the firmness of cooked pasta samples, underscoring the high quality of the CP-enriched pasta [24,25].

LF NMR is a method designed to study the dynamics of protons, that can be employed in numerous applications [26]. Recently, there has been increasing interest in the application of LF NMR for food analysis [27,28]. The main reason for that is the possibility to study different processes in model food systems, including gelatinization [29,30], retrogradation [31] or hydratation of starch [32], lipid oxidation [33], and enzymatic modification of proteins [34]. Moreover, it is useful in the analysis of complex food matrices, as proton fractions of water, lipids, or polysaccharides tend to form separate populations that relax at significantly different rates. This allows for observation of interactions that may occur as the product ages or is reformulated. Therefore, LF NMR has proved to be an useful tool in the quality design of emulsions [35,36], bread [37], dough [38], pâté [21,39], and many other food products [40]. Considering the advantages of CP, the importance of enriching food products, and the many changes induced by enrichment and the usefulness of the LF NMR technique in the analysis of food, this investigation was carried out to evaluate how an addition of various levels of CP influenced the nutritional value, consumer acceptance, textural properties, and water behavior on a molecular level of shortcake biscuits.

2. Results and Discussion

2.1. Consumer Study

The use of insects to enrich food may be negatively perceived by consumers [41]. It is extremely important to raise consumer awareness and identify potential health benefits [42–44]. For this reason, this study aimed to assess what level of replacement of wheat flour (WF) with CP in biscuits would be acceptable (Figure 1). There were no changes in the ratings of taste, texture, appearance, or the overall desirability of CP2 biscuits compared to reference biscuits (R); however, further increases in the replacement of WF with CP resulted in a significant reduction in the consumer acceptance scores awarded. In the case of the flavor evaluation, for both CP2 and CP6, the scores were significantly higher than for R. A small addition of CP significantly improved the flavor of the cookies. Commercial CP was used in this study, but in order to obtain CP, crickets were processed sequentially before being ground, including steaming, roasting, frying, and drying [45]. Technological treatment of insects can significantly improve the aroma of the resultant CP [46], and thus increase the consumer acceptance of such enhanced products. However, a 10% replacement of WF with CP resulted in an unpleasant, odd smell, which consumers indicated as undesirable. As reported by Grossmann et al. [47], most of the volatile odor-active compounds of crickets have been described as green, earthy or potato-mushroom, but have also been associated with a description of the smell of fat, sweat, cheese or popcorn. The volatile phenols present in crickets are responsible for the smell of smoke and feces. Therefore, too high a concentration of compounds present in crickets is unacceptable. With the increase in the conversion of WF to CP, biscuits more and more resembled wholemeal flour biscuits (see Section 2.2) and although such products are commonly considered to be more healthy [48], unfortunately this did not meet with growing marks in the consumer assessment of the appearance of biscuits. The evaluation of the texture of the biscuits has also changed. As in the case of flavor and appearance, texture scores also decreased with increasing WF to CP conversion (above 6%). Gluten proteins present in WF are responsible for creating the appropriate structure of cereal products [49]. Reducing its share in biscuits with the addition of CP resulted in an increase in their crispness and brittleness compared to biscuits without CP. Replacing WF with CP in the amount of 2% did not cause any significant changes in the taste assessment. Burt et al. suggest that the primary problem with the use of crickets in food production in Western cultures is a psychological one [50]; thus, on the basis of the obtained results, the 2% addition of CP could be fully acceptable, and the obtained shortcake biscuits could be successfully introduced to the market.

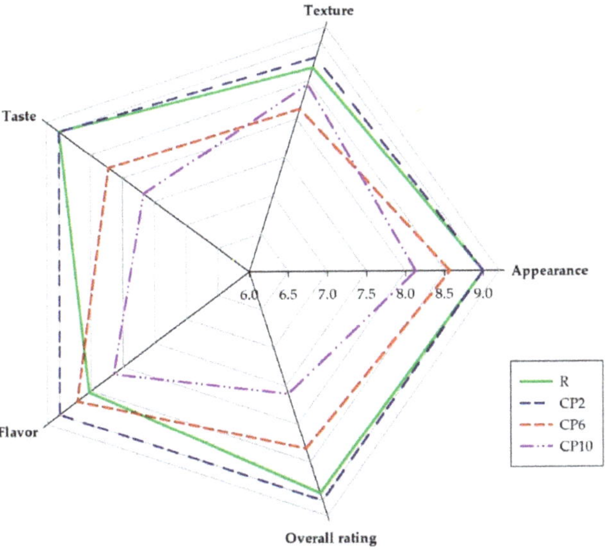

Figure 1. Results of consumer acceptance study. R—reference biscuits; CP2, CP6, and CP10—biscuits with 2%, 6%, and 10% of wheat flour replacement with CP, respectively.

2.2. Biscuits Appearance

The use of CP in the recipe of shortcake biscuits caused changes in the consumer assessment, including when assessing for their appearance. This may be due to the discoloration of the final products, readily visible to the naked eye (Figure 2).

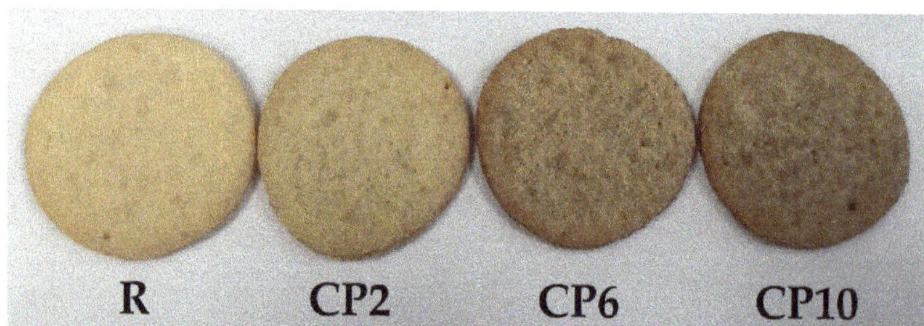

Figure 2. Biscuits with CP: R—reference biscuits; CP2, CP6, and CP10—biscuits with 2%, 6%, and 10% of wheat flour replacement with CP, respectively.

The color component analysis showed a significant darkening of the biscuits with CP (Table 1). The more WF was replaced with CP, the darker the biscuits became (lowering the L* value). The increase in the proportion of protein and the high-temperature process of baking the biscuits caused the formation of colored melanoid-forming products, among others in a Maillard reaction, but biscuits also can become darker due to the carbohydrate transformation including caramelization [51,52]. It was observed that the color of biscuits with the addition of CP gradually shifted in the red and blue directions (increase in red saturation (a*) and a decrease in yellow saturation (b*)). This phenomenon contributed to the initial increase in the whiteness index of the biscuits (CP2), that was neglected at higher CP content due to a more substantial lightness decrease. Similar changes were observed by Zielińska and Pankiewicz [53] in cookies enriched with *Tenebrio molitor*, as well as in other cereal products enriched with CP [20,24,54].

Total color difference analysis (ΔE) confirmed that the color changes caused by the addition of CP are significant. According to Mokrzycki and Tatol [55], the higher the ΔE value, the easier it is to observe the color difference, and untrained people can spot slight differences above ΔE = 2.0 and clear differences above ΔE = 3.5. This was reflected by the reduced appearance scores in the consumer analysis of biscuits containing elevated amounts of CP (6 and 10%) and similar ones for and CP2 (Figure 1).

Table 1. Color parameters of cricket powder and biscuits.

Parameter	CP	R	CP2	CP6	CP10
L*	52.04 ± 0.70	75.53 ± 0.28 [a]	73.90 ± 0.28 [b]	65.98 ± 0.29 [c]	63.29 ± 0.16 [d]
a*	6.02 ± 0.20	3.12 ± 0.05 [d]	4.17 ± 0.08 [c]	4.96 ± 0.02 [b]	5.17 ± 0.06 [a]
b*	14.65 ± 1.77	25.19 ± 0.93 [a]	22.86 ± 0.37 [b]	22.42 ± 0.53 [b]	20.47 ± 0.12 [c]
ΔE	-	-	3.03	10.11	13.28
WI	49.49	64.74	65.05	58.96	57.65

Mean values in biscuits samples with the same letters in the row ([a–d]) were not significantly different (α = 0.05). CP—cricket powder; R—reference biscuits; WI—whiteness index; CP2, CP6, and CP10—biscuits with 2%, 6%, and 10% of wheat flour replacement with CP, respectively.

2.3. Nutritional Value

Insects are widely described as a good source of protein, fat and minerals [56–58], so the use of CP can improve the nutritional value of a biscuit recipe. An increase in protein, fat, and ash content was observed, along with an increase in the conversion of WF to CP (Table 2). The consequence of the observed increases in their content was a gradual reduction in carbohydrate content. Additionally, an increase in the energy value of the cookies was observed. The most desirable biscuit, CP2, had more energy, fat, and protein than an average commercial biscuit. The 50 g portion of CP2 biscuits (about seven pieces) had realized 12% of the reference intake for energy and fat and, 8% for protein, and 14% for carbohydrates, being a nutritionally attractive sweet snack [59]. The moisture content did not differ statistically significantly.

Table 2. Proximate composition and energy value.

Parameter	R	CP2	CP6	CP10
Moisture (%)	1.75 ± 0.49 [a]	1.65 ± 0.21 [a]	2.08 ± 0.31 [a]	1.75 ± 0.19 [a]
Protein (%)	6.08 ± 0.08 [d]	7.80 ± 0.24 [d]	9.24 ± 0.18 [b]	10.30 ± 0.09 [a]
Fat (%)	14.7 ± 0.4 [d]	16.2 ± 0.1 [b]	16.8 ± 0.2 [b]	17.5 ± 0.4 [a]
Ash (%)	1.01 ± 0.12 [b]	1.03 ± 0.03 [b]	1.09 ± 0.15 [b]	1.35 ± 0.09 [a]
Carbohydrates [1] (%)	76.5 ± 1.14 [a]	73.3 ± 1.03 [c]	70.8 ± 1.01 [b]	69.1 ± 1.15 [d]
Energy value [2] (kcal/100 g)	454.5 [d]	462.4 [b]	461.5 [b]	466.8 [a]

[1] The carbohydrate content was estimated by subtracting the average content of ash, fat, and protein from 100%. [2] Energy value was calculated based on the average moisture, protein, fat, and carbohydrate content. Mean values with the same letters in the row ([a–d]) were not significantly different (α = 0.05). R—reference biscuits; CP2, CP6, and CP10—biscuits with 2%, 6%, and 10% of wheat flour replacement with CP, respectively.

The content of minerals: Ca, Mg, K, Na, Cu, Fe, Mn, and Zn is presented in Table 3. As CP is an important source of minerals, its addition to biscuits increased their elemental profile (except for Na) [19]. The content of most minerals was higher in biscuits with CP addition than in a commercial sample. The most significant differences between CP2 and a control sample were noticed for Ca (23%), Fe (12%), Mn (14%), and Zn (16%). For Mg and K the content changed by 6% and 7%, respectively. However, a nutritional claim on "source of mineral" could only apply to Mn in CP6 or Cu, Mn, and Zn in CP10 which were scored significantly lower in a sensory test [59,60]. The content of Na was comparable (302–323 mg/100 g) in all biscuits and resulted from the salt (sodium chloride) addition to the biscuit dough. Generally, the worldwide intake of Na is above nutritional recommendations, so it is suggested to lower its content in food products [61]. On the other hand, salt plays an extremely important role in sensory attributes of food products, so it is added to most of food categories. All products under the study, despite delivering 10–11% of adequate intake (AI) for Na in 50 g portion, would fulfill the clearly defined and rigorous latest British targets for salt reduction (maximum of 340 mg of Na/100 g in a category of biscuits) [62]. Moreover, the portion of CP2 biscuits provided 10% of nutrient reference value/adequate intake (NRV/AI) for Mn, 4% for Zn, 2% for Ca, Cu, Fe, and K, and 1% for Mg. In general, bakery confectionary products are not regarded as a source of minerals, so those developed with CP addition can be regarded as a healthier option than commercial ones.

Table 3. Mineral composition (expressed as mg per 100 g of biscuits).

Mineral	NRV/AI (mg/Day)	R (mg/100)	CP2 (mg/100 g)	CP6 (mg/100 g)	CP10 (mg/100 g)
Ca	800	31.4 ± 1.8 [d]	38.5 ± 2.0 [c]	53.2 ± 0.9 [b]	67.0 ± 3.2 [a]
Mg	375	10.4 ± 0.1 [d]	11.1 ± 0.2 [c]	13.6 ± 0.1 [b]	17.6 ± 1.0 [a]
K	3500	102.6 ± 1.0 [d]	109.8 ± 2.2 [c]	137.8 ± 4.4 [b]	152.3 ± 11.0 [a]
Na	1500	323.1 ± 10.3 [a]	310.2 ± 9.1 [b]	302.0 ± 8.3 [b]	310.6 ± 11.9 [b]
Cu	1	0.021 ± 0.001 [d]	0.044 ± 0.004 [c]	0.106 ± 0.007 [b]	0.196 ± 0.006 [a]
Fe	14	0.536 ± 0.017 [d]	0.602 ± 0.014 [c]	0.662 ± 0.027 [b]	0.786 ± 0.039 [a]
Mn	2	0.191 ± 0.004 [d]	0.216 ± 0.009 [c]	0.310 ± 0.010 [b]	0.365 ± 0.008 [a]
Zn	10	0.706 ± 0.003 [d]	0.819 ± 0.051 [c]	1.23 ± 0.08 [b]	1.61 ± 0.07 [a]

NRV—nutrient reference value (for Ca, Mg, Cu, Fe, Mn, Zn), AI—adequate intake (for K, Na); Mean values with the same letters in the row ([a–d]) were not significantly different (α = 0.05). R—reference biscuits; CP2, CP6, and CP10—biscuits with 2%, 6%, and 10% of wheat flour replacement with CP, respectively.

The literature data indicate that crickets are a good source of fat; therefore, a change in the fatty acid profile in finished products was expected. Udomsil et al. [63] indicated that in the fats of crickets, the most abundant are saturated fatty acids (SFA), mainly palmitic acid (C16:0) and stearic acid (C18:0), and of monounsaturated fatty acids (MUFA), oleic acid (C18:1). There are also polyunsaturated fatty acids (PUFA), such as linolenic acid (C18:3) and linoleic acid (C18:2). Importantly, other studies have shown that the fatty acid profile does not differ across the tissues of the cricket that are eaten [64]. The test results confirmed the expected changes in the fatty acid profile (Table 4). A slight increase in the share of MUFA and PUFA was observed along with the increase in the replacement of WF with CP. Unfortunately, due to the use of large amounts of baking margarine in the recipe of cookies (see Section 3.1) produced from vegetable oils in varying proportions (palm, rapeseed, sunflower), it cannot be concluded that the nutritional value of CP cookies in the context of fatty acids is improved. Nevertheless, it can be expected that, similar to other low-fat products (e.g., pasta or bread), it will be possible to improve the nutritional value of the biscuits.

Table 4. Fatty acid composition of biscuits enriched with CP (as a percentage of total fatty acids).

Fatty Acid	R	CP2	CP6	CP10
C 8:0	0.482 ± 0.002 [a]	0.477 ± 0.016 [b]	0.479 ± 0.013 [b]	0.487 ± 0.001 [a]
C 10:0	0.463 ± 0.005 [b]	0.453 ± 0.013 [b]	0.448 ± 0.009 [a]	0.448 ± 0.005 [a]
C 12:0	6.340 ± 0.002 [a]	6.179 ± 0.064 [b]	6.102 ± 0.085 [b]	6.100 ± 0.015 [b]
C 14:0	2.863 ± 0.004 [a]	2.837 ± 0.008 [b]	2.810 ± 0.013 [b]	2.803 ± 0.002 [b]
C 16:0	31.223 ± 0.019 [b]	31.629 ± 0.063 [a]	31.560 ± 0.033 [a]	31.343 ± 0.197 [b]
C 16:1	0.159 ± 0.045 [a]	0.130 ± 0.001 [b]	0.129 ± 0.003 [b]	0.133 ± 0.001 [b]
C 18:0	3.855 ± 0.029 [c]	3.956 ± 0.004 [b]	4.014 ± 0.027 [b]	4.044 ± 0.001 [a]
C 18:1	31.401 ± 0.024 [a]	31.218 ± 0.025 [b]	31.123 ± 0.080 [b]	31.146 ± 0.120 [b]
C 18:2	21.311 ± 0.032 [a]	21.288 ± 0.061 [a]	21.502 ± 0.049 [b]	21.655 ± 0.042 [c]
C 18:3	1.058 ± 0.021 [c]	1.055 ± 0.046 [c]	1.093 ± 0.006 [b]	1.126 ± 0.008 [a]
C 20:0	0.667 ± 0.010 [a]	0.603 ± 0.005 [b]	0.556 ± 0.007 [c]	0.548 ± 0.065 [c]
C 22:0	0.175 ± 0.016 [b]	0.175 ± 0.025 [b]	0.182 ± 0.030 [a]	0.170 ± 0.002 [b]
Σ SFA	46.069 ± 0.016	46.308 ± 0.131	46.151 ± 0.023	45.941 ± 0.152
Σ MUFA	31.561 ± 0.069	31.347 ± 0.024	31.253 ± 0.077	31.278 ± 0.118
Σ PUFA	22.370 ± 0.053	22.344 ± 0.107	22.596 ± 0.055	22.781 ± 0.034

Mean values with the same letters in the row ([a–c]) were not significantly different (α = 0.05). R—reference biscuits; CP2, CP6, and CP10—biscuits with 2%, 6%, and 10% of wheat flour replacement with CP, respectively.

The amino acid profile is presented in Table 5. Along with the increase in the amount of CP in the biscuit recipe, a higher content of all analyzed amino acids was observed, except for phenylalanine and methionine. It is well known that the major amino acids in cereal prolamins are proline and glutamine [65], which is in line with the results of our research. The lowest-content essential amino acid in grains, in particular wheat, is lysine, and next up is threonine [66]. It has been noticed that even a 2% incorporation of CP into biscuit formula led to an increase in the content of essential amino acids by 13.6%. In comparison to the control sample (R), the content of lysine in the samples CP2, CP6, CP10 increased by almost 40%, 83.5%, and 108.3%, respectively. Moreover, in the case of analyzed biscuits samples, the concentration of threonine increased by an average of 31.3%. It should be noted that the higher lysine and arginine contents led to increased susceptibility of flour to the progress of the Maillard reaction [67]. The drawback is that some of the Maillard reaction products (MRPs) are currently suspected to have deleterious health effects. The accumulation of MRPs in vivo has been implicated as a major pathogenic process in diabetic complications and other disorders, such as atherosclerosis, Alzheimer's disease, and normal aging [68].Thus, due to the possibility of the potentially harmful Maillard reaction compounds formation, it is worth noting to control their levels by the recipe's modification, e.g., adding functional ingredients and/or different flours sources, especially in cereal products such as cereal products biscuits, and bread [69].

Table 5. Amino acids profile expressed as mg per g of biscuits.

Amino Acid	R	CP2	CP6	CP10
Essential amino acids				
Histidine	1.48 ± 0.01 [d]	1.64 ± 0.03 [c]	1.90 ± 0.01 [b]	2.05 ± 0.02 [a]
Isoleucine	2.65 ± 0.17 [c]	2.78 ± 0.20 [c]	3.43 ± 0.08 [b]	3.70 ± 0.06 [a]
Leucine	4.91 ± 0.07 [d]	5.44 ± 0.21 [c]	6.37 ± 0.11 [b]	6.65 ± 0.07 [a]
Lysine	1.33 ± 0.05 [d]	1.86 ± 0.03 [c]	2.44 ± 0.05 [b]	2.77 ± 0.01 [a]
Cysteine	3.75 ± 0.08 [d]	4.36 ± 0.01 [a]	4.10 ± 0.05 [b]	3.93 ± 0.02 [c]
Methionine	1.12 ± 0.05 [c]	1.32 ± 0.14 [b]	1.51 ± 0.02 [a]	1.57 ± 0.03 [a]
Phenylalanine	3.31 ± 0.13 [c]	3.79 ± 0.11 [b]	4.04 ± 0.17 [ab]	4.30 ± 0.11 [a]
Tyrosine	2.27 ± 0.02 [d]	2.65 ± 0.03 [c]	3.12 ± 0.01 [b]	3.39 ± 0.01 [a]
Threonine	2.33 ± 0.02 [d]	2.58 ± 0.07 [c]	3.15 ± 0.01 [b]	3.45 ± 0.02 [a]
Tryptophan	0.102 ± 0.003 [d]	0.138 ± 0.003 [c]	0.356 ± 0.002 [b]	0.399 ± 0.005 [a]
Valine	2.93 ± 0.04 [d]	3.18 ± 0.09 [c]	4.07 ± 0.01 [b]	4.48 ± 0.04 [a]
Σ EAA *	26.18	29.74	34.49	36.69
Dispensable amino acids				
Alanine	1.23 ± 0.01 [d]	1.78 ± 0.04 [c]	2.30 ± 0.01 [b]	2.94 ± 0.01 [a]
Arginine	2.13 ± 0.06 [d]	2.48 ± 0.05 [c]	3.26 ± 0.02 [b]	3.79 ± 0.02 [a]
Aspartic acid	4.70 ± 0.11 [d]	6.05 ± 0.04 [c]	6.84 ± 0.05 [b]	8.89 ± 0.06 [a]
Glutamic acid	22.02 ± 0.12 [d]	23.19 ± 0.28 [c]	22.70 ± 0.08 [b]	24.79 ± 0.08 [a]
Glycine	2.14 ± 0.01 [d]	2.53 ± 0.04 [c]	3.16 ± 0.02 [b]	3.64 ± 0.02 [a]
Proline	6.86 ± 0.04 [d]	7.23 ± 0.03 [c]	7.87 ± 0.06 [b]	8.08 ± 0.02 [a]
Serine	3.76 ± 0.01 [d]	4.41 ± 0.12 [c]	4.73 ± 0.03 [b]	5.12 ± 0.01 [a]
Σ DAA *	42.84	47.67	50.86	57.25

* sums were calculated from the mean values. Mean values with the same letters in the row ([a–d]) were not significantly different ($\alpha = 0.05$). R—reference biscuits; CP2, CP6, and CP10—biscuits with 2%, 6%, and 10% of wheat flour replacement with CP, respectively.

2.4. Physical Properties

The physical properties of the obtained biscuits were analyzed by characterizing their dimensions, weight, and texture (Table 6). The weight and thickness of the biscuits obtained did not differ significantly across the variants ($\alpha = 0.05$); however, it was found that the addition of CP caused an increase in the diameter of the biscuits. Gluten proteins present in WF (replaced with CP) are responsible for the proper consistency and structure of the dough [70,71]. The observed increase in diameter may be caused by a reduction in the content of gluten proteins in the dough, which does not maintain its shape during preparation and baking. One of the parameters describing the quality of shortcake biscuits is the spread ratio. The larger the diameter to thickness ratio, the better the biscuit quality [72]. The overall spread ratio increased with the addition of CP and ranged from 6.30 for R to 7.75 for CP10. A significantly lower spread ratio in the case of R may result from a stronger binding by the action of gluten proteins, creating a dough with higher compactness. Literature data indicated that the addition of vegetable proteins, which bind water and other biopolymers, reduced the spread ratio and, on the other hand, increased the thickness of the biscuits [73–75]. According to Kulkarnia et al. [76], an increase in the biscuits spread ratio may indicate a poor connection of the protein and carbohydrate networks in the biscuits. These two components are important nutrients, but from a physical point of view, their interaction with one another can cause changes in the hardness of the biscuits. As expected, it was noted that replacing WF with CP resulted in a successive reduction in the hardness of the biscuits from 29.44 N for R to 24.50 N for CP10. The reduction in hardness can be explained by the uneven mixing process and the potential uneven distribution of the added ingredients, which may result in limiting the availability of water for proteins, which should be hydrated during the preparation of the dough. Too little water or additional dough ingredients such as fat and sugar prevent the proteins from being properly hydrated. The dough from which the biscuits are made is high in both sugar and fat and low in water, resulting in a dough with a sticky and consistent character and, consequently, increased hardness [77,78]. These results are in line with other studies by Ho and Abdul-Latif [74] and Chauhan et al. [79] who noted that replacing WF with other flours, and thus reducing the amount of gluten in the dough, also resulted in a reduction in the hardness of the biscuits.

Table 6. Physical properties of biscuits.

Parameter	R	CP2	CP6	CP10
Weight (g)	7.21 ± 0.30 [a]	7.55 ± 0.37 [a]	7.54 ± 0.45 [a]	7.46 ± 0.29 [a]
Diameter (cm)	4.54 ± 0.21 [b]	4.63 ± 0.30 [b]	4.79 ± 0.17 [ab]	4.88 ± 0.18 [a]
Thickness (cm)	0.72 ± 0.06 [a]	0.65 ± 0.05 [a]	0.65 ± 0.07 [a]	0.63 ± 0.06 [a]
Spread Ratio (–)	6.30 ± 0.12	7.12 ± 0.03	7.40 ± 0.08	7.75 ± 0.09
Firmness (N)	29.44 ± 3.07 [a]	25.44 ± 6.80 [ab]	25.22 ± 5.16 [ab]	24.50 ± 2.56 [b]

Mean values with the same letters in the row ([a–b]) were not significantly different (α = 0.05). R—reference biscuits; CP2, CP6, and CP10—biscuits with 2%, 6%, and 10% of wheat flour replacement with CP, respectively.

2.5. Water Behavior

Measurements of the relaxation parameters revealed two CPMG (Carr-Purcell-Meiboom-Gill) proton populations and one FID (free induction decay) proton population. This is expected for low moisture products that are rich in carbohydrates and fats. In fresh dough samples with water content significantly above 50%, up to three CPMG proton populations can be observed T_{21} (<10 ms), T_{22} (20–50 ms) and T_{23} (>100 ms), namely tightly, less tightly, and weakly bound water, respectively. As the water content in dough decreases below 50%, the T_{23} component disappears, as there is no longer an excess of water in the system. Moreover, T_{21} and T_{22} components tend to merge, forming one proton population [38]. This is not the case for shortcake biscuits, as both short T_{21} (Figure 3A) and long T_{22} (Figure 3B) components of spin–spin relaxation time could be observed. Shortcake biscuits are characterized by a very low water content <2%, so one can expect that it will be bound very "tightly", meaning that the T_{21} will correspond to the amount of water present in the system. Therefore, T_{22} will rather correspond to the amount of starch and fat in the system as those ingredients are present in large quantities and are the most proton abundant. This is in accordance with literature data, as for pure fat or fat in emulsion, relaxation times are estimated between 40–100 ms [36], whereas for pure starch the relaxation time may range between 40–180 ms (depending on water content) [80]. The presence of one spin–lattice relaxation time, T_1 (Figure 3C), is once again conditioned by the low amount of water in the system. Starches at hydration levels below 10% are characterized by only one component of spin–lattice relaxation time; above that value, when bulk water starts to be present in the system, a long component of relaxation time T_{12} can be separated [80].

A reduction in the value of short components of the spin–spin relaxation times T_{21} is observed in the samples containing CP, compared to the reference sample R. This indicates limiting the dynamics of water molecules bound to the polymer matrix. This phenomenon may be attributed to the inclusion of cricket proteins as the behavior of water in food is significantly affected by the solubility of proteins, which consists of hydrophobic (protein–protein) and hydrophilic (protein–solvent) interactions [81]. Literature data indicate that CP is hydrophilic in nature [82], which limits the amount of water hydrating the proteins and starch of WF [22]. This corresponds to changes in firmness (Table 6), as it has followed the same manner as T_{21}, suggesting that the addition of CP that causes decrease of water mobility results in softer texture of obtained biscuits, which were more fragile.

In contrast to short components, the long components of spin–spin relaxation time increased in samples where part of the WF was replaced with CP. This is the effect of an increase in fat content in samples containing more CP. The lack of a further increase in T_{22} with the increase in CP should be attributed to an overall lower amount of carbohydrates and fats. This is because of the fact that fat is a more proton-dense ingredient than starch, whereas a 10% replacement of WF with CP results in an over 5% reduction in sum of carbohydrates and fats.

From the point of view of the molecular properties of water, replacing a part of WF with CP reduces the binding of H_2O molecules with biopolymers. This is normally manifested by an increase in the value of spin–lattice T_1 relaxation times [83,84]; however, in shortcake biscuits, water molecules are present in relatively small quantities in comparison to starch or fat.

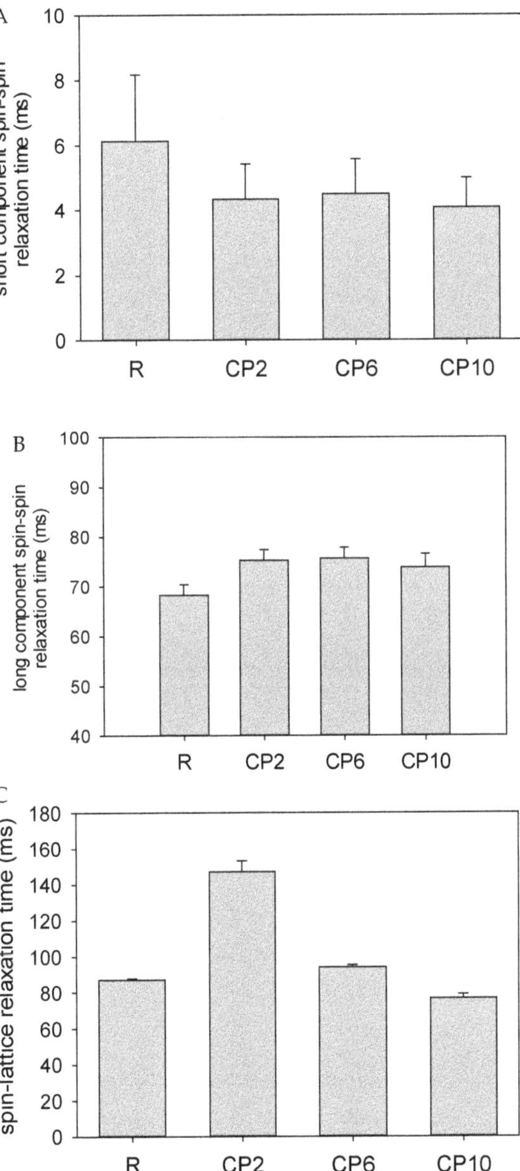

Figure 3. Results of relaxation times. R—reference biscuits; CP2, CP6, and CP10—biscuits with 2%, 6%, and 10% of wheat flour replacement by CP, respectively. (**A**)—results of short component spin–spin relaxation time. (**B**)—results of long component spin–spin relaxation time. (**C**)—results of spin–lattice relaxation time.

The smallest 2% replacement of WF with CP caused a significant increase in the T_1 value compared with the R. This should be normally interpreted as an increase in the amount of bulk water compared to bound one, but CP2 was characterized by the lowest water content, so one should assume that any water present will be completely bound. In the samples CP6 and CP10, the values were comparable with those observed for R. This result does not allow for an unambiguous interpretation of the effect of CP and the removal of part of the WF on quantitative changes in water binding in the recipe-modified cookie; however, these irregular changes in the values of spin-lattice relaxation times are confirmed by the results of the equilibrium analysis of the water activity (a_r) of the biscuits (Table 7). Taking into consideration the changes in water activity, water content, and spin–lattice relaxation time, it can be concluded that the sole implementation of CP in the recipe of WF shortcake biscuits causes interactions that decrease the binding of water at a molecular level. However, an increase in the replacement ratio of WF to CP reverses this effect. Although, due to low water content in the final product, this phenomenon was not reflected in texture analysis, it was noticed by consumers, as indicated by texture acceptance.

Table 7. Results of water activity in biscuits.

Parameter	R	CP2	CP6	CP10
water activity a_r (-)	0.3123 ± 0.0012 [b]	0.4098 ± 0.0012 [a]	0.2522 ± 0.0038 [c]	0.1940 ± 0.0008 [d]
transport rate V_D (s^{-1})	0.0296 ± 0.0022 [a]	0.0228 ± 0.0023 [c]	0.0260 ± 0.0031 [b]	0.0263 ± 0.0023 [b]

Mean values with the same letters in the row ([a-d]) were not significantly different ($\alpha = 0.05$). R—reference biscuits; CP2, CP6, and CP10—biscuits with 2%, 6%, and 10% of wheat flour replacement with CP, respectively.

A correlation was found between T_1 and a_r (Figure 4). The increase in the equilibrium water activity in the product determines the increase in the amount of bulk water compared to the amount of bound water. The mobility of the molecules of both water fractions is reflected in the values of the spin–spin relaxation time components. Linear correlations were found between the transport rate of water in the system (V_D) and the mobility of rotational movements of bulk and bound water molecules (Figure 5). As the translational movement rate of the water molecules in the product increases, the possibility of rotational movements of the water molecules in the bulk fraction decreases, and at the same time, the bound fraction molecules achieve a greater possibility of rotational movements around the water–polymer matrix bond.

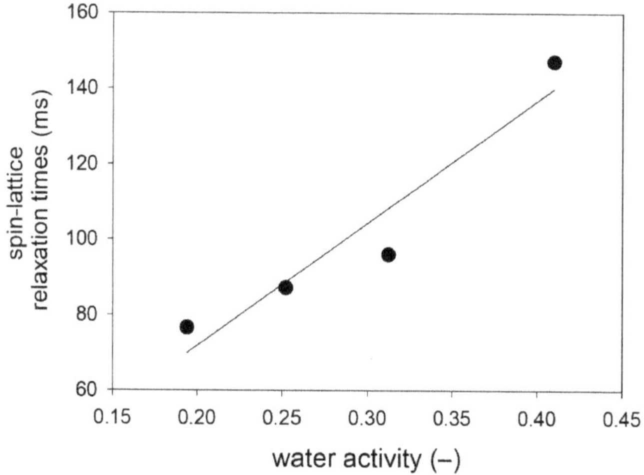

Figure 4. Correlation between water activity and spin–lattice relaxation times T_1 (Pearson r = 0.900; *p* = 0.050).

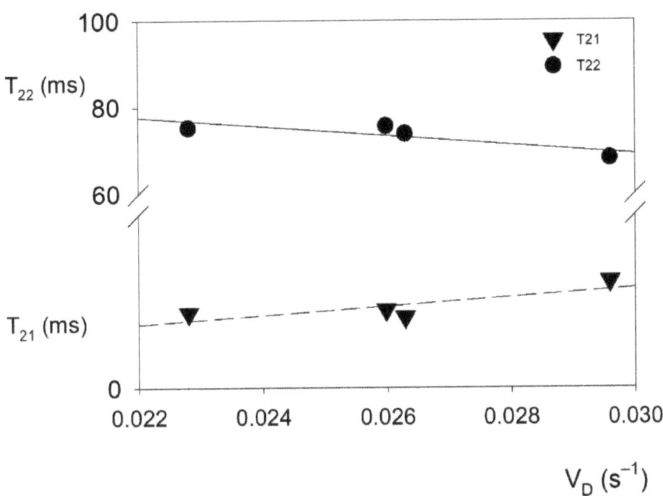

Figure 5. Linear correlation between transport rate (V_D) and short (T_{21}) and long components (T_{22}) of the spin–spin relaxation times. (Pearson $r = -0.849$; $p = 0.076$ and $r = 0.801$; $p = 0.100$ for T_{21} and T_{22}, respectively).

3. Materials and Methods

3.1. Shortcake Biscuits Manufacturing

The recipe for reference biscuits (denoted as R in the text) was as follows: 200 g wheat flour (type 500) (GoodMills Polska sp. z o.o., Grodzisk Wielkopolski, Poland), 64 g white sugar (Pfeifer & Langen Polska S.A., Środa Wielkopolska, Poland), 20 g brown sugar (Pfeifer & Langen Polska S.A., Środa Wielkopolska, Poland), 2 g milk powder (SM Mlekovita, Wysokie Mazowieckie, Poland), 2.5 g salt (Kopalnia Soli 'Kłodawa' S.A., Kłodawa, Poland), 2 g baking powder (Dr. Oetker Polska Sp. z o.o., Gdańsk, Poland), 80 g baking margarine (Upfield Polska sp. z o.o., Warsaw, Poland), and 44 g water. In the test samples, wheat flour was replaced with cricket powder (Crunchy Critters, Derby, UK) in three different quantities of 2%, 6%, and 10% (w/w) and denoted as CP2, CP6, and CP10, respectively. The amounts of other components were unchanged. The composition of CP (determined in a previously published study [16]) is: protein 42.0 ± 0.4 [%]; fat 29.1 ± 0.6 [%]; 3.6 ± 0.3 [%]; fiber 3.5 ± 0.02; and carbohydrate 21.8 ± 0.8 [%]. All the dry compounds were mixed together with the KitchenAid mixer (5KPM5EWH, KitchenAid, Greenville, OH, USA) for 3 min. Water was then added and mixing continued for another 1 min. The dough was rolled into 2 mm thick sheets, rounded shapes were cut with a Ø60 mm cookie cutter and placed in an aluminum tray. The position of biscuits on baking trays was the same for all variations. Biscuits were baked at 205 °C (MIWE Condo, MIWE Michael Wenz GmbH, Amstein, Germany) for 11 min, and then allowed to cool at room temperature for 15 min. The cool biscuits were packed in polypropylene pouches and stored at room temperature in darkness until analysis.

3.2. Consumer Acceptance

The rating of consumer acceptance was assessed by using the 9-point hedonic line scale (ranging from 1 "dislike very much" to 9 "like very much") [85]. In this study, sixty-five untrained panelists, aged between twenty-six and forty-five, were invited to participate. The study involved 29 men and 36 women, students and employees of the Poznań University of Life Sciences (Poznań, Poland), who expressed a voluntary willingness to participate. Consumers were asked to evaluate the appearance, flavor, taste, texture, and overall rating of analyzed biscuits.

3.3. Color Measurements

A Chroma Meter CR-410 (Konica Minolta Sensing Inc., Tokyo, Japan) was used for the color measurements of biscuits [20]. Differences in color were recorded in CIE L*a*b* scale in terms of lightness (L*) and color (a*—redness; b*—yellowness). Analysis was repeated 10 times for each sample. The total color difference (ΔE) and whiteness index (WI) was calculated, with the R values used as baseline for all CP variants:

$$\Delta E = \sqrt{\Delta L^{*2} + \Delta a^{*2} + \Delta b^{*2}}$$
$$WI = 100 - \sqrt{(100 - L^*)^2 + a^{*2} + b^{*2}}$$

3.4. Proximate Composition and Energy Value

Determination of the moisture was carried out in accordance with the AACC 44-19.01 method [86]. Total nitrogen content was determined by the Kjeldahl method according to ISO 20483 [87] and was used to calculate the protein content by multiplying the result by the conversion factor of 5.7, suitable for wheat [88] and recommended by Ritvanen et al. [89] for crickets. The fat content was determined (Soxhlet method) according to AACC 30-25.01 [90], and ash content according to AACC Method 08-12.01 [91]. Moreover, the proximate carbohydrate content was estimated by subtracting the total fat, protein, ash, and moisture content from 100%. The ash, carbohydrate, fat, and protein contents were presented on a dry weight basis. The energy value (EV) was calculated with the following formula:

$$EV \ (kcal/100 \ g) = 4 \times protein \ (\%) + 4 \times carbohydrate \ (\%) + 9 \times fat \ (\%)$$

3.5. Minerals Content

The concentrations of the minerals Ca, Cu, Fe, K, Mg, Mn, Na, and Zn were determined using flame atomic absorption spectroscopy (F-AAS; SpectrAA-800, Varian, Palo Alto, CA, USA) preceded by microwave mineralization with nitric acid [92]. The recommendations for Ca, Cu, Fe, Mg, Mn, and Zn were established at the level of Nutrient Reference Value (NRV) [93]. The contents of minerals were expressed in g/100 g of the sample.

3.6. Amino Acid Composition

Samples before analysis of the amino acid profile were subjected to acidic hydrolysis in 6 M HCl under nitrogen at 110 °C for 24 h with modifications as reported by Kwanyuen and Burton [94]. The contents of amino acids were determined as derivatives of phenylisothiocyanate (PITC) according to the procedure described by Polanowska et al. [95] Norleucine (500 nM) was added as internal standard. The tryptophan content was examined after alkaline hydrolysis of proteins in 4 M NaOH at 110 °C for 18 h under nitrogen according to the method proposed by Çevikkalp et al. [96] The analysis was performed using LC Agilent Technologies 1200 Rapid Resolution (Santa Clara, CA, USA) system equipped with a UV-Vis detector DAD 1260 (Agilent Technologies, Santa Clara, CA, USA) and a reversed-phase column Zorbax Eclipse Plus C18 (4.6 × 150 mm, 5 μm) (Agilent Technologies, Santa Clara, CA, USA).

3.7. Fatty Acid Composition Analysis

Fat was extracted from the biscuits using the standard procedure described by Folch et al. [97] and the fatty acid composition was determined according to the AOCS Official Method Ce 1 h-05 [98] according to the parameters described in detail previously [54] with an Agilent 7820A GC (Agilent Technologies, Santa Clara, CA, USA) equipped with a flame ionization detector (FID) and SLB-IL111 capillary column (Supelco, Bellefonte, PA, USA) (100 m, 0.25 mm, 0.20 μm). The results were expressed as a percentage of total fatty acids.

3.8. Texture Analysis

A TA.XTplus texture analyzer (Stable Micro Systems Co. Ltd., Godalming, UK) equipped with a load cell of 5 kg was used to determine the texture properties of biscuits. Hardness was determined by the bend test, using the 3-point bend rig HDP/3PB (Stable Micro Systems Co. Ltd., Godalming, UK) with the following settings: pre-test speed of 5 mm/s, test speed of 3 mm/s, post-test speed of 10 mm/s, distance of 5 mm and distance between the supports of 2 cm [99]. At least 10 biscuit measurements were taken for each batch. Each biscuit was compressed once and maximum force was recorded.

3.9. LF NMR Relaxometry

Biscuit samples were placed in $\varnothing 8$ mm measuring tubes. The height of the sample in the tube was set at 15 mm. After placing the samples, the tubes were closed with Parafilm® and measurements were made.

^{1}H NMR relaxation times (spin–lattice (T_1) and spin–spin (T_2)) were analyzed with a pulse NMR spectrometer PS15T operating at 15 MHz (Ellab, Poznań, Poland) at $21.0 \pm 0.5\,°C$. The inversion–recovery ($\pi - \tau - \pi/2$) [100] pulse sequence was applied for measurements of the T_1 relaxation times. Distances between RF pulses (τ) were changed within the range from 0.5 to 50 ms and the repetition time was from 15 s. Each time, 32 FID signals and 110 points from each FID signal were collected. Calculations of the spin–lattice relaxation time values were performed with the assistance of the CracSpin program [101].

Measurements of the spin–spin (T2) relaxation times were taken using the pulse train of the Carr–Purcell–Meiboom–Gill spin echoes ($\pi/2 - TE/2 - (\pi)_n$ [100]. The distance (TE) between π RF pulses ranged from 0.1 to 1.0 ms. The repetition time was 15 s. The number of spin echoes (n) amounted to 100. Five accumulation signals were employed. The calculations were performed by using the dedicated software by application of a non-linear least-square algorithm.

3.10. Water Activity

Rollers 1 cm thick and 2 cm in diameter, each cut from the tested product, were used for the measurements. The sample was placed in the measurement chamber. The analysis was performed by using water diffusion and activity analyzer ADA-7 (COBRABID, Poznań, Poland) with a sample temperature control panel. The analyzer is equipped with dedicated software to record temporary water activity during water evacuation process [34]. All measurements were performed at $21.0 \pm 0.2\,°C$. The duration of one measurement was set to 1000 s. Based on the obtained curves, the equilibrium value of water activity a_w in the product and the transport rate V_D were determined. All presented results are mean values ($n = 7$) and standard deviation.

3.11. Statistical Analysis

Each biscuit variant was analyzed in three samples, with triple measurement of each, unless stated otherwise. One-way analysis of variance (ANOVA) was carried out independently for each dependent variable. A post hoc Tukey HSD multiple comparison test was used to identify statistically homogeneous subsets at $\alpha = 0.05$. Moreover, the Pearson correlation coefficient was calculated between relaxation times and water activity parameters. Statistical analysis was performed with Statistica 13 software (Dell Software Inc., Round Rock, TX, USA).

4. Conclusions

Partial replacement of wheat flour with cricket powder in biscuits and other food products augmented their physical properties as well as their nutritional and functional values. A small (2%) addition of CP improved the ratings for flavor, texture, appearance, and the overall desirability of biscuits. However, further addition of CP (6% and 10%) resulted in significantly lower scores in consumer test. CP2 delivered 462 kcal, 7.8 g protein, 16.2 g fat, and 73 g carbohydrates in 100 g. Moreover, it had higher content of minerals:

Ca (↑23%), Zn (↑16%), Mn (↑14%), Fe (↑12%), K (↑7%), and Mg (↑6%) than the commercial (control) product.

Changes were also observed in the physical properties of the biscuits. Replacing wheat flour with cricket powder resulted in a successive reduction in the hardness of cookies from 29.44 N for R to 24.50 N for CP10. A decrease in the values of the short components of the T_{21} spin–spin relaxation times was also observed in the samples containing CP compared to the reference sample R, measured by LF NMR, which indicates a reduction in the dynamics of water molecules bound to the polymer matrix. Due to the increase in the fat content of CP biscuits as opposed to the short ones, the long spin–spin relaxation time components increased in samples where some flour was replaced by CP. Nevertheless, on the basis of the results, it was found that the obtained shortbreads with a 2% CP addition could be successfully marketed. Moreover, the use of products with such superior characteristics as edible insects poses a viable scenario for the future demands of growing human population.

Author Contributions: Conceptualization, P.Ł.K.; Data curation, P.J. and H.M.B.; Formal analysis, K.S.; Investigation, K.S., P.S., P.Ł.K., M.B.R., I.R., K.P., M.F., D.K., Ł.M., P.J. and H.M.B.; Methodology, P.Ł.K., I.R., K.P. and H.M.B.; Project administration, P.Ł.K.; Supervision, P.Ł.K.; Validation, P.Ł.K.; Visualization, H.M.B.; Writing—original draft, P.Ł.K., M.B.R., I.R., M.N., J.L., M.K., M.Ś., T.P. and H.M.B.; Writing—review and editing, P.Ł.K., I.R., M.N., J.L. and P.J. All authors have read and agreed to the published version of the manuscript.

Funding: This research received no external funding.

Institutional Review Board Statement: Not applicable.

Informed Consent Statement: Informed consent was obtained from all subjects involved in the study.

Data Availability Statement: The datasets generated during and/or analyzed during the current study are available from the corresponding author on reasonable request.

Conflicts of Interest: The authors declare no conflict of interest.

Sample Availability: Samples of cricket powder are commercially available on the market.

References

1. Metro, D.; Tardugno, R.; Papa, M.; Bisignano, C.; Manasseri, L.; Calabrese, G.; Gervasi, T.; Dugo, G.; Cicero, N. Adherence to the Mediterranean diet in a Sicilian student population. *Nat. Prod. Res.* **2018**, *32*, 1775–1781. [CrossRef]
2. Metro, D.; Papa, M.; Manasseri, L.; Gervasi, T.; Campone, L.; Pellizzeri, V.; Tardugno, R.; Dugo, G. Mediterranean diet in a Sicilian student population. Second part: Breakfast and its nutritional profile. *Nat. Prod. Res.* **2020**, *34*, 2255–2261. [CrossRef]
3. Cammilleri, G.; Vazzana, M.; Arizza, V.; Giunta, F.; Vella, A.; Lo Dico, G.; Giaccone, V.; Giofrè, S.V.; Giangrosso, G.; Cicero, N.; et al. Mercury in fish products: What's the best for consumers between bluefin tuna and yellowfin tuna? *Nat. Prod. Res.* **2018**, *32*, 457–462. [CrossRef]
4. Cena, H.; Calder, P.C. Defining a Healthy Diet: Evidence for the Role of Contemporary Dietary Patterns in Health and Disease. *Nutrients* **2020**, *12*, 334. [CrossRef]
5. Jakubowska, D.; Staniewska, K. Information on food fortification with bioactive compounds in observation andconsumer studies. *Pol. J. Nat. Sci.* **2015**, *30*, 307–318.
6. Betoret, E.; Rosell, C.M. Enrichment of bread with fruits and vegetables: Trends and strategies to increase functionality. *Cereal Chem.* **2020**, *97*, 9–19. [CrossRef]
7. Phongnarisorn, B.; Orfila, C.; Holmes, M.; Marshall, L. Enrichment of Biscuits with Matcha Green Tea Powder: Its Impact on Consumer Acceptability and Acute Metabolic Response. *Foods* **2018**, *7*, 17. [CrossRef]
8. Maner, S.; Sharma, A.K.; Banerjee, K. Wheat Flour Replacement by Wine Grape Pomace Powder Positively Affects Physical, Functional and Sensory Properties of Cookies. *Proc. Natl. Acad. Sci. India Sect. B Biol. Sci.* **2017**, *87*, 109–113. [CrossRef]
9. Statista Sugar Confectionary Retail Sales Value in Europe 2014–2018. Available online: https://www.statista.com/statistics/1134398/sugar-confectionary-retail-sales-value-in-europe/ (accessed on 12 July 2021).
10. Cookie Report—TOP Agency. Available online: https://topagency.com/report/cookie-report/ (accessed on 12 July 2021).
11. Statista Confectionery: Weekly Consumption UK 2006–2019. Available online: https://www.statista.com/statistics/284521/weekly-household-consumption-of-confectionery-in-the-united-kingdom-uk/ (accessed on 12 July 2021).
12. Hooda, S.; Jood, S. Organoleptic and nutritional evaluation of wheat biscuits supplemented with untreated and treated fenugreek flour. *Food Chem.* **2005**, *90*, 427–435. [CrossRef]
13. Tyagi, S.K.; Manikantan, M.R.; Oberoi, H.S.; Kaur, G. Effect of mustard flour incorporation on nutritional, textural and organoleptic characteristics of biscuits. *J. Food Eng.* **2007**, *80*, 1043–1050. [CrossRef]

14. De Nogueira, A.C.; Steel, C.J. Protein enrichment of biscuits: A review. *Food Rev. Int.* **2018**, *34*, 796–809. [CrossRef]
15. FAO. *Assessing the Potential of Insects as Food and Feed in Assuring Food Security*; Food and Agriculture Organization of the United Nations: Rome, Italy, 2012.
16. Montowska, M.; Kowalczewski, P.Ł.; Rybicka, I.; Fornal, E. Nutritional value, protein and peptide composition of edible cricket powders. *Food Chem.* **2019**, *289*, 130–138. [CrossRef]
17. Skotnicka, M.; Karwowska, K.; Kłobukowski, F.; Borkowska, A.; Pieszko, M. Possibilities of the Development of Edible Insect-Based Foods in Europe. *Foods* **2021**, *10*, 766. [CrossRef]
18. Zielińska, E.; Pankiewicz, U.; Sujka, M. Nutritional, Physiochemical, and Biological Value of Muffins Enriched with Edible Insects Flour. *Antioxidants* **2021**, *10*, 1122. [CrossRef] [PubMed]
19. Zielińska, E.; Baraniak, B.; Karaś, M.; Rybczyńska, K.; Jakubczyk, A. Selected species of edible insects as a source of nutrient composition. *Food Res. Int.* **2015**, *77*, 460–466. [CrossRef]
20. Pauter, P.; Różańska, M.; Wiza, P.; Dworczak, S.; Grobelna, N.; Sarbak, P.; Kowalczewski, P.Ł. Effects of the replacement of wheat flour with cricket powder on the characteristics of muffins. *Acta Sci. Pol. Technol. Aliment.* **2018**, *17*, 227–233. [CrossRef] [PubMed]
21. Walkowiak, K.; Kowalczewski, P.Ł.; Kubiak, P.; Baranowska, H.M. Effect of cricket powder addition on ^{1}H NMR mobility and texture of pork pate. *J. Microbiol. Biotechnol. Food Sci.* **2019**, *9*, 191–194. [CrossRef]
22. Kowalczewski, P.; Walkowiak, K.; Masewicz, Ł.; Bartczak, O.; Lewandowicz, J.; Kubiak, P.; Baranowska, H. Gluten-Free Bread with Cricket Powder—Mechanical Properties and Molecular Water Dynamics in Dough and Ready Product. *Foods* **2019**, *8*, 240. [CrossRef]
23. Smarzyński, K.; Sarbak, P.; Musiał, S.; Jeżowski, P.; Piątek, M.; Kowalczewski, P.Ł. Nutritional analysis and evaluation of the consumer acceptance of pork pâté enriched with cricket powder-preliminary study. *Open Agric.* **2019**, *4*, 159–163. [CrossRef]
24. Duda, A.; Adamczak, J.; Chełmińska, P.; Juszkiewicz, J.; Kowalczewski, P. Quality and Nutritional/Textural Properties of Durum Wheat Pasta Enriched with Cricket Powder. *Foods* **2019**, *8*, 46. [CrossRef]
25. Bruneel, C.; Pareyt, B.; Brijs, K.; Delcour, J.A. The impact of the protein network on the pasting and cooking properties of dry pasta products. *Food Chem.* **2010**, *120*, 371–378. [CrossRef]
26. Małyszek, Z.; Lewandowicz, J.; Le Thanh-Blicharz, J.; Walkowiak, K.; Kowalczewski, P.Ł.; Baranowska, H.M. Water Behavior of Emulsions Stabilized by Modified Potato Starch. *Polymers* **2021**, *13*, 2200. [CrossRef]
27. Kamal, T.; Cheng, S.; Khan, I.A.; Nawab, K.; Zhang, T.; Song, Y.; Wang, S.; Nadeem, M.; Riaz, M.; Khan, M.A.U.; et al. Potential uses of LF-NMR and MRI in the study of water dynamics and quality measurement of fruits and vegetables. *J. Food Process. Preserv.* **2019**, *43*, e14202. [CrossRef]
28. Ezeanaka, M.C.; Nsor-Atindana, J.; Zhang, M. Online Low-field Nuclear Magnetic Resonance (LF-NMR) and Magnetic Resonance Imaging (MRI) for Food Quality Optimization in Food Processing. *Food Bioprocess Technol.* **2019**, *12*, 1435–1451. [CrossRef]
29. Lewandowicz, J.; Baranowska, H.M.; Szwengiel, A.; Le Thanh-Blicharz, J. Molecular structure vs. Functional properties of waxy and normal corn starch. In Proceedings of the 12th International Conference on Polysaccharides-Glycoscience, Prague, Czech Republic, 19–21 October 2016; Rapkova, R., Copikova, J., Sarka, E., Eds.; Czech Chemical Society: Prague, Czech Republic, 2016; pp. 53–57.
30. Lewandowicz, J.; Ostrowska-Ligeza, E.; Baranowska, H.M. Gelatinization of Starch: A Comparative Study of Viscographic, Differential Scanning Calorimetry and Low Field NMR Analyses. In Proceedings of the 16th International Conference on Polysaccharides-Glycoscience, Prague, Czech Republic, 4–6 November 2020; Rapkova, R., Copikova, J., Sarka, E., Eds.; Czech Chemical Society: Prague, Czech Republic, 2020; pp. 9–14.
31. Sikora, M.; Krystyjan, M.; Dobosz, A.; Tomasik, P.; Walkowiak, K.; Masewicz, Ł.; Kowalczewski, P.Ł.; Baranowska, H.M. Molecular Analysis of Retrogradation of Corn Starches. *Polymers* **2019**, *11*, 1764. [CrossRef] [PubMed]
32. Masewicz, L.; Pers, K.; Le Thanh-Blicharz, J.; Lewandowicz, J.; Baranowska, H.M. The effect of degree of substitution on dynamics of molecules of hydration water in acetylated distarch adipate powders. In Proceedings of the 13th International Conference on Polysaccharides-Glycoscience, Prague, Czech Republic, 9–10 November 2017; Rapkova, R., Copikova, J., Sarka, E., Eds.; Czech Chemical Society: Prague, Czech Republic, 2017; pp. 29–32.
33. Resende, M.T.; Osheter, T.; Linder, C.; Wiesman, Z. Proton Low Field NMR Relaxation Time Domain Sensor for Monitoring of Oxidation Stability of PUFA-Rich Oils and Emulsion Products. *Foods* **2021**, *10*, 1385. [CrossRef] [PubMed]
34. Stangierski, J.; Rezler, R.; Baranowska, H.M.; Poliszko, S. Effect of enzymatic modification on chicken surimi. *Czech J. Food Sci.* **2012**, *30*, 404–411. [CrossRef]
35. Baranowska, H.M.; Rezler, R. Water binding analysis of fat-water emulsions. *Food Sci. Biotechnol.* **2015**, *24*, 1921–1925. [CrossRef]
36. Le Thanh-Blicharz, J. *The Influence of the Physicochemical Properties and Structure of Modified Starches on Their Functionality in the Formation and Stabilization of Food Emulsions*; Wydawnictwo Uniwerystetu Przyrodniczego w Poznaniu: Poznań, Poland, 2018; ISBN 978-83-7160-903-9.
37. Kowalczewski, P.Ł.; Walkowiak, K.; Masewicz, Ł.; Smarzyński, K.; Le Thanh-Blicharz, J.; Kačániová, M.; Baranowska, H.M. LF NMR spectroscopy analysis of water dynamics and texture of gluten-free bread with cricket powder during storage. *Food Sci. Technol. Int.* **2021**, 108201322098791. [CrossRef]
38. Li, J.; Hou, G.G.; Chen, Z.; Gehring, K. Effects of endoxylanases, vital wheat gluten, and gum Arabic on the rheological properties, water mobility, and baking quality of whole-wheat saltine cracker dough. *J. Cereal Sci.* **2013**, *58*, 437–445. [CrossRef]

39. Baranowska, H.M.; Masewicz, Ł.; Kowalczewski, P.Ł.; Lewandowicz, G.; Piątek, M.; Kubiak, P. Water properties in pâtés enriched with potato juice. *Eur. Food Res. Technol.* **2018**, *244*, 387–393. [CrossRef]

40. Colnago, L.A.; Wiesman, Z.; Pages, G.; Musse, M.; Monaretto, T.; Windt, C.W.; Rondeau-Mouro, C. Low field, time domain NMR in the agriculture and agrifood sectors: An overview of applications in plants, foods and biofuels. *J. Magn. Reson.* **2021**, *323*, 106899. [CrossRef] [PubMed]

41. Castro, M.; Chambers, E. Consumer Avoidance of Insect Containing Foods: Primary Emotions, Perceptions and Sensory Characteristics Driving Consumers Considerations. *Foods* **2019**, *8*, 351. [CrossRef] [PubMed]

42. Piha, S.; Pohjanheimo, T.; Lähteenmäki-Uutela, A.; Křečková, Z.; Otterbring, T. The effects of consumer knowledge on the willingness to buy insect food: An exploratory cross-regional study in Northern and Central Europe. *Food Qual. Prefer.* **2018**, *70*, 1–10. [CrossRef]

43. House, J. Consumer acceptance of insect-based foods in the Netherlands: Academic and commercial implications. *Appetite* **2016**, *107*, 47–58. [CrossRef] [PubMed]

44. Mishyna, M.; Chen, J.; Benjamin, O. Sensory attributes of edible insects and insect-based foods—Future outlooks for enhancing consumer appeal. *Trends Food Sci. Technol.* **2020**, *95*, 141–148. [CrossRef]

45. Melgar-Lalanne, G.; Hernández-Álvarez, A.; Salinas-Castro, A. Edible Insects Processing: Traditional and Innovative Technologies. *Compr. Rev. Food Sci. Food Saf.* **2019**, *18*, 1166–1191. [CrossRef]

46. Seo, H.; Kim, H.R.; Cho, I.H. Aroma Characteristics of Raw and Cooked *Tenebrio molitor* Larvae (Mealworms). *Food Sci. Anim. Resour.* **2020**, *40*, 649–658. [CrossRef]

47. Grossmann, K.K.; Merz, M.; Appel, D.; De Araujo, M.M.; Fischer, L. New insights into the flavoring potential of cricket (*Acheta domesticus*) and mealworm (*Tenebrio molitor*) protein hydrolysates and their Maillard products. *Food Chem.* **2021**, *364*, 130336. [CrossRef]

48. Hazard, B.; Trafford, K.; Lovegrove, A.; Griffiths, S.; Uauy, C.; Shewry, P. Strategies to improve wheat for human health. *Nat. Food* **2020**, *1*, 475–480. [CrossRef]

49. Blanshard, J.M.V. Elements of cereal product structure. In *Food Structure*; Elsevier: Amsterdam, The Netherlands, 1988; pp. 313–330.

50. Burt, K.G.; Kotao, T.; Lopez, I.; Koeppel, J.; Goldstein, A.; Samuel, L.; Stopler, M. Acceptance of Using Cricket Flour as a Low Carbohydrate, High Protein, Sustainable Substitute for All-Purpose Flour in Muffins. *J. Culin. Sci. Technol.* **2020**, *18*, 201–213. [CrossRef]

51. Kowalski, S.; Lukasiewicz, M.; Juszczak, L.; Kutyła-Kupidura, E.M. Dynamics of 5-hydroxymethylfurfural formation in shortbreads during thermal processing. *Czech J. Food Sci.* **2013**, *31*, 33–42. [CrossRef]

52. Zanoni, B.; Peri, C.; Bruno, D. Modelling of browning kinetics of bread crust during baking. *LWT-Food Sci. Technol.* **1995**, *28*, 604–609. [CrossRef]

53. Zielińska, E.; Pankiewicz, U. Nutritional, Physiochemical, and Antioxidative Characteristics of Shortcake Biscuits Enriched with Tenebrio molitor Flour. *Molecules* **2020**, *25*, 5629. [CrossRef] [PubMed]

54. Kowalczewski, P.Ł.; Gumienna, M.; Rybicka, I.; Górna, B.; Sarbak, P.; Dziedzic, K.; Kmiecik, D. Nutritional Value and Biological Activity of Gluten-Free Bread Enriched with Cricket Powder. *Molecules* **2021**, *26*, 1184. [CrossRef]

55. Mokrzycki, W.; Tatol, M. Color difference Delta E—A survey. *Mach. Graph. Vis.* **2011**, *20*, 383–411.

56. Van Huis, A. New Sources of Animal Proteins: Edible Insects. In *New Aspects of Meat Quality*; Elsevier: Amsterdam, The Netherlands, 2017; pp. 443–461.

57. Van Huis, A. Nutrition and health of edible insects. *Curr. Opin. Clin. Nutr. Metab. Care* **2020**, *23*, 228–231. [CrossRef]

58. Jantzen da Silva Lucas, A.; Menegon de Oliveira, L.; da Rocha, M.; Prentice, C. Edible insects: An alternative of nutritional, functional and bioactive compounds. *Food Chem.* **2020**, *311*, 126022. [CrossRef]

59. Regulation (EU) No 1169/2011 of the European Parliament and of the Council of 25 October 2011 on the Provision of Food Information to Consumers, Amending Regulations (EC) No 1924/2006 and (EC) No 1925/2006 of the European Parliament and of the Council. Available online: https://eur-lex.europa.eu/legal-content/EN/ALL/?uri=CELEX%3A32011R1169 (accessed on 12 July 2021).

60. Regulation (EC) No 1924/2006 of the European Parliament and of the Council of 20 December 2006 on Nutrition and Health Claims Made on Foods. Available online: https://eur-lex.europa.eu/legal-content/en/ALL/?uri=CELEX%3A32006R1924 (accessed on 12 July 2021).

61. World Health Organization. Salt Reduction. Available online: https://www.who.int/news-room/fact-sheets/detail/salt-reduction (accessed on 12 July 2021).

62. Public Health England. *Salt Reduction Targets for 2024*; 2020. Available online: https://www.gov.uk/government/publications/salt-reduction-targets-for-2024 (accessed on 12 July 2021).

63. Udomsil, N.; Imsoonthornruksa, S.; Gosalawit, C.; Ketudat-Cairns, M. Nutritional Values and Functional Properties of House Cricket (*Acheta domesticus*) and Field Cricket (*Gryllus bimaculatus*). *Food Sci. Technol. Res.* **2019**, *25*, 597–605. [CrossRef]

64. Grapes, M.; Whiting, P.; Dinan, L. Fatty acid and lipid analysis of the house cricket, *Acheta domesticus*. *Insect Biochem.* **1989**, *19*, 767–774. [CrossRef]

65. Martínez-Villaluenga, C.; Peñas, E.; Hernández-Ledesma, B. Pseudocereal grains: Nutritional value, health benefits and current applications for the development of gluten-free foods. *Food Chem. Toxicol.* **2020**, *137*, 111178. [CrossRef]

66. Mathai, J.K.; Liu, Y.; Stein, H.H. Values for digestible indispensable amino acid scores (DIAAS) for some dairy and plant proteins may better describe protein quality than values calculated using the concept for protein digestibility-corrected amino acid scores (PDCAAS). *Br. J. Nutr.* **2017**, *117*, 490–499. [CrossRef] [PubMed]

67. Wu, T.; Taylor, C.; Nebl, T.; Ng, K.; Bennett, L.E. Effects of chemical composition and baking on in vitro digestibility of proteins in breads made from selected gluten-containing and gluten-free flours. *Food Chem.* **2017**, *233*, 514–524. [CrossRef] [PubMed]

68. Delgado-Andrade, C. Maillard reaction products: Some considerations on their health effects. *Clin. Chem. Lab. Med.* **2014**, *52*. [CrossRef]

69. Lund, M.N.; Ray, C.A. Control of Maillard Reactions in Foods: Strategies and Chemical Mechanisms. *J. Agric. Food Chem.* **2017**, *65*, 4537–4552. [CrossRef]

70. Zheng, B.; Zhao, H.; Zhou, Q.; Cai, J.; Wang, X.; Cao, W.; Dai, T.; Jiang, D. Relationships of protein composition, gluten structure, and dough rheological properties with short biscuits quality of soft wheat varieties. *Agron. J.* **2020**, *112*, 1921–1930. [CrossRef]

71. Chevallier, S.; Colonna, P.; Buléon, A.; Della Valle, G. Physicochemical Behaviors of Sugars, Lipids, and Gluten in Short Dough and Biscuit. *J. Agric. Food Chem.* **2000**, *48*, 1322–1326. [CrossRef] [PubMed]

72. Mudgil, D.; Barak, S.; Khatkar, B.S. Cookie texture, spread ratio and sensory acceptability of cookies as a function of soluble dietary fiber, baking time and different water levels. *LWT* **2017**, *80*, 537–542. [CrossRef]

73. Nanyen, D. Nutritional Composition, Physical and Sensory Properties of Cookies from Wheat, Acha and Mung Bean Composite Flours. *Int. J. Nutr. Food Sci.* **2016**, *5*, 401. [CrossRef]

74. Ho, L.-H.; Abdul Latif, N.W.B. Nutritional composition, physical properties, and sensory evaluation of cookies prepared from wheat flour and pitaya (*Hylocereus undatus*) peel flour blends. *Cogent Food Agric.* **2016**, *2*, 1136369. [CrossRef]

75. Zaker, A.; Genitha, T.R.; Hashmi, S.I. Effects of Defatted Soy Flour Incorporation on Physical, Sensorial and Nutritional Properties of Biscuits. *J. Food Process. Technol.* **2012**, *3*, 1–4. [CrossRef]

76. Kulkarni, A.S.; Joshi, D.C. Effect of replacement of wheat flour with pumpkin powder on textural and sensory qualities of biscuit. *Int. Food Res. J.* **2013**, *20*, 587.

77. Gaines, C.S. Influence of chemical and physical modification of soft wheat protein on sugar-snap cookie dough consistency, cookie size, and hardness. *Cereal Chem.* **1990**, *67*, 73–77.

78. Pareyt, B.; Delcour, J.A. The Role of Wheat Flour Constituents, Sugar, and Fat in Low Moisture Cereal Based Products: A Review on Sugar-Snap Cookies. *Crit. Rev. Food Sci. Nutr.* **2008**, *48*, 824–839. [CrossRef] [PubMed]

79. Chauhan, A.; Saxena, D.C.; Singh, S. Physical, textural, and sensory characteristics of wheat and amaranth flour blend cookies. *Cogent Food Agric.* **2016**, *2*, 1125773. [CrossRef]

80. Walkowiak, K.; Lewandowicz, J.; Masewicz, L.; Le Thanh-Blicharz, J.; Baranowska, H.M. Application of empirical model of relationship of spin-lattice relaxation times and values of critical hydration for analysis of potato starch aerogels. In Proceedings of the 14th International Conference on Polysaccharides-Glycoscience, Prague, Czech Republic, 7–9 November 2018; Rapkova, R., Hinkova, A., Copikova, J., Sarka, E., Eds.; Czech Chemical Society: Prague, Czech Republic, 2018; pp. 90–92.

81. Zayas, J.F. Solubility of Proteins. In *Functionality of Proteins in Food*; Springer: Berlin/Heidelberg, Germany, 1997; pp. 6–75.

82. Bassett, F. Comparison of Functional, Nutritional, and Sensory Properties of Spray-Dried and Oven-Dried Cricket (*Acheta domesticus*) Powder. Master's Thesis, Brigham Young University, Provo, UT, USA, 2018.

83. Bosmans, G.M.; Lagrain, B.; Ooms, N.; Fierens, E.; Delcour, J.A. Biopolymer Interactions, Water Dynamics, and Bread Crumb Firming. *J. Agric. Food Chem.* **2013**, *61*, 4646–4654. [CrossRef]

84. Berendsen, H.J.C. Specific Interactions of Water with Biopolymers. In *Water in Disperse Systems*; Springer: Boston, MA, USA, 1975; pp. 293–330.

85. Villanueva, N.D.; Petenate, A.J.; Da Silva, M.A.A. Performance of three affective methods and diagnosis of the ANOVA model. *Food Qual. Prefer.* **2002**, *11*, 363–370. [CrossRef]

86. AACC 44-19.01 Moisture–Air-Oven Method, Drying at 135 degrees. In *AACC International Approved Methods*; AACC International: Washington, DC, USA, 2009.

87. *ISO 20483:2013 Cereals and Pulses-Determination of the Nitrogen Content and Calculation of the Crude Protein Content-Kjeldahl Method*; International Organization for Standardization: Geneva, Switzerland, 2013.

88. Tkachuk, R. Note on the nitrogen-to-protein conversion factor for wheat flour. *Cereal Chem.* **1966**, *43*, 223–225.

89. Ritvanen, T.; Pastell, H.; Welling, A.; Raatikainen, M. The nitrogen-to-protein conversion factor of two cricket species-*Acheta domesticus* and *Gryllus bimaculatus*. *Agric. Food Sci.* **2020**, *29*. [CrossRef]

90. AACC Crude Fat in Wheat, Corn, and Soy Flour, Feeds, and Mixed Feeds. *AACC Approved Methods of Analysis*. 2009. Available online: http://methods.aaccnet.org/summaries/30-25-01.aspx (accessed on 12 July 2021). [CrossRef]

91. AACC Ash in Farina and Semolina. *AACC Approved Methods of Analysis*. 2009. Available online: http://methods.aaccnet.org/summaries/08-12-01.aspx (accessed on 12 July 2021). [CrossRef]

92. Rybicka, I.; Gliszczyńska-Świgło, A. Minerals in grain gluten-free products. The content of calcium, potassium, magnesium, sodium, copper, iron, manganese, and zinc. *J. Food Compos. Anal.* **2017**, *59*, 61–67. [CrossRef]

93. European Food Safety Authority. Dietary Reference Values for nutrients Summary report. *EFSA Support. Publ.* **2017**, *14*, e15121. [CrossRef]

94. Kwanyuen, P.; Burton, J.W. A Modified Amino Acid Analysis Using PITC Derivatization for Soybeans with Accurate Determination of Cysteine and Half-Cystine. *J. Am. Oil Chem. Soc.* **2010**, *87*, 127–132. [CrossRef]

95. Polanowska, K.; Grygier, A.; Kuligowski, M.; Rudzińska, M.; Nowak, J. Effect of tempe fermentation by three different strains of Rhizopus oligosporus on nutritional characteristics of faba beans. *LWT* **2020**, *122*, 109024. [CrossRef]
96. Çevikkalp, S.A.; Löker, G.B.; Yaman, M.; Amoutzopoulos, B. A simplified HPLC method for determination of tryptophan in some cereals and legumes. *Food Chem.* **2016**, *193*, 26–29. [CrossRef]
97. Folch, J.; Lees, M.; Sloane, S.G.H.; Stanley, G.H.S.; Sloane Stanley, G.H. A simple method for the isolation and purification of total lipides from animal tissues. *J. Biol. Chem.* **1957**, *226*, 497–509. [CrossRef]
98. AOCS Official Method Ce 1h-05. Determination of cis-, trans-, Saturated, Monounsaturated and Polyunsaturated Fatty Acids in Vegetable or Non-Ruminant Animal Oils and Fats by Capillary GLC. 2009. Available online: https://www.aocs.org/attain-lab-services/methods/methods/search-results?method=111777 (accessed on 12 July 2021).
99. Canali, G.; Balestra, F.; Glicerina, V.; Pasini, F.; Caboni, M.F.; Romani, S. Influence of different baking powders on physico-chemical, sensory and volatile compounds in biscuits and their impact on textural modifications during soaking. *J. Food Sci. Technol.* **2020**, *57*, 3864–3873. [CrossRef]
100. Brosio, E.; Gianferri, R.R. An analytical tool in foods characterization and traceability. In *Basic NMR in Foods Characterization*; Research Signpost: Kerala, India, 2009; pp. 9–37.
101. Weglarz, W.P.; Haranczyk, H. Two-dimensional analysis of the nuclear relaxation function in the time domain: The program CracSpin. *J. Phys. D Appl. Phys.* **2000**, *33*, 1909–1920. [CrossRef]

 *molecules*

Article

Nutritional Value and Biological Activity of Gluten-Free Bread Enriched with Cricket Powder

Przemysław Łukasz Kowalczewski [1,*], **Małgorzata Gumienna** [1], **Iga Rybicka** [2], **Barbara Górna** [1], **Paulina Sarbak** [3], **Krzysztof Dziedzic** [1] and **Dominik Kmiecik** [1]

1   Department of Food Technology of Plant Origin, Poznań University of Life Sciences, 31 Wojska Polskiego St., 60-624 Poznań, Poland; malgorzata.gumienna@up.poznan.pl (M.G.); barbara.gorna@up.poznan.pl (B.G.); krzysztof.dziedzic@up.poznan.pl (K.D.); dominik.kmiecik@up.poznan.pl (D.K.)
2   Department of Technology and Instrumental Analysis, Poznań University of Economics and Business, Al. Niepodległości 10, 61-875 Poznań, Poland; iga.rybicka@ue.poznan.pl
3   Students' Scientific Club of Food Technologists, Poznań University of Life Sciences, 31 Wojska Polskiego St., 60-624 Poznań, Poland; paulina.sarbak@onet.pl
*   Correspondence: przemyslaw.kowalczewski@up.poznan.pl; Tel.: +48-61-848-7297

Abstract: Cricket powder, described in the literature as a source of nutrients, can be a valuable ingredient to supplement deficiencies in various food products. Work continues on the implementation of cricket powder in products that are widely consumed. The aim of this study was to obtain gluten-free bread with a superior nutritional profile by means of insect powder addition. Gluten-free breads enriched with 2%, 6%, and 10% of cricket (*Acheta domesticus*) powder were formulated and extensively characterized. The nutritional value, as well as antioxidant and β-glucuronidase activities, were assessed after simulated in vitro digestion. Addition of cricket powder significantly increased the nutritional value, both in terms of the protein content (exceeding two-, four-, and seven-fold the reference bread (RB), respectively) and above all mineral compounds. The most significant changes were observed for Cu, P, and Zn. A significant increase in the content of polyphenolic compounds and antioxidant activity in the enriched bread was also demonstrated; moreover, both values additionally increased after the digestion process. The total polyphenolic compounds content increased about five-fold from RB to bread with 10% CP (BCP10), and respectively about three-fold after digestion. Similarly, the total antioxidant capacity before digestion increased about four-fold, and after digestion about six-fold. The use of CP also reduced the undesirable activity of β-glucuronidase by 65.9% (RB vs. BCP10) in the small intestine, down to 78.9% in the large intestine. The influence of bread on the intestinal microflora was also evaluated, and no inhibitory effect on the growth of microflora was demonstrated, both beneficial (*Bifidobacterium* and *Lactobacillus*) and pathogenic (*Enterococcus* and *Escherichia coli*). Our results underscore the benefits of using cricket powder to increase the nutritional value and biological activity of gluten-free food products.

Keywords: *Acheta domesticus*; antioxidant activity; β-glucuronidase; edible insect; gut microbiome; insect protein; in vitro digestion

Citation: Kowalczewski, P.Ł.; Gumienna, M.; Rybicka, I.; Górna, B.; Sarbak, P.; Dziedzic, K.; Kmiecik, D. Nutritional Value and Biological Activity of Gluten-Free Bread Enriched with Cricket Powder. *Molecules* 2021, 26, 1184. https://doi.org/10.3390/molecules26041184

Academic Editor: Francisco J. Barba

Received: 5 February 2021
Accepted: 20 February 2021
Published: 23 February 2021

Publisher's Note: MDPI stays neutral with regard to jurisdictional claims in published maps and institutional affiliations.

1. Introduction

Bakery products, especially breads, are a crucial part of the everyday human diet. Commercially available gluten-free (GF) breads are often disregarded by people on a GF diet, because of their comparatively inferior characteristics: taste, aroma, texture, artificial ingredients, or nutritional value [1]. The technological and sensory challenges of GF products appear because of the absence of gluten, responsible for formulating dough in gluten-containing (wheat, barley, rye) products [2,3]. The overall poorer nutritional quality results from the artificial food additives and the fact that the most popular GF raw materials are corn, rice, and GF starch, generally not rich in vitamins or minerals [4]. Gluten-free bakery products are called the Achilles' heel of a GF diet, and numerous studies have been conducted to improve their attractiveness [1,5]. On the other hand, the food market offers various natural food products and ingredients which can improve the nutritional value of this diet.

The use of insects as food is well established, particularly in Africa, Latin America, and Asia [6,7], and according to FAO/WHO data, more than 1900 species of insects are edible worldwide, including crickets, meal larvae, ants, grasshoppers and flies [8]. To introduce edible insects to a wider consumption is one of the latest nutritional trends [9]. Edible insects are extremely rich in nutrients such as protein, fat, vitamins, and minerals; therefore, their consumption is recommended [8,10,11]. Insects can thus act as an enriching additive for food production [12–14]. Both the breeding method and the age of insects significantly affect their nutritional value [15–17]. Most consumers, however, do not accept eating entire insects; therefore, powder is the most preferable form [18,19]. Previous studies have shown that cricket (*Acheta domesticus*) powder (CP) is a rich source of protein and minerals [17,20]. GF products often show a deficiency of minerals and proteins [21]; therefore, CP is an interesting additive that can increase the nutritional value of GF products. Edible insects, apart from basic nutrients, can also provide compounds with bioactive activity [22–24]. Among the most frequently mentioned compounds are those with antioxidant activity, of particular importance in the prevention of broadly understood oxidative stress, associated with many diseases [25,26]. Literature reports describe the effect of in vitro digestion on the stability and bioconversion of some antioxidative compounds in the human gastrointestinal model [27–30].

Stull et al. [31] indicated that the consumption of crickets can improve the condition of the intestines, reduce inflammation, and positively affect the growth of the intestinal microflora, hence it is worth investigating the effect of using CP to fortify food for people suffering from intestinal diseases, e.g., celiac disease. It can be assumed that the use of crickets in GF products will, on the one hand, increase the nutritional value, and on the other hand, will allow for obtaining products with new, attractive biological properties. Only scarce data are available regarding the effects of the digestion of bioactive compounds from insects, especially in the GF bread matrix. Therefore, the aim of this work was to comprehensively characterize the biological activity of GF breads enriched with CP. Such products are not available on the food market and similar studies have never been conducted. For this study, we prepared GF breads with 2%, 6%, and 10% substitution of starch by CP. The analyses included: nutritional value (moisture, ash, protein, fat, carbohydrates, dietary fiber, microelements, and macroelements), color, in vitro digestion, effects on intestinal microflora, polyphenols content, and antioxidant activity.

2. Results and Discussion

2.1. Proximate Composition and Energy Value

GF foods often display an inappropriate nutritional profile, deficient in many nutrients [4,32]. Montowska et al. [20] showed that CP is a rich source of protein and other substances, including minerals. In the present study, CP increased the content of protein, fat, and dietary fiber in breads enriched with it (Table 1). Despite the widespread recognition of insects as a very good source of protein [10], a significant, but not spectacular, increase in its content was noted. Replacing starch with CP at the amount of 2%, 6%, and 10% (BCP2, BCP6, and BCP10, respectively) resulted in a two, four, and seven-fold increase in protein content, respectively. The Kjeldahl method measures nitrogen and has been validated for protein determination in food using a specific conversion factor for various products considering that all nitrogen present is in the form of protein. An incorrect nitrogen-to-protein conversion factor results in an overestimated protein content in edible insects [33]. Due to the exoskeleton of arthropods, composed of, inter alia, chitin, glucosamine polysaccharides, and nitrogen-rich N-acetylglucosamine [34], the use of an appropriate conversion factor is essential [35]. The fat content increased by 23%, 59%, and 105% for BCP2, BCP6, and BCP10, respectively, compared to reference bread (RB). Insects are rich in fat in their early stages of development [36,37], whereas CP was prepared from adult crickets. In addition to macronutrients, insects are also a source of dietary fiber, mainly insoluble [38], which resulted in three-fold increase in its content in CP-enriched breads. Importantly, despite the differences in the content of individual macronutrients in breads, including carbohydrates, no significant change in the energy value was observed.

Table 1. Proximate composition and energy value of obtained breads.

Parameter		RB	BCP2	BCP6	BCP10
Moisture [%]		51.14 ± 2.13 [a]	50.20 ± 2.61 [a]	51.94 ± 1.94 [a]	50.45 ± 2.05 [a]
Protein [%]		1.23 ± 0.21 [d]	2.56 ± 0.19 [c]	5.85 ± 0.34 [b]	8.48 ± 0.53 [a]
Fat [%]		0.78 ± 0.09 [c]	0.96 ± 0.11 [c]	1.24 ± 0.07 [b]	1.60 ± 0.12 [a]
Fiber [%]	SDF	1.64 ± 0.08 [b]	1.79 ± 0.15 [b]	2.00 ± 0.13 [a]	2.06 ± 0.11 [a]
	IDF	0.44 ± 0.06 [d]	0.73 ± 0.11 [c]	1.14 ± 0.04 [b]	1.48 ± 0.13 [a]
	TDF	2.08 ± 0.02 [d]	2.52 ± 0.28 [c]	3.14 ± 0.14 [b]	3.54 ± 0.17 [a]
Ash [%]		1.10 ± 0.07 [c]	1.85 ± 0.08 [b]	1.91 ± 0.08 [ab]	2.01 ± 0.05 [a]
Carbohydrates [1] [%]		43.67 ± 1.07 [a]	41.91 ± 1.22 [b]	35.92 ± 2.01 [c]	33.92 ± 1.29 [c]
Energy value [2] [kcal/100 g]		190.78 ± 7.11 [a]	191.56 ± 4.28 [a]	184.52 ± 6.62 [a]	191.08 ± 8.03 [a]

[1] The carbohydrate content was estimated by subtracting the average content of ash, fat, fiber, and protein from 100%. [2] Energy value was calculated based on the average moisture, protein, fat, fiber, and carbohydrate content. Mean values with the same letters in the row ([a–d]) were not significantly different (α = 0.05). RB—reference bread; BCP2, BCP6, BCP10—breads with starch replaced with cricket powder (CP) at 2%, 6%, and 10%, respectively; IDF—insoluble dietary fiber; SDF—soluble dietary fiber; TDF—total dietary fiber.

The addition of CP changed the content of most of the minerals in the analyzed breads, although the degree of these changes varied among the assessed minerals. Crickets are a good source of minerals. According to the data reported by Ghosh et al. [17], they contain significant amounts of calcium, magnesium, and iron. Phytic acid present in insects can chelate minerals, including iron [39], rendering them effectively indigestible. The content of minerals, expressed in mg for a 100 g edible portion, and the values of mineral requirements, population reference intakes (PRIs) and adequate intakes (AIs), are presented in Table 2. The percentage of provided dietary reference intakes (DRIs) was calculated for 100 g portion (two regular or four thin slices) of bread. The percentage of DRIs and AI for Ca, Fe, K, and Mg increased from about 1% to 2% (portion of control bread) to between 3% and 4% (BCP10). The content of Na was at the same level in all analyzed breads (approximately 300 mg/100 g) and resulted from their recipe and added salt. The most desirable improvement in the mineral profile was obtained for Cu, P, Mn, and Zn. BCP10 could be regarded as an important source of Cu (23% of DRI) and P (13% of DRI), whereas a portion of RB provided only 8% and 5% of DRI for these minerals, respectively. The content of Zn increased from 0.40 mg in RB (4% of DRI) to 1.08 mg in BCP10 (11% of DRI), and that of Fe increased from 0.24 to 0.59 mg (2% to 4% of DRI).

Table 2. Minerals in gluten-free (GF) breads enriched with CP.

Mineral	PRI/AI [mg/day]	RB mg/100 g	RB [%PRI/AI[1]]	BCP2 mg/100 g	BCP2 [%PRI/AI]	BCP6 mg/100 g	BCP6 [%PRI/AI]	BCP10 mg/100 g	BCP10 [%PRI/AI]
Ca	950	17.0 ± 1.9 [a]	2	18.8 ± 2.0 [a]	2	25.8 ± 1.8 [b]	3	28.4 ± 2.9 [b]	3
Mg	300	3.48 ± 0.28	1	4.51 ± 0.28	2	7.50 ± 0.41	3	9.53 ± 0.10	3
K	3500	31.8 ± 0.5	1	40.0 ± 1.9	1	68.0 ± 5.0	2	94 ± 2	3
Na	1500	304 ± 10 [a]	20	293 ± 14 [a]	20	324 ± 19 [a]	22	317 ± 9 [a]	21
P	550	28.8 ± 1.1	5	36.7 ± 1.3	7	49.5 ± 0.5	9	71.4 ± 3.3	13
Cu	10	0.08 ± 0.00	8	0.12 ± 0.01	12	0.18 ± 0.00	18	0.23 ± 0.02	23
Fe	16	0.24 ± 0.01	2	0.29 ± 0.01	2	0.39 ± 0.02	2	0.59 ± 0.08	4
Mn	3	0.02 ± 0.00	1	0.06 ± 0.00	2	0.13 ± 0.00	4	0.21 ± 0.01	7
Zn	10	0.40 ± 0.05 [a]	4	0.48 ± 0.02 [a]	5	0.85 ± 0.08	9	1.08 ± 0.07	11

Mean values with the same letters in the row ([a, b]) were not significantly different (α = 0.05). RB—reference bread; BCP2, BCP6, BCP10—breads with starch replaced with CP at 2%, 6%, and 10%, respectively; PRI—population reference intakes; AI—adequate intakes.

In addition to protein and minerals, CP is also a source of fat. Montowska et al. [20] reported that the fat content of commercial CPs ranges from 23.6% to 29.1%. According to Kim et al. [40], the main fatty acids in CP are palmitic acid (C16:0), oleic acid (C18:1), and linoleic acid (C18:2). Fats present in the dough affect the nutritional value of bread, but also their derivatives (hydroperoxides) are responsible for the formation of volatile compounds in breads [41].

The addition of CP changed the fatty acid profile in the prepared bread (Table 3). In all samples, the main fatty acid was oleic acid (C18:1), which constituted from 54.97% to 68.72% of all fatty acids. For oleic acid, a decrease in the share with an increase in CP addition was observed. Linoleic acid (C18:2) was also characterized by a high content, with its share increasing from 18.56% to 25.29% along with an increase in the CP addition. The increase in the proportion was also characteristic of palmitic acid (C16:0), which content increased from 4.04% in the RB bread to 8.80% in the BCP10 bread. Changes in the share of individual fatty acids also manifested in the content of individual groups of fatty acids. Along with the increase in CP addition, an increase in the content of saturated fatty acids (SFA) and polyunsaturated fatty acids (PUFA) and a decrease in monounsaturated fatty acids (MUFA) were observed. This phenomenon can be explained, as the main source of fat in the RB sample was the oleic-rich rapeseed oil. CP is characterized by a high content of linoleic acid (C18:2; 35%), palmitic acid (C16:0; 25.52%), and stearic acid (C18:0; 7.76%) [42]. In addition, Ghosh et al. [17] indicated that crickets contain more PUFA than MUFA, which is consistent with the observed changes in the fat acid profile in CP-enriched breads. The increase in CP content in the samples significantly increased the share of these acids in the pool of fatty acids.

Table 3. Fatty acid composition of GF breads enriched with CP [%].

Fatty Acid	RB	BCP2	BCP6	BCP10
C16:0	4.04 ± 0.03 [a]	5.21 ± 0.01 [b]	7.37 ± 0.18 [c]	8.80 ± 0.11 [d]
C16:1	0.24 ± 0.03 [ab]	0.18 ± 0.03 [a]	0.28 ± 0.01 [b]	0.31 ± 0.01 [b]
C18:0	0.99 ± 0.03 [a]	1.97 ± 0.22 [b]	2.76 ± 0.06 [c]	3.40 ± 0.08 [d]
C18:1	68.72 ± 1.24 [a]	64.43 ± 0.38 [b]	58.59 ± 0.06 [c]	54.97 ± 0.24 [d]
C18:2	18.56 ± 0.56 [a]	20.59 ± 0.11 [b]	23.58 ± 0.12 [c]	25.29 ± 0.03 [d]
C18:3	6.00 ± 0.54 [a]	5.87 ± 0.16 [b]	5.48 ± 0.08 [c]	5.44 ± 0.05 [d]
C20:1	1.46 ± 0.06 [a]	1.77 ± 0.08 [b]	1.74 ± 0.21 [c]	1.53 ± 0.01 [d]
C22:0	N/D	N/D	0.20 ± 0.03 [a]	0.27 ± 0.06 [a]
SFA	5.03 [a]	7.18 [b]	10.33 [c]	12.47 [d]
MUFA	70.42 [a]	66.38 [b]	60.61 [c]	56.81 [d]
PUFA	24.56 [a]	26.46 [b]	29.06 [c]	30.73 [d]

Mean values with the same letters in the row ([a–d]) were not significantly different ($\alpha = 0.05$). RB—reference bread; BCP2, BCP6, BCP10—breads with starch replaced with CP at 2%, 6%, and 10%, respectively; N/D—not detected; SFA, saturated fatty acids; MUFA, monounsaturated fatty acids; PUFA, polyunsaturated fatty acids. Calculated based on the mean content.

2.2. Characteristics of Bread Color

Replacing starch with CP changed the color of the resulting breads. The bread crumb was increasingly darker the more starch was replaced with CP. The obtained breads are presented in Figure 1.

Noticeable with the naked eye, color changes were then analyzed using a colorimeter. The results of instrumental color analysis are presented in Table 4. A significant decrease in crumb lightness was observed due to the addition of CP. There was a clear decrease in crumb lightness due to CP addition, by 16.4% for BCP2, 27.3% for BCP6, and 33.2% for BCP10. The darker color of bread is perceived by consumers as more desirable, as they associate it with healthier, whole-grain bread [43]. Therefore, it can be concluded that a color change to a darker one will be well received by consumers. There was also a significant increase in the value of the red saturation parameter (a*), with a slight decrease in yellow saturation (b*). The color of the crumb may depend not only on the ingredients used, but also on the conditions of the technological process in which reactions resulting in a color change may occur, i.e., caramelization and Maillard reactions [44,45]. Both reactions depend on the temperature, the content of reducing sugars, and amino groups, and can occur simultaneously during the baking process. The total color difference (ΔE) ranged from 13.8 to 27.5, signifying very large differences from RB without CP additive, and as reported by Mokrzycki and Tatol [46], differences exceeding 2 may already be noticed by an observer inexperienced in color assessment.

RB

BCP2

BCP6

BCP10

Figure 1. GF breads with CP: RB—reference bread; BCP2, BCP6, BCP10—breads with starch replaced with CP at 2%, 6%, and 10%, respectively.

Table 4. The results of color analysis.

Parameter	RB	BCP2	BCP6	BCP10
L*	78.38 ± 0.84 [a]	65.50 ± 0.71 [b]	56.98 ± 1.12 [c]	52.36 ± 1.17 [d]
A*	−3.16 ± 0.10 [d]	1.43 ± 0.53 [c]	3.53 ± 0.31 [b]	4.38 ± 0.43 [a]
B*	16.86 ± 1.72 [a]	14.72 ± 1.66 [b]	12.43 ± 1.08 [c]	12.19 ± 0.61 [c]
ΔE	−	13.84	22.85	27.49

Mean values with the same letters in the row ([a–d]) were not significantly different (α = 0.05). RB—reference bread; BCP2, BCP6, BCP10—breads with starch replaced with CP at 2%, 6%, and 10%, respectively.

2.3. Total Phenolic Compounds and Antioxidative Activity

Chronic oxidative stress can cause a variety of diseases [47,48]. Reactive oxygen species are involved in the oxidation of lipids, proteins, and nucleic acids, which can lead to changes in cells and even cell death. Oxidative stress can be reduced by providing antioxidant compounds to the diet. Plants are a widely reported source of antioxidants [49–52]. Edible insects, in addition to basic nutrients, also provide biologically active compounds, including antioxidants, but also anti-nutritional compounds, such as phytic acid, saponins, oxalates, and tannins. Those undesirable compounds may adversely affect health after prolonged consumption, so their levels in food products should be monitored [53,54]. Table 5 shows the results of the antioxidant activity as well as the total polyphenol content. With the increase in the amount of starch replaced with CP, the polyphenol content in bread increased by 336% (RB vs. BCP10). The analyzed antioxidant activity also increased due to the addition of CP. However, providing antioxidants in food will not have a beneficial effect on our body. Similar to other nutrients, antioxidants must first be released from the food matrix, initially by grinding the food mechanically and then chemically and enzymatically. They can then be absorbed by the digestive tract, especially in the upper part of the small intestine [55]. A significant increase in the content of polyphenolic compounds derived from RB and breads with CP addition was observed after the digestion process. As in the case of total polyphenols content (TPC), the highest value of 6.2 mg/g was recorded for BCP10 vs. 1.9 mg/g for RB. The total antioxidant activity of the breads after digestion increased significantly for each analyzed bread type as well. The highest TEAC value of 42.79 mg/g was recorded for BCP10, which was also the largest increase in activity caused by the digestive process (by as much as 2009%). The increase in the antioxidant capacity due to CP addition results from the presence of active compounds in it, but also from the method of the CP preparation. According to Zielińska et al. [23], thermal treatment of insects may significantly increase their biological activity. Similarly, the enzymatic hydrolysis process, analogous to the digestive process in the human gastrointestinal tract, may cause an additional increase in activity [22], also observed herein. Furthermore, the influence of the intestinal microflora may increase the antioxidant potential of the digested products [55,56]. The possible impacts of these metabolic processes taking place mainly in the large intestine on the CP nutritional properties cannot be ignored.

Table 5. Antioxidant activity and total polyphenol content during digestion in the gastrointestinal tract model.

The Stages of Digestion	Total Polyphenols (mg Gallic Acid/g)				Antioxidant Activity (mg Trolox/g)			
	RB	BCP2	BCP6	BCP10	RB	BCP2	BCP6	BCP10
Before digestion	0.224 ± 0.008	0.246 ± 0.006	0.400 ± 0.090	0.977 ± 0.014	0.478 ± 0.071	1.296 ± 0.047	1.699 ± 0.262	2.029 ± 0.036
2h at pH 2.0 "after stomach"	0.851 ± 0.040	1.192 ± 0.067	2.054 ± 0.070	2.780 ± 0.024	0.199 ± 0.047	0.965 ± 0.055	1.516 ± 0.069	2.001 ± 0.182
pH 7.4 "in small intestine"	0.904 ± 0.027 [a]	0.993 ± 0.048 [a]	1.964 ± 0.025	2.314 ± 0.115	3.914 ± 0.675	5.617 ± 0.497	9.323 ± 0.234	15.398 ± 0.306
pH 7.4 "in small intestine" with fecal flora	0.588 ± 0.063	0.993 ± 0.073	2.040 ± 0.097	2.833 ± 0.236	3.583 ± 0.284	5.810 ± 0.142	11.059 ± 0.169	16.226 ± 0.308
2 h at pH 7.4 "after small intestine"	0.651 ± 0.062	0.876 ± 0.039	2.599 ± 0.025	3.166 ± 0.151	4.617 ± 0.799 [a]	5.621 ± 0.028 [a]	11.617 ± 0.581	18.597 ± 0.318
pH 8.0 "in large intestine"	0.656 ± 0.031	1.246 ± 0.016	2.773 ± 0.093	4.026 ± 0.118	5.524 ± 0.389	8.823 ± 0.488	15.585 ± 0.426	23.840 ± 0.600
18 h at pH 8.0 "after large intestine"	1.934 ± 0.084	2.339 ± 0.049	4.265 ± 0.421	6.236 ± 0.610	7.475 ± 0.288	17.204 ± 0.566	30.625 ± 0.693	42.791 ± 0.922

Mean values with the same letters in the row ([a]) were not significantly different ($\alpha = 0.05$). RB—reference bread; BCP2, BCP6, BCP10—breads with starch replaced with CP at 2%, 6%, and 10%, respectively.

2.4. β-Glucuronidase Activity

The level of β-glucuronidase (β-Glu) activity in body fluids is considered a potential biomarker in the diagnosis of certain intestinal pathological conditions [57]. Therefore, the search for potent β-Glu inhibitors in the human intestinal microflora has attracted increasing attention over the years [58], due to its role in colon carcinogenesis In particular, work is underway to discover natural dietary inhibitors of this enzyme. To date, the main strategy for reducing or eliminating the gastrointestinal toxicity caused by bacterial β-Glu is by administration of antibiotics [59,60], but plant food and herbal medicines are a promising source of bacterial inhibitors as well [58,61,62]. To the best of our knowledge, there are no published reports on the β-Glu inhibitory activity of GF bread with insects. Our results indicate a significant ability to reduce the activity of β-Glu by CP (Table 6). The use of 6% and 10% CP in the bread recipe resulted in a decrease in the activity of β-Glu at the stage of adding the intestinal microflora to the digestive process by 63.5% and 65.9%, respectively. The β-Glu activity is also effectively inhibited in the subsequent stages of the digestive process in the large intestine. After 18 h of BCP6 and BCP10 digestion, after large intestine, the β-Glu activity was reduced by 70.6% and 78.9%, respectively. The use of food additives that inhibit β-Glu activity may also be of key importance in the treatment of certain diseases. As β-Glu plays a key role in reducing the effectiveness of anticancer drugs [63], its inhibition with food ingredients may aid the treatment of certain diseases in a non-pharmacological manner. It therefore seems that CP has the potential to be used as a new β-Glu inhibitor, and further studies of the biological activity of CP may deepen our understanding of the mechanisms underlying the beneficial effects observed in the current experiments.

Table 6. The results of β-glucuronidase activity after bread digestion.

The Stages of Digestion	β-Glucuronidase Activity (mg/g of Soluble Nitrogen)			
	RB	BCP2	BCP6	BCP10
pH 7.4 "in small intestine" with fecal flora	0.572 ± 0.004	0.695 ± 0.005	0.209 ± 0.001	0.195 ± 0.001
2 h at pH 7.4 "after small intestine"	0.599 ± 0.017	0.649 ± 0.008	0.243 ± 0.001	0.211 ± 0.001
pH 8.0 "in large intestine"	0.648 ± 0.007 [a]	0.648 ± 0.013 [a]	0.254 ± 0.001	0.205 ± 0.001
18 h at pH 8.0 "after large intestine"	1.051 ± 0.010	1.281 ± 0.012	0.309 ± 0.002	0.222 ± 0.001

Mean values with the same letters in the row ([a]) were not significantly different ($\alpha = 0.05$). RB—reference bread; BCP2, BCP6, BCP10—breads with starch replaced with CP at 2%, 6%, and 10%, respectively.

2.5. Effect on Intestinal Microflora

Antioxidant compounds very often also show antimicrobial activity. The polyphenols, present mainly in plants [64,65], show a strong antimicrobial activity against human pathogens, but may also adversely affect the growth of the beneficial intestinal microflora. The exact mechanisms of the antimicrobial action of phenolic compounds are not yet fully understood, as they are added to food products for preservation [64,66,67]. After the in vitro digestion process, no inhibitory effect of ingredients derived from bread with CP was observed on the growth of microorganisms, either beneficial (*Bifidobacterium* and *Lactobacillus*) or pathogenic (*Enterococcus* and *Escherichia coli*) (Table 7). Literature data indicate that some physiological functions of bacteria, such as tolerance of pH changes in the gastrointestinal tract, growth, temperature, and availability of substrates necessary for growth, have a decisive impact on the survival of a specific group of microorganisms in the human gastrointestinal tract [68,69]. Pectin, starch, and sugar were used to prepare the bread dough, which allows easy access to nutrients for the microflora. Nevertheless, in the case of BCP10, it was noticed that the growth of the microflora was slightly slowed from the very beginning of the digestive process. Therefore, it can be assumed that a small addition of CP does not impede microfloral growth, but concentration of the antimicrobial compounds increased with an increasing portion of CP in the bread recipe. This hypothesis requires further research for full explanation.

Table 7. Quantitative changes in the intestinal microflora during digestion of the analyzed breads [log 10 cfu/mL].

Microorganisms	RB			BCP2		
	pH 7.4 [1]	2 h pH 7.4 [2]	18 h pH 8.0 [3]	pH 7.4 [1]	2 h pH 7.4 [2]	18 h pH 8.0 [3]
Bifidobacterium	7.471 ± 0.154	8.883 ± 0.028	10.123 ± 0.114	7.537 ± 0.045	9.007 ± 0.075	10.382 ± 0.114
Lactobacillus	7.543 ± 0.035	8.875 ± 0.025	10.181 ± 0.256	7.576 ± 0.081	8.995 ± 0.025	10.254 ± 0.152
Enterococcus	6.576 ± 0.081	6.880 ± 0.032	8.605 ± 0.069	6.465 ± 0.094	7.203 ± 0.038	9.228 ± 0.162
E. coli	6.894 ± 0.027	8.851 ± 0.009	9.488 ± 0.206	7.465 ± 0.094	8.928 ± 0.051	10.477 ± 0.020
Microorganisms	BCP6			BCP10		
	pH 7.4 [1]	2 h pH 7.4 [2]	18 h pH 8.0 [3]	pH 7.4 [1]	2 h pH 7.4 [2]	18 h pH 8.0 [3]
Bifidobacterium	7.941 ± 0.032	8.922 ± 0.070	12.300 ± 0.031	6.511 ± 0.028	7.594 ± 0.070	9.339 ± 0.153
Lactobacillus	7.672 ± 0.049	8.841 ± 0.059	12.385 ± 0.006	6.578 ± 0.049	7.560 ± 0.059	9.272 ± 0.178
Enterococcus	7.244 ± 0.029	8.484 ± 0.814	9.253 ± 0.010	6.021 ± 0.029	6.655 ± 0.814	9.145 ± 0.044
E. coli	7.961 ± 0.097	8.657 ± 0.069	10.466 ± 0.005	6.500 ± 0.097	7.554 ± 0.069	8.724 ± 0.172

[1] "in small intestine" with fecal flora; [2] "after small intestine"; [3] "after large intestine". RB—reference bread; BCP2, BCP6, BCP10—breads with starch replaced with CP at 2%, 6%, and 10%, respectively.

2.6. Principal Component Analysis

Principal component analysis (PCA) of the proximate composition (ash, carbohydrates, fat, protein and total dietary fiber (TDF) contents), saturated fatty acids (SFA), mono- and polyunsaturated fatty acids (MUFA, PUFA), as well as total polyphenols content (TPC), antioxidative activity (TEAC), and β-glucuronidase was performed to analyze the main factors determining the properties of the analyzed GF breads enriched with CP. The first two principal factors accounted for 99% (F1 = 94.36% and F2 = 4.64%) of the total variation. The projection of cases on the factor plane showed significant differences between the properties of the individual analyzed bread variants, with the smallest differences observed for BCP6 and BCP10, whereas RB differed significantly from breads enriched with CP (Figure 2). The loadings plot (Figure 2A) shows that factor 1 was mainly correlated with MUFA (r = 0.999), carbohydrates content (r = 0.991) and activity of β-glucuronidase (r = 0.895). It was also strongly negatively correlated with the SFA (r = −0.999), PUFA (r = −0.998), TDF (r = −0.998), TEAC (r = −0.998), protein content (r = −0.996), fat content (r = −0.988), and TPC (r = −0.979). The score plot (Figure 2B) shows data divided into three groups. Each group placed in a different quadrant on the score plot. The first and second analyzed samples (RB and BCP2) are on the right side of the Y axis, but at a large distance from each other. These samples showed a low content of TPC (below 2.34 mg gallic acid/g), high activity of β-glucuronidase (above 1.051 mg/g of soluble nitrogen), and low content of SFA (below 7.18%). TEAC was the value that had the greatest impact on their diversity and the distance between them. On the left side of the Y axis and under the X axis, the third group is located; it includes the BCP6 and BCP10 samples. These two samples showed a very low activity of β-glucuronidase (below 0.309 mg/g of soluble nitrogen), higher TEAC value (above 30.625 mg Trolox/g) and higher protein content (above 5.85%). Because the only variable in the recipe was the equivalent replacement of starch with CP, it can therefore be clearly stated that such replacement improved the nutritional value of the bread, and changed its biological activity, which is crucial in products for people with intestinal diseases.

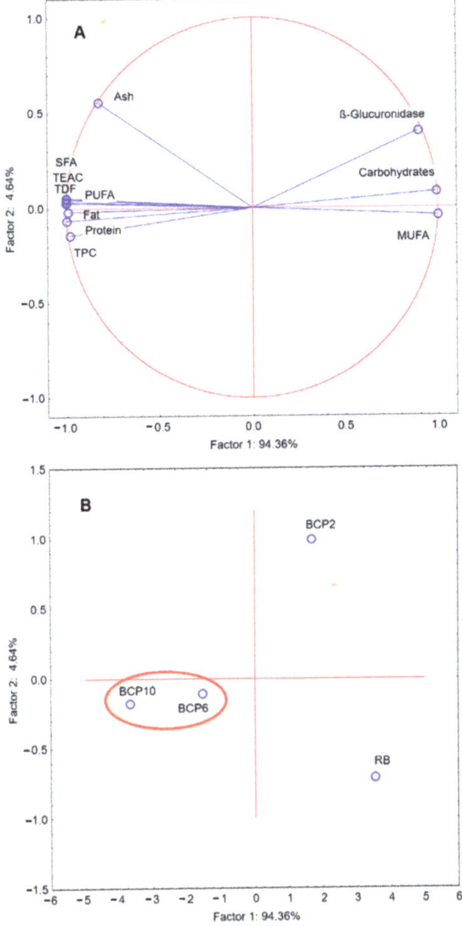

Figure 2. Principal component analysis (PCA) of the loadings plot (**A**) and the score plot (**B**) of data from ash, carbohydrates, fat, protein, and total dietary fiber (TDF) contents, saturated fatty acids (SFA), mono- and polyunsaturated fatty acids (MUFA, PUFA), and total polyphenols content (TPC), antioxidative (TEAC) and β-glucuronidase activity. RB—reference bread; BCP2, BCP6, BCP10—breads with starch replaced with CP at 2%, 6%, and 10%, respectively.

3. Materials and Methods

3.1. Bread Production

Bread was prepared according to the method previously described [70]. The reference bread (RB) consisted of: 200 g corn starch, 50 g potato starch, 4.25 g guar gum, 4.25 g pectin, 15 g yeast, 5 g sugar, 4.25 g salt, 7.5 g rapeseed oil, and 275 g demineralized water. The analyzed breads were obtained by replacing the total starch by the amount of 2% (BCP2), 6% (BCP6), and 10% (BCP10) of commercially available cricket (*Acheta domesticus*) powder (Crunchy Critters, Derby, UK), produced from adult crickets. Dough was prepared using a straight dough method (mixing time: 8 min, 70 rpm). The finished dough was placed in a fermentation chamber for 20 min (temperature 35 °C, relative humidity 85%), punched, divided into equal pieces (280 g), and fermentation continued for 15 min more. The dough was then baked (at 230 °C for 30 min), cooled, and packed in polypropylene pouches. Bread production was performed three times and then the samples were standardized.

3.2. Nutritional Value and Minerals Content

The total nitrogen was determined by Kjeldahl method, according to ISO 20,483 [71], and was used to calculate protein content (P) by multiplying by a nitrogen conversion factor equaling 5.09. The ash content was determined according to ISO 2171 [72] and the total fat content (F) was determined according to AACC 30-25.01 [73]. The moisture content was analyzed according to AACC 44-19.01 [74]. The content of dietary fiber (TDF), both soluble (SDF) and insoluble (IDF), was determined by the enzymatic method in accordance with the AOAC 991.43 method [75]. The proximate carbohydrate content (C) was estimated by subtracting the total ash, fat, fiber, protein, and moisture content from 100%. Moreover, the energy value (EV) was calculated with the following formula [76]:

$$EV\ [kcal/100\ g] = 4 \times (P + C) + 2 \times TDF + 9 \times F$$

For mineral content determination, about 1 g of each sample was weighted into Teflon vessels and 7 mL of HNO_3 (65%) and 1 mL of H_2O_2 (30%) were added [77]. The digestion was carried out according to the conditions: temperature of 210 °C, ramp time of 15 min, hold time of 15 min, pressure of 800 psi and power of 900 to 1050 W (CEM 6, Mars, CEM Corporation, Matthews, NC, USA). After cooling, the digests were transferred to polymethylopropylene flasks and diluted to 50 mL with demineralized water. For each bread, three digestions were prepared.

The concentrations of the most common minerals—Ca, Cu, Fe, K, Mg, Mn, and Zn—were determined by microwave plasma atomic emission spectrometry using Agilent MP-AES 4210 (Agilent Technologies, Melbourne, Australia) [78]. At least two calibration curves were prepared for the measurement. The content of P was determined using the molybdate-ascorbic acid colorimetric method AOAC 995.11 [79] transformed into microwell plate measurements (PowerWave XS2, Biotek Instruments, Winooski, VT, USA). For each digest, three reactions were performed. The values of population reference intake (PRI) and adequate intake (AI) were established at the level of EFSA recommendations [80]: PRIs for Ca, Cu, Fe, Mg, Mn, and Zn, and AIs for K and Na.

The fatty acid composition of the lipids in the bread was determined according to the AOCS Official Method Ce 1 h-05 [81]. The lipids were extracted by the traditional Folch method [82] and then fatty acid methyl esters (FAME) were analyzed with an Agilent 7820A GC (Agilent Technologies, Santa Clara, CA, USA) equipped with a flame ionization detector (FID) and SLB-IL111 capillary column (Supelco, Bellefonte, PA, USA) (100 m, 0.25 mm, 0.20 µm). The detailed parameters of the analysis were described earlier [83]. The results are presented as a percentage of total fatty acids.

3.3. Color Measurements

The color of the crumb was measured using a Chroma Meter CR-410 (Konica Minolta Sensing Inc., Tokyo, Japan) [84]. Differences in color were recorded in CIE L*a*b* scale in terms of lightness (L*) and color (a*—redness; b*—yellowness). Color measurement was repeated 15 times for each sample. Additionally, the total color difference (ΔE) was calculated using the following formula:

$$\Delta E = \sqrt{\Delta L^{*2} + \Delta a^{*2} + \Delta b^{*2}}$$

3.4. In Vitro Digestion Process

The in vitro digestion was conducted in a glass bioreactor equipped with 4 inlets, allowing the introduction of the pH electrode, programming of the active acidity, dosage of biochemical agents, and appropriate media as well as collection of analytical samples. Samples for the in vitro digestion process were prepared by taking 23 g of products and dissolving them in demineralized water to a volume of 230 mL. During the process, the total polyphenol content, antioxidation potential, and β-glucuronidase activity were determined.

The bioreactor was thermally stable and the reactions were carried out at 37 °C. The conditions of the process in the bioreactor were designed in such a way as to comprise the following stages of the model: the 'stomach', the 'small intestine', and the 'large intestine' (Figure 3). The parameters of the digestion process were selected on the basis of our previous investigations [27,30]. *Stomach stage*: the digestion process was started by simulating digestion in the stomach. In this process, 60,000 U pepsin (Sigma-Aldrich, St. Louis, MO, USA) suspended in 2 mL 0.1 M HCl was added to the sample. The pH was then adjusted to 2.0 and the digestion continued for 4 h. *Small intestine stage*: after 4 h of digestion, the pH was adjusted to 6.0 by the addition of 1 M $NaHCO_3$, followed by the addition of 10 mL of a pancreatic-intestinal extract composed of 0.02 g of pancreatic extract (Sigma-Aldrich, St. Louis, MO, USA) and 0.12 g of bile salt (Sigma-Aldrich, St. Louis, MO, USA) dissolved in 10 mL 0.1 M $NaHCO_3$. The pH was then adjusted to 7.4 with 1 M $NaHCO_3$, and human intestinal microflora prepared according to the method described by Knarreborg et al. [85] was added at a total count of about 10^6 cfu/mL. Intestinal digestion was carried out for 2 h. *Large intestine stage*: after the digestion process in the small intestine, the pH was adjusted to 8.0 by adding 2 M $NaHCO_3$ and the fermentation was continued for another 18 h. A nitrogen stream was passed through the bioreactor to ensure anaerobic conditions

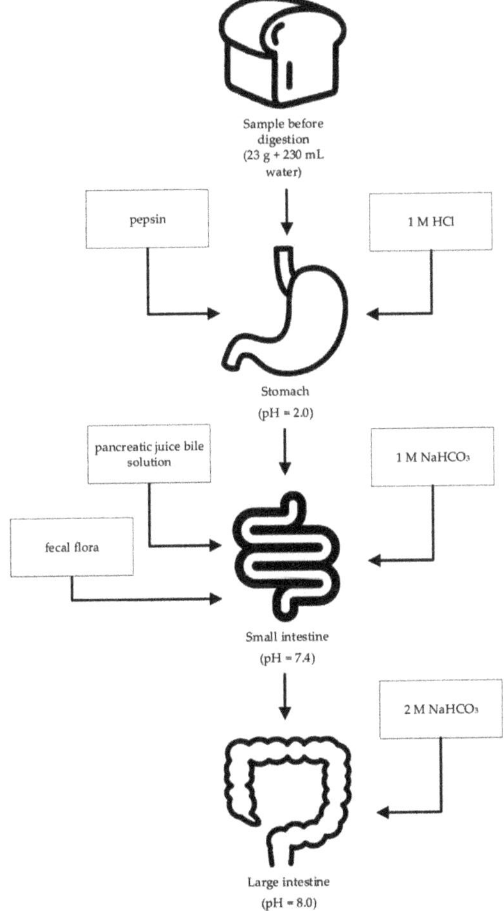

Figure 3. Gastrointestinal tract model.

3.5. Extraction Process of Antioxidants

The extraction process was carried out using a 70% solution of acetone employing a single extraction of polyphenols from the examined samples. In this process, 3 mL of the digested material was mixed with 7 mL of acetone ($\geq$99%) to obtain 70% concentration of acetone in the solution. The mixture was then shaken (shaker type KL-942) for 1 h at room temperature. Afterwards, the samples were centrifuged by $1700\times g$ and 300 µL of the supernatant was used for total polyphenols and antioxidative activity analysis.

3.6. Total Polyphenols Content

The total polyphenols content was measured using the modified Folin–Ciocalteu method, and its values were estimated from a standard curve of gallic acid. All results were corrected for the presence of phenols in the pancreatin/bile salts mixture. The results were expressed as equivalents of gallic acid in mg per g of digested products [86].

3.7. Antioxidant Activity

The antioxidative activity (TEAC) was determined against the ABTS reagent (2,2′-azinobis-(3-ethylbenzothiazoline-6-sulphonic acid) according to the method described by Re et al. [87]. The results of the TEAC assay were expressed as mg Trolox/g of the examined extract.

3.8. β-Glucuronidase Activity

Determination of β-glucuronidase (EC 3.2.1.31) was based on the Kapnoor and Mulimani methodology [88]. The test was prepared by withdrawing from the bioreactor 200 µL of the contents, to which phosphate buffer (pH 7.0) and NaCl (0.1 M) were added. The samples were shaken for 1 h, then centrifuged to give the supernatant. Then, 200 mL substrate (1 mg/mL β-glucuronidase suspended in phosphate buffer, pH 6.7) was added to 200 mL of the supernatant, and incubated for 2.5 h at 40 °C. The reaction was stopped with sodium carbonate. The absorbance was measured at 420 nm. The total content of β-glucuronidase was expressed as mM/g of nitrogen soluble.

3.9. Effect on Intestinal Microflora

To control the influence of the conditions prevailing in the gastrointestinal tract on the growth of microorganisms, control inoculations were made after 2 h from the moment (pH 7.4, small intestine) of introducing the microorganisms into the environment and at the moment of the termination of the digestion process (after 21 h). The intestinal microflora isolated from the faecalis of a mature person was introduced into the experimental model. The determined groups of microorganisms included: *Entrobacteriaceae* (MacConkey selective medium—Sigma Aldrich, Saint Louis, MO, USA), *Lactobacillus* (MRS agar medium—Sigma Aldrich, Saint Louis, MO, USA), *Enteroccocus* (substrate—agar with kanamycin, esculin, and sodium azide), and *Bifidobacterium* (Garche medium—Sigma Aldrich, Saint Louis, MO, USA). Inoculated media were incubated in anaerobic conditions depending on the determined group of microorganisms for the period of 48 to 72 h at 37 °C [27]. The number of viable bacterial cells was determined using Koch's plate method.

3.10. Statistical Analysis

Statistical analysis of the data was performed with Statistica 13 (Dell Software Inc., USA) software. For every test, three independent measurements were taken, unless stated otherwise. All measurements were studied using one-way analysis of variance independently for each dependent variable. Post hoc Tukey honest significant difference (HSD) multiple comparison tests were used to identify statistically homogeneous subsets at $\alpha = 0.05$. Principal component analysis (PCA) was performed using selected data obtained in the analyses. The results were presented in a two-dimensional system (biplot) obtained by plotting the observations and variables on the plane formed by the calculated principal components.

4. Conclusions

To improve the available product selection, as well as the nutritional properties and the consumer appeal of the available gluten-free breads, cricket powder can be used as an effective fortifier. The obtained bread enriched with CP was characterized by a higher content of protein, polyunsaturated fatty acids, and minerals desirable in nutrition. The more starch that was replaced with CP, the greater the observed increases in the content of said nutrients. Other valuable benefits of CP enrichment included the increased antioxidant activity of the bread and the decreased β-glucuronidase activity. Simulated digestion further underscored the observed superiority of the CP-enriched breads. Overall, cricket powder is a promising raw material for the production of gluten-free functional foods. Further studies on the biological activities of CP and products enriched with it could improve our knowledge about the mechanisms underlying the beneficial effects reported herein.

Author Contributions: Conceptualization, P.Ł.K.; investigation, P.Ł.K., M.G., I.R., B.G., P.S., K.D. and D.K.; methodology, P.Ł.K., M.G. and I.R.; project administration, P.Ł.K.; supervision, P.Ł.K.; visualization, P.Ł.K.; writing—original draft, P.Ł.K., M.G. and I.R.; writing—review and editing, P.Ł.K. and D.K. All authors have read and agreed to the published version of the manuscript.

Funding: The authors received no financial support for the research, authorship, and/or publication of this article. The APC was funded by MDPI.

Institutional Review Board Statement: Not applicable.

Informed Consent Statement: Not applicable.

Data Availability Statement: The datasets generated during and/or analyzed during the current study are available from the corresponding author on reasonable request.

Acknowledgments: The authors thank Olga Bartczak and Krzysztof Smarzyński (Students' Scientific Club of Food Technologists, Poznań University of Life Sciences, Poznań, Poland) for their help with the preparation of bread and analyses. Marcin Nowicki (University of Tennessee, Knoxville, USA) is gratefully acknowledged for their copyediting and critical reading of this manuscript.

Conflicts of Interest: On behalf of all authors, the corresponding author states that there is no conflict of interest.

Sample Availability: Samples of breads and CP are available from the corresponding author.

References

1. Rybicka, I.; Doba, K.; Bińczak, O. Improving the sensory and nutritional value of gluten-free bread. *Int. J. Food Sci. Technol.* **2019**, *54*, 2661–2667. [CrossRef]
2. Xhakollari, V.; Canavari, M.; Osman, M. Factors affecting consumers' adherence to gluten-free diet, a systematic review. *Trends Food Sci. Technol.* **2019**, *85*, 23–33. [CrossRef]
3. Pruska-Kędzior, A.; Kędzior, Z.; Gorący, M.; Pietrowska, K.; Przybylska, A.; Spychalska, K. Comparison of rheological, fermentative and baking properties of gluten-free dough formulations. *Eur. Food Res. Technol.* **2008**, *227*, 1523–1536. [CrossRef]
4. Vici, G.; Belli, L.; Biondi, M.; Polzonetti, V. Gluten free diet and nutrient deficiencies: A review. *Clin. Nutr.* **2016**, *35*, 1236–1241. [CrossRef] [PubMed]
5. Naqash, F.; Gani, A.; Gani, A.; Masoodi, F.A. Gluten-free baking: Combating the challenges—A review. *Trends Food Sci. Technol.* **2017**, *66*, 98–107. [CrossRef]
6. Raheem, D.; Carrascosa, C.; Oluwole, O.B.; Nieuwland, M.; Saraiva, A.; Millán, R.; Raposo, A. Traditional consumption of and rearing edible insects in Africa, Asia and Europe. *Crit. Rev. Food Sci. Nutr.* **2019**, *59*, 2169–2188. [CrossRef]
7. Ghosh, S.; Jung, C.; Meyer-Rochow, V.B. What Governs Selection and Acceptance of Edible Insect Species? In *Edible Insects in Sustainable Food Systems*; Springer International Publishing: Cham, Switzerland, 2018; pp. 331–351.
8. van Huis, A. Potential of Insects as Food and Feed in Assuring Food Security. *Annu. Rev. Entomol.* **2013**, *58*, 563–583. [CrossRef] [PubMed]
9. Rossman, S. 2019 Food Trends: Will Cricket Powder, Edible Insects Enter US Diet? Available online: https://eu.usatoday.com/story/news/investigations/2018/12/21/2019-food-trends-cricket-powder-edible-insects-enter-us-diet/2351371002/ (accessed on 29 June 2019).
10. FAO. Edible Insects—Future prospects for food and feed security. In *FAO Forestry Paper*; Food and Agriculture Organization of the United Nations: Rome, Italy, 2013; Volume 171, ISBN 9789251075951.

11. van Huis, A. Edible insects are the future? *Proc. Nutr. Soc.* **2016**, *75*, 294–305. [CrossRef]
12. Duda, A.; Adamczak, J.; Chełmińska, P.; Juszkiewicz, J.; Kowalczewski, P. Quality and Nutritional/Textural Properties of Durum Wheat Pasta Enriched with Cricket Powder. *Foods* **2019**, *8*, 46. [CrossRef]
13. Zielińska, E.; Pankiewicz, U. Nutritional, Physiochemical, and Antioxidative Characteristics of Shortcake Biscuits Enriched with Tenebrio molitor Flour. *Molecules* **2020**, *25*, 5629. [CrossRef]
14. Kowalczewski, P.Ł.; Walkowiak, K.; Masewicz, Ł.; Smarzyński, K.; le Thanh-Blicharz, J.; Kačániová, M.; Baranowska, H.M. LF NMR spectroscopy analysis of water dynamics and texture of Gluten-Free bread with cricket powder during storage. *Food Sci. Technol. Int.* **2021**. [CrossRef] [PubMed]
15. Meyer-Rochow, V.B.; Ghosh, S.; Jung, C. Farming of insects for food and feed in South Korea: Tradition and innovation. *Berl. Munch. Tierarztl. Wochenschr.* **2019**, *132*, 236–244. [CrossRef]
16. Kouřimská, L.; Adámková, A. Nutritional and sensory quality of edible insects. *NFS J.* **2016**, *4*, 22–26. [CrossRef]
17. Ghosh, S.; Lee, S.-M.; Jung, C.; Meyer-Rochow, V.B. Nutritional composition of five commercial edible insects in South Korea. *J. Asia Pac. Entomol.* **2017**, *20*, 686–694. [CrossRef]
18. Hartmann, C.; Shi, J.; Giusto, A.; Siegrist, M. The psychology of eating insects: A cross-cultural comparison between Germany and China. *Food Qual. Prefer.* **2015**, *44*, 148–156. [CrossRef]
19. Kauppi, S.-M.; Pettersen, I.N.; Boks, C. Consumer acceptance of edible insects and design interventions as adoption strategy. *Int. J. Food Des.* **2019**, *4*, 39–62. [CrossRef]
20. Montowska, M.; Kowalczewski, P.Ł.; Rybicka, I.; Fornal, E. Nutritional value, protein and peptide composition of edible cricket powders. *Food Chem.* **2019**, *289*, 130–138. [CrossRef]
21. Rybicka, I. The Handbook of Minerals on a Gluten-Free Diet. *Nutrients* **2018**, *10*, 1683. [CrossRef]
22. Zielińska, E.; Baraniak, B.; Karaś, M. Identification of antioxidant and anti-inflammatory peptides obtained by simulated gastrointestinal digestion of three edible insects species (Gryllodes sigillatus, Tenebrio molitor, Schistocerca gragaria). *Int. J. Food Sci. Technol.* **2018**, *53*, 2542–2551. [CrossRef]
23. Zielińska, E.; Baraniak, B.; Karaś, M. Antioxidant and Anti-Inflammatory Activities of Hydrolysates and Peptide Fractions Obtained by Enzymatic Hydrolysis of Selected Heat-Treated Edible Insects. *Nutrients* **2017**, *9*, 970. [CrossRef]
24. Zielińska, E.; Baraniak, B.; Karaś, M.; Rybczyńska, K.; Jakubczyk, A. Selected species of edible insects as a source of nutrient composition. *Food Res. Int.* **2015**, *77*, 460–466. [CrossRef]
25. Klimova, B.; Kuca, K.; Maresova, P. Global View on Alzheimer's Disease and Diabetes Mellitus: Threats, Risks and Treatment Alzheimer's Disease and Diabetes Mellitus. *Curr. Alzheimer Res.* **2018**, *15*, 1277–1282. [CrossRef] [PubMed]
26. Incalza, M.A.; D'Oria, R.; Natalicchio, A.; Perrini, S.; Laviola, L.; Giorgino, F. Oxidative stress and reactive oxygen species in endothelial dysfunction associated with cardiovascular and metabolic diseases. *Vascul. Pharmacol.* **2018**, *100*, 1–19. [CrossRef] [PubMed]
27. Gumienna, M.; Lasik, M.; Czarnecki, Z. Bioconversion of grape and chokeberry wine polyphenols during simulated gastrointestinal in vitro digestion. *Int. J. Food Sci. Nutr.* **2011**, *62*, 226–233. [CrossRef]
28. Rios, L.Y.; Bennett, R.N.; Lazarus, S.A.; Rémésy, C.; Scalbert, A.; Williamson, G. Cocoa procyanidins are stable during gastric transit in humans. *Am. J. Clin. Nutr.* **2002**, *76*, 1106–1110. [CrossRef] [PubMed]
29. Petri, N.; Tannergren, C.; Holst, B.; Mellon, F.A.; Bao, Y.; Plumb, G.W.; Bacon, J.; O'Leary, K.A.; Kroon, P.A.; Knutson, L.; et al. Absorption/Metabolism of Sulforaphane and Quercetin, and Regulation of Phase II Enzymes, in Human Jejunum in Vivo. *Drug Metab. Dispos.* **2003**, *31*, 805–813. [CrossRef] [PubMed]
30. Gumienna, M.; Goderska, K.; Nowak, J.; Czarnecki, Z. Changes in the antioxidative activities of bean products and intestinal microflora in the model of the gastrointestinal tract "in vitro". *Polish J. Environ. Stud.* **2006**, *15*, 29–32.
31. Stull, V.J.; Finer, E.; Bergmans, R.S.; Febvre, H.P.; Longhurst, C.; Manter, D.K.; Patz, J.A.; Weir, T.L. Impact of Edible Cricket Consumption on Gut Microbiota in Healthy Adults, a Double-blind, Randomized Crossover Trial. *Sci. Rep.* **2018**, *8*, 10762. [CrossRef]
32. Melini, V.; Melini, F. Gluten-Free Diet: Gaps and Needs for a Healthier Diet. *Nutrients* **2019**, *11*, 170. [CrossRef]
33. Jonas-Levi, A.; Martinez, J.-J.J.I. The high level of protein content reported in insects for food and feed is overestimated. *J. Food Compos. Anal.* **2017**, *62*, 184–188. [CrossRef]
34. Chapman, R.F. *The Insects: Structure and Function*; Cambridge University Press: Cambridge, UK, 1998; ISBN 9780521570480.
35. Ritvanen, T.; Pastell, H.; Welling, A.; Raatikainen, M. The nitrogen-to-protein conversion factor of two cricket species—Acheta domesticus and Gryllus bimaculatus. *Agric. Food Sci.* **2020**, *29*. [CrossRef]
36. Tzompa-Sosa, D.A.; Yi, L.; van Valenberg, H.J.F.; van Boekel, M.A.J.S.; Lakemond, C.M.M. Insect lipid profile: Aqueous versus organic solvent-based extraction methods. *Food Res. Int.* **2014**, *62*, 1087–1094. [CrossRef]
37. Rumpold, B.A.; Schlüter, O.K. Nutritional composition and safety aspects of edible insects. *Mol. Nutr. Food Res.* **2013**, *57*, 802–823. [CrossRef]
38. Kinyuru, J.N.; Mogendi, J.B.; Riwa, C.A.; Ndung'u, N.W. Edible insects—A novel source of essential nutrients for human diet: Learning from traditional knowledge. *Anim. Front.* **2015**, *5*, 14–19. [CrossRef]
39. Udousoro, I.I.; Udo, E.S.; Udoh, A.P.; Udoanya, E.E. Proximate and antinutrients compositions, and health risk assessment of toxic metals in some edible vegetables. *Niger. J. Chem. Res.* **2018**, *23*, 51–62.

40. Kim, D.-H.; Kim, E.-M.; Chang, Y.-J.; Ahn, M.-Y.; Lee, Y.-H.; Park, J.J.; Lim, J.-H. Determination of the shelf life of cricket powder and effects ofstorage on its quality characteristics. *Korean J. Food Preserv.* **2016**, *23*, 211–217. [CrossRef]

41. Bahrami, N.; Yonekura, L.; Linforth, R.; da Silva, M.C.; Hill, S.; Penson, S.; Chope, G.; Fisk, I.D. Comparison of ambient solvent extraction methods for the analysis of fatty acids in non-starch lipids of flour and starch. *J. Sci. Food Agric.* **2014**, *94*, 415–423. [CrossRef]

42. Osimani, A.; Milanović, V.; Cardinali, F.; Roncolini, A.; Garofalo, C.; Clementi, F.; Pasquini, M.; Mozzon, M.; Foligni, R.; Raffaelli, N.; et al. Bread enriched with cricket powder (Acheta domesticus): A technological, microbiological and nutritional evaluation. *Innov. Food Sci. Emerg. Technol.* **2018**, *48*, 150–163. [CrossRef]

43. Sandvik, P.; Nydahl, M.; Kihlberg, I.; Marklinder, I. Consumers' health-related perceptions of bread—Implications for labeling and health communication. *Appetite* **2018**, *121*, 285–293. [CrossRef] [PubMed]

44. Ronda, F. Effects of polyols and nondigestible oligosaccharides on the quality of sugar-free sponge cakes. *Food Chem.* **2005**, *90*, 549–555. [CrossRef]

45. Purlis, E.; Salvadori, V.O. Modelling the browning of bread during baking. *Food Res. Int.* **2009**, *42*, 865–870. [CrossRef]

46. Mokrzycki, W.; Tatol, M. Color difference Delta E—A survey. *Mach. Graph. Vis.* **2011**, *20*, 383–411.

47. Alfadda, A.A.; Sallam, R.M. Reactive Oxygen Species in Health and Disease. *J. Biomed. Biotechnol.* **2012**, *2012*, 1–14. [CrossRef] [PubMed]

48. Collin, F. Chemical Basis of Reactive Oxygen Species Reactivity and Involvement in Neurodegenerative Diseases. *Int. J. Mol. Sci.* **2019**, *20*, 2407. [CrossRef]

49. Rovná, K.; Ivanišová, E.; Žiarovská, J.; Ferus, P.; Terentjeva, M.; Kowalczewski, P.Ł.; Kačániová, M. Characterization of Rosa canina Fruits Collected in Urban Areas of Slovakia. Genome Size, iPBS Profiles and Antioxidant and Antimicrobial Activities. *Molecules* **2020**, *25*, 1888. [CrossRef]

50. Ražná, K.; Sawinska, Z.; Ivanišová, E.; Vukovic, N.; Terentjeva, M.; Stričík, M.; Kowalczewski, P.Ł.; Hlavačková, L.; Rovná, K.; Žiarovská, J.; et al. Properties of Ginkgo biloba L.: Antioxidant Characterization, Antimicrobial Activities, and Genomic MicroRNA Based Marker Fingerprints. *Int. J. Mol. Sci.* **2020**, *21*, 3087. [CrossRef] [PubMed]

51. Kumar, S.; Sharma, S.; Vasudeva, N. Review on antioxidants and evaluation procedures. *Chin. J. Integr. Med.* **2017**. [CrossRef] [PubMed]

52. Alkadi, H. A Review on Free Radicals and Antioxidants. *Infect. Disord. Drug Targets* **2020**, *20*, 16–26. [CrossRef] [PubMed]

53. Kunatsa, Y.; Chidewe, C.; Zvidzai, C.J. Phytochemical and anti-nutrient composite from selected marginalized Zimbabwean edible insects and vegetables. *J. Agric. Food Res.* **2020**, *2*, 100027. [CrossRef]

54. Musundire, R.; Zvidzai, C.J.; Chidewe, C.; Samende, B.K.; Manditsera, F.A. Nutrient and anti-nutrient composition of Henicus whellani (Orthoptera: Stenopelmatidae), an edible ground cricket, in south-eastern Zimbabwe. *Int. J. Trop. Insect Sci.* **2014**, *34*, 223–231. [CrossRef]

55. Velderrain-Rodríguez, G.R.; Palafox-Carlos, H.; Wall-Medrano, A.; Ayala-Zavala, J.F.; Chen, C.-Y.O.; Robles-Sánchez, M.; Astiazaran-García, H.; Alvarez-Parrilla, E.; González-Aguilar, G.A. Phenolic compounds: Their journey after intake. *Food Funct.* **2014**, *5*, 189–197. [CrossRef]

56. Keppler, K.; Humpf, H.-U. Metabolism of anthocyanins and their phenolic degradation products by the intestinal microflora. *Bioorg. Med. Chem.* **2005**, *13*, 5195–5205. [CrossRef] [PubMed]

57. Waszkiewicz, N.; Szajda, S.D.; Konarzewska-Duchnowska, E.; Zalewska-Szajda, B.; Gałązkowski, R.; Sawko, A.; Nammous, H.; Buko, V.; Szulc, A.; Zwierz, K.; et al. Serum β-glucuronidase as a potential colon cancer marker: A preliminary study. *Postepy Hig. Med. Dosw.* **2015**, *69*, 436–439. [CrossRef]

58. Awolade, P.; Cele, N.; Kerru, N.; Gummidi, L.; Oluwakemi, E.; Singh, P. Therapeutic significance of β-glucuronidase activity and its inhibitors: A review. *Eur. J. Med. Chem.* **2020**, *187*, 111921. [CrossRef] [PubMed]

59. Ahmad, S.; Hughes, M.A.; Lane, K.T.; Redinbo, M.R.; Yeh, L.-A.; Scott, J.E. A High Throughput Assay for Discovery of Bacterial β-Glucuronidase Inhibitors. *Curr. Chem. Genom.* **2011**, *5*, 13–20. [CrossRef]

60. LoGuidice, A.; Wallace, B.D.; Bendel, L.; Redinbo, M.R.; Boelsterli, U.A. Pharmacologic Targeting of Bacterial β-Glucuronidase Alleviates Nonsteroidal Anti-Inflammatory Drug-Induced Enteropathy in Mice. *J. Pharmacol. Exp. Ther.* **2012**, *341*, 447–454. [CrossRef] [PubMed]

61. Wei, B.; Yang, W.; Yan, Z.-X.; Zhang, Q.-W.; Yan, R. Prenylflavonoids sanggenon C and kuwanon G from mulberry (Morus alba L.) as potent broad-spectrum bacterial β-glucuronidase inhibitors: Biological evaluation and molecular docking studies. *J. Funct. Foods* **2018**, *48*, 210–219. [CrossRef]

62. Liew, S.Y.; Sivasothy, Y.; Shaikh, N.N.; Mohd Isa, D.; Lee, V.S.; Choudhary, M.I.; Awang, K. β-Glucuronidase inhibitors from Malaysian plants. *J. Mol. Struct.* **2020**, *1221*, 128743. [CrossRef]

63. Bhatt, A.P.; Pellock, S.J.; Biernat, K.A.; Walton, W.G.; Wallace, B.D.; Creekmore, B.C.; Letertre, M.M.; Swann, J.R.; Wilson, I.D.; Roques, J.R.; et al. Targeted inhibition of gut bacterial β-glucuronidase activity enhances anticancer drug efficacy. *Proc. Natl. Acad. Sci. USA* **2020**, *117*, 7374–7381. [CrossRef]

64. Kačániová, M.; Terentjeva, M.; Galovičová, L.; Ivanišová, E.; Štefániková, J.; Valková, V.; Borotová, P.; Kowalczewski, P.Ł.; Kunová, S.; Felšöciová, S.; et al. Biological Activity and Antibiofilm Molecular Profile of Citrus aurantium Essential Oil and Its Application in a Food Model. *Molecules* **2020**, *25*, 3956. [CrossRef]

65. Daglia, M. Polyphenols as antimicrobial agents. *Curr. Opin. Biotechnol.* **2012**, *23*, 174–181. [CrossRef]

66. Aziz, M.; Karboune, S. Natural antimicrobial/antioxidant agents in meat and poultry products as well as fruits and vegetables: A review. *Crit. Rev. Food Sci. Nutr.* **2016**, *58*, 1–26. [CrossRef]
67. Kačániová, M.; Galovičová, L.; Valková, V.; Tvrdá, E.; Terentjeva, M.; Žiarovská, J.; Kunová, S.; Savitskaya, T.; Grinshpan, D.; Štefániková, J.; et al. Antimicrobial and antioxidant activities of Cinnamomum cassia essential oil and its application in food preservation. *Open Chem.* **2021**, *19*, 214–227. [CrossRef]
68. Buriti, F.C.A.; Castro, I.A.; Saad, S.M.I. Viability of Lactobacillus acidophilus in synbiotic guava mousses and its survival under in vitro simulated gastrointestinal conditions. *Int. J. Food Microbiol.* **2010**, *137*, 121–129. [CrossRef] [PubMed]
69. Trastoy, R.; Manso, T.; Fernández-García, L.; Blasco, L.; Ambroa, A.; del Molino, M.L.P.; Bou, G.; García-Contreras, R.; Wood, T.K.; Tomás, M. Mechanisms of Bacterial Tolerance and Persistence in the Gastrointestinal and Respiratory Environments. *Clin. Microbiol. Rev.* **2018**, *31*. [CrossRef]
70. Kowalczewski, P.; Walkowiak, K.; Masewicz, Ł.; Bartczak, O.; Lewandowicz, J.; Kubiak, P.; Baranowska, H. Gluten-Free Bread with Cricket Powder—Mechanical Properties and Molecular Water Dynamics in Dough and Ready Product. *Foods* **2019**, *8*, 240. [CrossRef]
71. ISO. *20483:2013 Cereals and Pulses—Determination of the Nitrogen Content and Calculation of the Crude Protein Content—Kjeldahl Method*; International Organization for Standardization: Geneva, Switzerland, 2013.
72. ISO. *ISO 2171:2007 Cereals, Pulses and by-Products—Determination of Ash Yield by Incineration*; International Organization for Standardization: Geneva, Switzerland, 2007.
73. AACC. Crude Fat in Wheat, Corn, and Soy Flour, Feeds, and Mixed Feeds. *AACC Int. Approv. Methods* **2009**. [CrossRef]
74. AACC. 44-19.01 Moisture—Air-Oven Method, Drying at 135 degrees. In *AACC International Approved Methods*; AACC International: Saint Paul, MN, USA, 2009.
75. AOAC. *AOAC Official Methods of Analysis*, 18th ed.; AOAC International: Gaithersburg, MD, USA, 2012.
76. Manzi, P.; Aguzzi, A.; Pizzoferrato, L. Nutritional value of mushrooms widely consumed in Italy. *Food Chem.* **2001**, *73*, 321–325. [CrossRef]
77. Škrbić, B.; Živančev, J.; Mrmoš, N. Concentrations of arsenic, cadmium and lead in selected foodstuffs from Serbian market basket: Estimated intake by the population from the Serbia. *Food Chem. Toxicol.* **2013**, *58*, 440–448. [CrossRef] [PubMed]
78. Ozbek, N.; Akman, S. Method development for the determination of calcium, copper, magnesium, manganese, iron, potassium, phosphorus and zinc in different types of breads by microwave induced plasma-atomic emission spectrometry. *Food Chem.* **2016**, *200*, 245–248. [CrossRef]
79. AOAC. *AOAC Official Method 995.11 Phosphorus (Total) in Foods*; AOAC International: Rockville, MD, USA, 2016.
80. European Food Safety Authority. Dietary Reference Values for nutrients Summary report. *EFSA Support. Publ.* **2017**, *14*, e15121. [CrossRef]
81. AOCS AOCS Official Method Ce 1h-05. Determination of cis-, trans-, Saturated, Monounsaturated and Polyunsaturated Fatty Acids in Vegetable or Non-ruminant Animal Oils and Fats by Capillary GLC. *Off. Methods Recomm. Pract. Am. Oil Chem. Soc.* **2009**, *2005*, 1–29.
82. Folch, J.; Lees, M.; Sloane, S.G.H.; Stanley, G.H.S.; Stanley, G.H.S. A simple method for the isolation and purification of total lipides from animal tissues. *J. Biol. Chem.* **1957**, *226*, 497–509. [CrossRef]
83. Fedko, M.; Kmiecik, D.; Siger, A.; Kulczyński, B.; Przeor, M.; Kobus-Cisowska, J. Comparative characteristics of oil composition in seeds of 31 Cucurbita varieties. *J. Food Meas. Charact.* **2020**, *14*, 894–904. [CrossRef]
84. Kowalczewski, P.; Różańska, M.; Makowska, A.; Jeżowski, P.; Kubiak, P. Production of wheat bread with spray-dried potato juice: Influence on dough and bread characteristics. *Food Sci. Technol. Int.* **2019**, *25*, 223–232. [CrossRef] [PubMed]
85. Knarreborg, A.; Simon, M.A.; Engberg, R.M.; Jensen, B.B.; Tannock, G.W. Effects of Dietary Fat Source and Subtherapeutic Levels of Antibiotic on the Bacterial Community in the Ileum of Broiler Chickens at Various Ages. *Appl. Environ. Microbiol.* **2002**, *68*, 5918–5924. [CrossRef]
86. Singleton, V.L.; Orthofer, R.; Lamuela-Raventós, R.M. Analysis of total phenols and other oxidation substrates and antioxidants by means of folin-ciocalteu reagent. In *Methods in Enzymology*; Academic Press: Cambridge, MA, USA, 1999; Volume 299, pp. 152–178.
87. Re, R.; Pellegrini, N.; Proteggente, A.; Pannala, A.; Yang, M.; Rice-Evans, C. Antioxidant activity applying an improved ABTS radical cation decolorization assay. *Free Radic. Biol. Med.* **1999**, *26*, 1231–1237. [CrossRef]
88. Kapnoor, S.; Mulimani, V.H. Production of α-Galactosidase by Aspergillus oryzae through solid-state fermentation and its application in soymilk Galactooligosaccharide hydrolysis. *Brazilian Arch. Biol. Technol.* **2010**, *53*, 211–218. [CrossRef]

MDPI
St. Alban-Anlage 66
4052 Basel
Switzerland
Tel. +41 61 683 77 34
Fax +41 61 302 89 18
www.mdpi.com

Molecules Editorial Office
E-mail: molecules@mdpi.com
www.mdpi.com/journal/molecules